EASTERN
OCEAN
CHINA
Luçon
Sooloo Sea
Celebes Sea
Straits of Macassar
Borneo
Part inhabited by the Biayos
Minda Nao
Pelew Islands
New Guinea
Spice Islands
Celebes
Ceram I.
Banda I.
Timor
Timorland
Flores
Cumbava
Balli
Lamboc
Madura I.
Sunda
Terra Australis
Gulf of Carpentaria
Endeavour Straits
Arnheims Land
Dimens Land
I. of Hainan
Paracel
the Triangles
Cambodia
Comorin Bay
Sooloo
Sooloo Archipelago
Basilan
Ternate
Tidor
Batsina
Gilolo
I. Waygiou
Amboyna
Boero
Macassar
Gunong Tellu Bay
Lant I.
Arou Isles
Great Bay
Yap
Martyres
Sayavedra
St. Juan
Baganga
Mysory I.
Evening I.
Current I.
Wesel I.
Bales
C. Blanco
Sandalwood
Aru
110
115
120
125
130
135

£12-

British & Indian
Armies in the East Indies
1685–1935

Army Gold Medal for Java, 1811. Reverse.
Courtesy of the National Army Museum.

BRITISH & INDIAN ARMIES IN THE EAST INDIES 1685–1935

Alan Harfield

with a foreword by
Lieutenant-General Sir John Worsley, KBE, CB, MC
President of the British Association
for Cemeteries in South Asia

PICTON PUBLISHING (CHIPPENHAM) LTD
1984

First published by Picton Publishing (Chippenham) Ltd 1984
ISBN 0902633 95 3

Photoset in AM International Plantin
Text paper supplied by Howard Smith Papers, Bristol
Printed in Great Britain by Picton Print
Citadel Works, Bath Road, Chippenham, Wilts

PP41369

Contents

List of Illustrations and Maps

Foreword

by Lieutenant-General Sir John Worsley, KBE, CB, MC

The story of the part played by British and Indian soldiers in the late eighteenth, nineteenth and early twentieth centuries in what we now know as Malaysia, and in Singapore, Sumatra, Java and some adjacent islands is not well know. This is a pity, because it is certainly not without interest, nor lacking examples of gallantry and fortitude in adversity. The reason is, perhaps, partly that events in the area have been overshadowed by military exploits on a somewhat larger scale, during the same period on the sub-continent of India and in Burma and China; but I think chiefly because, hitherto, it is a story which has been told only in many separate, unconnected accounts.

Alan Harfield has done something of great value by, for the first time, bringing together the best of these separate accounts in one book. He has added material from many other sources, some not available in the United Kingdom, and some entirely original detail not previously published, together with many interesting illustrations, some obtained from museum collections and others from recent photographs, including a number taken by his wife or himself, and he has done all this with considerable skill.

The book is the outcome of an enormous amount of most careful and detailed research, and a remarkable degree of perseverence, over a period of many years, in the face of setbacks which would have deterred a less dedicated man. As I have suggested, I believe it meets a long-felt need. In addition to its value to those who are interested in the military history of the area concerned, particularly in the days when our activities there were under the direction and control of the Honourable East India Company, there are details which cannot fail to be helpful to collectors of medals; and there is much for those whose interest is in graves and memorials.

John Worsley July 1984

Preface

The idea for this book was given to me in 1970 when, on joining a new unit in Singapore, during an interview with the then Commander Singapore Area, Brigadier D. W. V. P. O'Flaherty, DSO (who was himself a keen military historian), he suggested that I might undertake research on the early days of the defences of Singapore.

I commenced work on this project and was able to carry out a great deal of research before leaving Singapore at the end of 1971 when the British forces withdrew from the State. On arrival in England I continued with the research and was given advice by Mr W. Y. Carman who suggested that as there was a lack of general reference books on the military in the Straits Settlements, I make that the subject of a reference book. I was encouraged throughout the whole project by Mr Carman and it was in a fairly advanced state when, in January 1974, I moved to Cyprus to take up a new appointment. Four chapters had been written and all the notes were ready for the completion of the book. My intention was to finish writing the book during my tour of duty in Cyprus. It was with complete dismay that I learned in March 1974 that the vessel on which my baggage had been travelling had sunk in Malta Harbour. The salvage of the baggage was not completed until October 1974 by which time the whole of my work on the book had been destroyed by sea water. Once again with the encouragement of Mr Carman, and of course my wife, I decided to start the whole book again and at that time I was fortunate enough to be selected for employment with the Royal Brunei Malay Regiment as a Loan Service Officer and this allowed me to be located back in the area about which I wished to write. With the help of my wife I was able to retrieve most of the information that had been lost on the ship at Malta. After one false start and thirteen years later the reference book has been finally completed.

It is a subject which still requires many years work but I hope that this book will enable the military historian to have a background from which to work and to this end I have written each chapter as a complete unit in order that it may be taken and developed by using the references that I have given as a starting point for further research.

My most sincere and grateful thanks are extended to Mr W. Y. Carman and to the many people who have helped me with my research over the past thirteen years. I cannot possibly list them all but I would like to extend special thanks to Mr P. B. Boyden, BA, of the National Army Museum and to the most helpful staff of the Reading Room; Tan Sri Dato' Mubin Sheppard, CMG, MBE, the Editor of the *Journal of the Malaysian Branch of the Royal Asiatic Society;* the National Library and the National Archives of Singapore; Mr Anthony Farrington of the India Office Library and Records; Mr Edward Edwards McKinnon without whose help the chapter dealing with Fort Marlborough could not have been written; Major J. M. A. Tamplin, TD of the Army Museum Ogilby Trust; Mrs Kathleen Clark and finally the staff of the Public Record Office, Kew.

My grateful thanks is also given to my wife, June, who has lived with this project for thirteen years and has, during all that time, encouraged me, helped with much of the research, acted as photographer, proof reader and without whose help this book would never have been completed.

A.G.H. June 1984

Author's Notes

Throughout this reference book I have used the contemporary spelling of place names relative to the document, or period, that is being covered. Two examples are Singapore which has been spelt as Sincapore, Sincapoor, and also as Singapura, and the island of Pulau Penang, as it is now known, is alternatively referred to as Pulo Pinang, Prince of Wales Island, Prince of Wales's Island, Pulau Penang and on occasion as Penang.

Likewise with the military ranks I have used the contemporary spelling so that Matross can appear in the text as Matross or Mattross and Golundauze appears in the following alternative spellings Golundauze, Golundaze, Golun Dauze and Golundaz.

GLOSSARY

Amboynese A soldier recruited on the island of Amboyna by the HEIC to serve in the Corps of Amboynese.

Centinell, Centinel, Central A private soldier recruited for the purposes of guard duties at a fort, factory or a plantation.

Golundauze, Golundaze, Golun Dauze, Golundaz Native artillery or native artilleryman.

Gun Laskar, Gun Lasker A labourer employed by the artillery. Can also be a general labourer or a camp follower.

Havildar Sepoy non-commissioned officer equating to the rank of sergeant.

Jemedar, Jemidar, Jemmadar An Indian junior officer equivalent to a Second Lieutenant.

Lieutenant (Fireworker) A Third or Fourth Lieutenant of Artillery; an officer junior to a Second Lieutenant of Artillery.

Matross, Mattross Hobson-Jobson refer to a Matross as 'an inferior class of soldier in the artillery.' Matrosses were apprentices to gunners who assisted in the loading, firing and sponging of the guns and acted as a guard to the guns.

Naik, Naique An Indian Corporal.

Sepoy, Seapoy A private soldier, usually disciplined in the European style.

Subedar, Subedah, Soubadar An Indian Lieutenant.

Temenggong, Toomongong, Tumunggung A Malay minister, usually in charge of defence, or police and markets.

Topaz, Topass A name used for dark-skinned or half-caste claimants of Portuguese descent who were, during the seventeenth and eighteenth centuries, employed as soldiers.

CHAPTER ONE

The Spice Islands and Sumatra

In order to appreciate the need for a military presence in the East Indies it is first essential to see the reason for the development of trade in that area. It was only due to the increased importance of the trading stations and factories and the need for some form of defence against other European nations and hostile local inhabitants, during the eighteenth and nineteenth centuries, that the military force began to develop.

The need to develop trading posts in the Far East stems from the requirement for spices which were, at that time, the beginning of the seventeenth century, of great value and importance to the European population as partial preservatives for food. Fresh meat was only available for a few months each year and meat of that time would be almost unacceptable to the present-day palate. Spices, such as cloves, nutmeg, cinnamon and ginger were therefore a source of wealth and power for those who were adventurous enough to journey to the East to secure them. The Dutch and Portuguese were already established in the East and trading in spices, so it was not surprising that Queen Elizabeth I agreed to sign the Charter for the Merchant Venturers of London to send an expedition to the East Indies for the 'Honour of this our realm of England' and for the advancement of trade of merchandise. The very wording of the petition was calculated to stir the imagination and this at a time when the country had no shortage of proud and patriotic seamen.

The actual establishment of English trading posts in the East Indies can be related to that expedition, organised by the Merchant Venturers of London, which departed from England during 1601. The expedition, commanded by John Lancaster, consisted of five ships manned by about 400 men. Because of the high rate of sickness and death, ships were generally over-manned. The expedition reached Acheen, in northern Sumatra, after eighteen months. After negotiations with the local ruler, Lancaster sailed east and established the first British trading factory in the East Indies at Bantam[1] on the island of Java. William Starkey was made the chief factor of this trading post and nine other men were left to man the post. The fleet returned to London arriving in the Thames on 11 September 1603. The ships had brought back 1,030,000 pounds of pepper which was sold for a good profit and the leader of the expedition, John Lancaster was knighted.[2] A second expedition set out in 1604 and relieved the men who had been left at Bantam. During the next twenty years trading continued in the area and other factories were set up at Jacarta in Java, Macassar in the Celebes and at Amboyna in the Moluccas. A number of smaller posts were also established in the area generally known as the Spice Islands. Unlike the Dutch, the English East India Company manned its factories with as few Europeans as possible and in some of the minor factories only two or three Englishmen were employed and were responsible not only for the trading with the local ruler but also the safe keeping, or defence, of the factory. The foothold in the area was very precarious and the British suffered a serious setback when, in 1623 at Amboyna, the Dutch arrested and tortured the factors and senior merchants on false charges of plotting against the Dutch Trading

Company. The Dutch Governor-General Jan Pieterszoon Coen[3] had returned to Europe and the Dutch governor said that a Javanese soldier employed by the English factory had informed him that the English were conspiring to take over the Dutch fort and kill off the garrison stationed there. A number of the English were arrested and tortured and, confessing to the plot, ten of them were executed by beheading. Despite this the East India Company continued to hold on to the factory at Bantam until the end of the century when they were forced to relinquish, temporarily, this last foothold in the Javanese cinnamon trade.

The Honourable East India Company in the early days of the East Indies settlements, had an establishment of civil servants without any military presence. In 1623 there were in the Amboyna establishment eighteen Englishmen of whom twelve were factors, two clerks, a barber-surgeon, a tailor and two servants. The factors lived in a type of collegiate society sharing a house which would usually be hired from the local Sultan. The houses were adopted from the local style, made of wood, raised above the ground on piles and with a palm-leaf thatch. Captain William Dampier visiting a Company post in 1688 describes the house as follows:

> '... There is a handsome dining-room in the middle, and at each end convenient apartments for the Merchants, Factors and Servants belonging to the Company to live in, and at each end of it there are smaller Houses for other uses, such as Kitchen, Storehouse etc., running in Line from the great House towards the River, Making two Wings, and a square Court open on to the River. In this square space, near to the Banks of the River, there stands a Flag Staff, purposely for the hoysting of the English colours on all occasions ...'[4]

The Dutch turned their factories into a combined factory and military establishment in the early seventeenth century but the British continued to man unfortified trading posts until the end of the century, and even then they were usually under-manned.

British factories were established in Borneo and Sumatra but these posts set up for the black pepper trade were generally abandoned within a decade. There were various reasons for these evacuations but the reason for the evacuation of the factory at Banjarmasin was given by Captain Alexander Hamilton as being forcibly expelled by the local Sultan. Hamilton records that '... their factory was not half finished before they began to domineer over the natives, who passed in their boats up and down the river, and, very imprudently, would needs search one of the king's boats which was carrying a lady of quality down the river, which so provoked the king that he swore revenge and accordingly gathered an army ...' Whilst he was gathering his army the Company departed from the factory and made no further attempt to re-establish trading in that area.[5]

British factories were then concentrated onto the west coast of Sumatra and the largest of these was at Bencoolen which was established in 1685 under the command of Ralph Ord.[6] This factory remained operative until 1825. Benkulen, or Bencoolen (both forms of spelling have been used for this settlement) was a fortified factory although it is recorded that the original fort was a very '... precarious work ...' The building of York Fort, as it was to become known, proceeded extremely slowly due to the death of the engineer in charge only a few months after the work had started. The fort was situated on the banks of the Benkulen river and is said to have looked very impressive from the sea. There is no trace of the establishment of gunners or guards for this fort. The voyager William Dampier, author of *Voyages and Discoveries*, served as a gunner at Benkulen for a short time and commented that it was '... the most irregular piece ...' of defence work he had ever seen. He recommended to the Governor that the fort, which consisted of earthworks and was subject to erosion during the monsoon season, be faced with either brick or stone. Dampier did what he could under the circumstances to effect repairs and made the bastions as regular as he could; only four had been completed and these he repaired, and also staked out the fifth bastion, but with the limited resources of the garrison this was never built. The defence work was manned in the early years by a mixed force of locally enlisted men and young soldiers recruited and sent out direct from England. Generally the British soldiers filled the posts of officers and gunners. The infantry were recruited from Topazes. The definition Topaz, or Topass, is given in Hobson-Jobson as a name '... used in the 17th and 18th centuries for dark skinned or halfe caste claiments of Portuguese descent, and the Christian profession ...'[7]

At this time it was considered unwise to recruit the local Malays as they generally formed an alliance with the local Sultan and frequently made sorties against the British. Although at the outset of the formation of the settlement at Benkulen, the forces were small in number, sickness, probably malaria, and the rigours of a dissipated life led by some Europeans in the East Indies, reduced still further the strength of the Garrison. Of eleven British and nineteen Topazes only eleven were fit enough to take any part in the defence duties for the factory. Sickness, and subsequent death, was a continuing problem on the West Coast. As early as 1685 the following report was sent which illustrates this problem. '... All our servants are Sick & dead, & at this Minute [there is] not a Cooke to gett victualls ready for those that sett at the Compas table, and Such have been our straites yt wee many times have fasted. Ye Sick lyes Neglected, some cry for remedies, but none [are] to bee had: those that could eate have none to Cooke ym victuals, ... soe yt wee now have not liveing to bury ye dead, & if one is sick ye other will not watch, for hee says that better one [than] two dies, Soe that people dies & noe notice taken thereof...'[8] Despite these conditions the East India Company never lacked recruits both for civil and military posts.

With the problems of the unreliability of local inhabitants, and the cost of sending reinforcements from England to such a far outpost as Benkulen, the East India Company headquarters suggested that slaves from Madagascar might be enlisted in the Company's military service for the defence of the Sumatran settlements. This, the Directors stated, would maintain a balance in the case of need against the forces of the local Sultans. They considered that '... five of them will not cost us so much to maintain as one English soldier ...'[9] They were to be used as guards under British officers and were to be armed with '... lances, darts, and swords and other weapons of India, but teach them not ye use of firearms ...'[10] The instruction was repeated in a later directive and was further qualified with the following statement '... some of ye blacks that speak English if in want of soldiers you may arm and may make them keep guard to ease your English soldiers; but trust them not too much; neither ever arm or exercise them above ten of them to thirty English soldiers ...'[11] The experiment seems to have been short-lived as the 'slaves' proved to be poor soldiers.

The conditions in the factories in West Sumatra were very poor with malaria, cholera and dysentery taking a heavy toll. Living conditions were bad compared with factory life in India. Food was not only scarce but also expensive and labour costs for servants were high. There were very few of the diversions, taken for granted in the factories in India, to those serving on the west coast of Sumatra. The Military had little to occupy their time between duties, unlike the civilian staff who were involved in the organisation and running of the factory.

York Fort had cost the Company approximately a quarter of a million pounds sterling in

View of Guning Bengko from York Fort, 1982. Courtesy of E. Edwards McKinnon.

HEIC sword hilt excavated at the site of York Fort, 1982.
Courtesy of E. Edwards McKinnon.

repairs. It was later decided to abandon the site and in 1714 work commenced on a new fort, which was to be named Fort Marlborough and which was erected on an elevated and secure site approximately two miles south of York Fort. Fort Marlborough was a formidable defence work and its outer ramparts and gateway still stand 270 years later. Fort Marlborough came into being during the Deputy Governorship of Joseph Collet between the years 1712 and 1716. Collet wrote home saying that he had '. . . six or seven garrisons under my command . . .' and that Fort Marlborough and Bental, a factory about ninety miles north of Fort Marlborough, had 'forty or fifty guns each . . .'[12] In the Bill of Mortality for Bental (also spelt Bantall in some documents) dated 28 January 1713, of the fourteen deaths recorded nine were military and were shown on the document as follows:

Name	*Station*	*Time of their decease*
William Mathews	Gunners Mate	August ye 22, 1712
George Lockhart	Surgeon	November ye 19
Augustine Derosario	Topass Centinell	February ye 6
Alexander Gamble	Ensign	March ye 13
Stephen Oliver	Ensign	May ye 2, 1713
John Sone	Gunner	Do ye 6
James Horsly	Gunners Mate	Ditto ye 21
William Newton	European Centinell	August ye 7
Charles Wickstead	Ensigne	September ye 6th[13]

The requirement for locally enlisted soldiers continued to be a problem and it is evident from letters written by Governor Collet that the Company had commenced its policy of recruiting Bugis for the guarding of the factories and Fort Marlborough. John Crawfurd, in his book *A Descriptive Dictionary of the Indian Islands and Adjacent Countries* states that Bugis is the Malay name given to the natives of the south-western part of the Celebes, although they refer to themselves as 'Wugi'.[14] In a letter written on 23 September 1712 Governor Collet records '. . . If I go riding I have a Horse Guard attending me and a Union Flag carry'd before me. I have also a foot Guard of Buggess (Bugis) soldiers who generally keep way with me. If I walk I have 4 men with Blunderbusses go before and a Guard of Buggess to bring up the rear . . .'[15]

The British Resident at Bantal, Henry White, had problems with the Sultan of that area and also two Rajas in the territory around Bantal and requested military assistance from Governor Collet. He hoped to raise a force in order to mount an expedition against the Malays. In his reply to Henry White, Governor Collet commented that he was unable to provide much help saying '. . . our military are reduced so thin that we have not sufficient to do duty but must be oblig'd to arm some of the slaves . . .'[16] The Resident at Bantal had asked for a military force of thirty Europeans and a hundred Bugis but in a second letter, dated 23 October 1713, Collet informs him that he will '. . . send two Europe ships, and Ensign, Sergeant and 12 soldiers beside the six soldiers you will receive by the President (a ship) with full power to command a peace, by Military Execution if it can't be done on better Terms . . .'[17] A satisfactory solution was negotiated without this very limited force having to take to the field.

The areas around the factories were generally a mixture of swamp and jungle. The pepper crop was the responsibility of the local Sultan, or Raja, and was usually not in the immediate vicinity of the factory or fort. Collet wrote to a friend in England and commented that the Fort, Marlborough, was situated in a swamp which no doubt accounted for the high rate of sickness and mortality. The Malay town adjoining the fort consisted of some seven to eight hundred houses and on the other side of the fort he noted that, near to the sea, is the small '. . . village inhabited by the Company's slaves . . .' Rice was grown near to the fort and the area was considered to produce a good crop.

In Sumatra, at this time, life was very restrictive for the European military personnel as not only were the living conditions very primitive, but also the climate was continually hot and humid. The humidity was such that clothing and equipment quickly became covered with mildew and required constant airing and cleaning. Trading ships visited very infrequently and therefore the small population became very insular and petty arguments were frequent and often assumed proportions which led to general disaffection within the settlement. The climatic conditions were conducive to malaria, denghi fever and many other tropical ailments which were constantly depleting the military force. With such conditions it was little wonder that many of the European personnel took to drinking alcohol in an attempt to keep the numerous fevers under some sort of control. The Europeans in the main settlement at Benkulen divided themselves naturally into two factions, these being the civilians who were employed by the Company, and the military. At times the military suffered under the command of some of the Governors who were reluctant to spend money and time on defence and used any excuse to attack the military organisation and in some instances individuals. Any officer who was found to be indulging in drink to any excess was an open target in this type of settlement. Governor Collet, writing in 1712, commented that '. . . Here are but 4 White Women . . . All the People that have dyed here since my arrivall which are but few, have apparently been destroy'd by Drinking or Women. There has not one of our private centinells been buried these 10 months which I ascribe to their not having the opportunity of Drinking so deep as their betters . . .'[18] Collet's task was a difficult one, he was responsible for the annual pepper trade and to help him achieve this end he ordered the construction of the Fort, which he named after John Churchill, first Duke of Marlbrough (1650–1722), and wrote '. . . I have layd the foundation of the strongest Fortification in India . . .'[19]

The Governors had, during the first decade of the eighteenth century, been granted direct control over the military within the Settlement and this authority continued to be exercised during the period leading up to the building of Fort Marlborough.

The document authorising Governor James Cross to exercise these powers was signed in London and dated 12 April 1706 and reads:

> '. . . To James Cross, Governor of our Fort called York Fort at Bencoolen on the West Coast of the island of Sumatra in the East Indies . . . we do hereby require all our Officers and souldiers, [*sic*] ffactors, [*sic*] writers and other Servants on the Island of Sumatra to yield due obedience and submit to the said James Cross . . .'[20]

NOTES TO CHAPTER ONE

1. Bantam situated at the western end of the island of Java was formerly an independent kingdom. It covered an area of 2,568 square miles of mountainous and volcanic country. In the seventeenth and eighteenth centuries the population consisted mainly of the Sunda nation and, in the coastal area, a mixture of Sunda, Malay and Javanese. The estimated population in 1814 was recorded as being 231,604. The main spice crop was, at that time, cinnamon.
2. Sir John Lancaster became a proprietor of the Honourable East India Company and, although he was responsible for the planning of the next few voyages after his return, he did not take part in any further expeditions. He died in 1618.
3. Jan Pieterszoon Coen was born in January 1587 at the town of Hoorn, a port on the Zuider Zee. He was known to the British as John Peter Sacone and was greatly feared by the British who were resident in his territory in the East Indies.

4. George Woodcock, *The British in the Far East,* (London, 1969), p6.
5. John Crawfurd, *A Descriptive Dictionary of the Indian Islands and Adjacent Countries,* (reprinted, Kuala Lumpur, Malaya, 1971), pp 36 and 37.
6. Ralph Ord was an ex-soldier who had also been employed, in Madras in 1678, as a teacher for European children. He was provisionally appointed factor in Madras in September 1682 and took up the appointment in Benkulen as the factor and founder of the newly-established factory in 1685. Some sources record that he died in 1686 due to his being poisoned by the Dutch. His death at Benkulen is confirmed but as yet no trace of his grave has been found.
7. Colonel H. Yule and A. C. Burnell, *Hobson-Jobson,* (London, 1903), pp 933 and 934.
8. Letter from Benjamin Bloome and Joshua Charlton, dated October 1685.
9. A. Wright and T. H. Reid, *The Malay Peninsula,* (London, 1912), p37.
10. John Bastin, *The British in West Sumatra (1685–1825),* (Kuala Lumpur, Malaya, 1965), pxviii.
11. Wright and Reid, op cit, p37.
12. Letter from Joseph Collet to Mrs M. Quincy, 19 September 1712. (Private letter books of Joseph Collet).
13. Although the date of the document is 'January 28th 1713' the list covers a period before and after that date. From an examination of Burial Registers from Sumatra it is noted that such registers are usually dated and then continue in an open form until the next visiting ship was due to return to London. The document was then closed and despatched with the ship's master. A copy was retained in the Settlement.
14. Crawfurd, op cit, pp 74 and 75.
15. Bastin, op cit, p 44.
16. Ibid, p48.
17. Ibid, p49.
18. Ibid, p45.
19. Ibid, p55.
20. India Office Library and Records, H/Misc/24.

CHAPTER TWO

Fort Marlborough, 1714–1825

The move of the main portion of the settlement at Bencoolen was brought about by the '... unhealthiness of the old situation ...'[1] and in consequence there was a need to build a new fortification. Fort Marlborough was planned and the foundations laid out in 1714 '... on a spot two or three miles distant ...' from the site of the then existing and dilapidated York Fort at Pasar Bengkulu. The site of the new fort was on high ground which afforded a good defendable area but Marsden comments that this advantage was '... counterbalanced by its want of the vicinity of a river, so necessary for the ready and plentiful supply of provisions ...'[2] The fort was of the general design used by the military in the Far East with strong defensive positions, a moat and capable of withstanding a siege, although this was later to be found wanting. The construction of the main portions of the fort, which was by then named Fort Marlborough, was completed during 1716. However before the fort could be fully completed the local Malays who had become incensed by the treatment that they were receiving from the Europeans at the settlement rose and attacked the settlement and the fort in 1719 and because of the numbers involved, forced the small garrison to quit the fort and take refuge on board the Company's ships that were anchored off-shore at that time. There was a fear that as the Company's garrison and civil staff had been forced to virtually evacuate the Settlement the Dutch would move in and take the settlement at Bencoolen. Fortunately before that situation developed peace was made with the local ruler and as supplies also arrived from Madras the garrison re-established itself ashore once again and took possession of the partially completed fort. This brought pressure on the military to complete the construction of the fort so as to prevent a similar situation occurring in the future.

The garrison at Fort Marlborough had little to occupy them other than drills and guard duties. The area around the fort was a mixture of swamp and jungle and the close proximity of the swamp accounted for the high rate of sickness and mortality suffered by the military stationed at the garrison. The life expectancy of the European officers and men serving at Benkulen during the late seventeenth and early eighteenth centuries, living under the prevailing conditions, was short and the Bombay proverb that 'two monsoons' was the life expectancy of the European was even more true for those posted to the west coast of Sumatra. Life in the fort was, for the European, very restrictive as, not only were the living conditions primitive when compared to those experienced by officers of the Company who were serving in India, but the climate was continually hot and humid which caused clothing and equipment to quickly become covered in mildew and require constant airing and cleaning. The conditions were conducive to malaria, dengue fever and many other tropical ailments which constantly depleted the military force at Fort Marlborough. Trading vessels visited infrequently and therefore the life of the small military community was very restrictive and it has been recorded that petty arguments were frequent and often assumed proportions which led to general disaffection within the garrison.

The Company gradually expanded along the west coast of Sumatra and in 1752 a settlement was established at Natal and shortly afterwards at Tappanuli (sometimes spelt Tapanuli) and these new settlements brought the Company once again into dispute with the Dutch who claimed that these two settlements infringed into their territory in Sumatra.[3] By 1754 the Company had established Residencies in the following areas which were all under the jurisdiction of 'Fort Marlborough', the list is in order of location (north to south, along the west coast):

Moco Moco Residency	(also spelt Moko Moko)
Cattown Residency	
Laye Residency	(also spelt Lais)
Tallo Residency	
Manna Residency	
Cawoor Residency	
Croee Residency	(also spelt Krui)[4]

These out-stations all needed a small guard which was, when required, provided from the main garrison at Fort Marlborough.[5]

In 1760 the position of the Settlement changed when the Directors issued the following order which was dated 22 October 1760:

> '... We have determined upon rendering our settlements on the West Coast of Sumatra independant of Fort St George, and erecting Fort Marlborough into a Presidency in the like manner as our other Presidencys [*sic*] ...'[6]

It was perhaps an unfortunate time to have issued such a document as it was during 1760 that the French, under Comte d'Estaing, attacked and destroyed all the Company's settlements along the west coast of Sumatra, including Bencoolen. The garrison at Fort Marlborough surrendered and were permitted to embark on the Company ships that were anchored off-shore.[7] The re-possession of the settlements was secured through the Treaty of Paris in 1763 and within a short time the west coast settlements were once again controlled by the Honorable East India Company. On their return they found '... many of the pepper gardens in ruins, the people restless and in many cases unwilling to recommence cultivation of the vines ...'[8] In addition the Sultan of Moko Moko had ordered the destruction of the vines in his territories and also the gardens in the '... Benkulen-Silebar region were ruined ...'

With the change of status there were other internal changes within the administration of the Settlement. The office of 'President and Governor' also carried the title of Commander-in-Chief. The document appointing Richard Wyatt to Governorship dated 15 January 1766 reads:

> '... To Richard Wyatt ... to be President and Governor of and for all our affairs on the Island of Sumatra in the East Indies and also to be our Commander-in-Chief of our Fort Marlborough at Bencoolen on the said island and all the Territories thereunto belonging and of all and singular the Towns, Forts, Factories and Settlements, Territories, Countries and Jurisdictions thereof ...'[9]

With the authority delegated to the Governor and President the control of the military forces in Sumatra also included the granting of commissions. The first of the local 'Fort Marlborough' commissions were granted in 1767 to the following Europeans, with the format of the commission being:

> '... Richard Wyatt, Governor Fort Marlborough, Commander-in-Chief of all the Forces employed in the Service of the United Company of Merchants of England

trading in the East Indies on the West Coast of Sumatra ... appoint you to be Lieutenant of Infantry ...
[the commission was addressed and dated]'
'... Neville Verier [dated] 7 July 1767 ...'[10]

During that year three other local commissions were granted as follows:

'... to be Second Lieutenant of the Train of Artillery.
Thomas Oldham 8 July 1767 ...'
'... to be Ensign of Infantry.
John Bourne 17 July 1767 ...'
'... to be Lieutenant Fireworker.
Thomas Scrivener 17 July 1767 ...'[11]

The garrison of Fort Marlborough was increased during the following year with more officers being recruited in India and in Britain to fill the vacancies. Unfortunately the burial records for Bencoolen are not available for this period and it is therefore impossible to see whether any of the commissions were granted to replace officers who may have died. The following commissions were granted by 'The Governor and Commander-in-Chief' to fill the vacancies as shown:

Establishment vacancy	*Appointed*	*Date*
Ensign of Infantry	Samuel Reynolds	1 May 1768
Lieutenant Fireworker	Alexander Milne	1 June 1768
Captain of Foot	John Hutcheon	17 September 1768
Captain/Lieutenant of Foot	John Burnham	17 September 1768
Lieutenant of Infantry	John Twells	17 September 1768
Ensign of Infantry	Henry Edwards	17 September 1768
Lieutenant Fireworker	Robert Stratford Byrne	17 September 1768

In addition to his commission of a Lieutenant of Infantry John Twells was granted an additional commission, also dated 17 September 1768, appointing him as '... Adjutant for all the Forces [on the west coast of Sumatra] ...'[12]

Local commissions were granted by 'The Governor' until 1785 when Fort Marlborough was reduced and became a '... Residency subordinate to the Supreme Government in Calcutta ...'[13] although a few commissions were granted after that date by the Deputy Governor. The list of local commissions granted during the period 1769 to 1800 is given at Annexure A to this Chapter.

In addition to the sepoys that had been recruited in India, the 'Fort Marlborough Corps' also had European recruits in the ranks. Discipline and living conditions for the soldiers was hard and justice was, when required, swift. A typical example of this can be seen through the 'Execution Warrant' that was issued in December 1771. The warrant reads:

'... Whereas Joseph Gordon, James Eagle and George Duberry, private Centinels in this Garrison were tried by a General Courte [*sic*] Martial held at this place the 11th day of December 1771, by Virtue of a Warrant from us, & have been found guilty of a breach of part of the Fifth Article of Fifteenth Section of the Articles of War, and have been sentenced by the Court that one of the three above named be shot to death and to which of their Lots, this sentence shall fall to be determined with a chance of a Die on the Drum Head of which Sentence we have thought proper to approve. We do therefore by the Virtue of the Power deputed unto us, order and direct you Mussenden Johnston Esq., Major Commandant of the forces here, to cause the said Sentence to be put in Execution on Wednesday morning the 18th Instant at seven-o-Clock at the usual place

of Execution near the Gallows for so doing this shall be your sufficient warrant.
Given under our hand & Broadseal.

Richard Wyatt
Joseph Darvall
Robert Hay
Robert Nairne
John Walton . . .'[14]

With the loss of the burial registers for this period it is not possible to ascertain which of the three soldiers lost in the throw of the dice and was subsequently executed.

The question of discipline within the garrison was one which required constant review and in the case of officers they were, like their counterparts serving in the British Army, subject to Court Martial. Two officers were dealt with at one Court Martial in 1774. The court was made up of the following officers who were serving on the garrison strength at that time

'. . . President of the General Court Martial - Captain David Pilmore
Members - Captain John Hutcheon
Captain John Burnham
Captain/Lieutenant Oldham
Lieutenant John Bourne
Lieutenant Archibald Ramsay
Ensign Andrew Dingwell
Ensign Richard Alexander Powell
Ensign John Swann . . .'[15]

The two officers were Captain/Lieutenant Samuel Reynolds, who was accused of misappropriation and was censured and Lieutenant Fireworker William Maitland who was accused of being '. . . Drunk on Duty . . .' but was acquitted.

In the following year one of the members of that Court Martial was himself court martialled for disobedience of orders, this was Lieutenant John Bourne. The President on this occasion was Captain Thomas Oldham who on 28 December 1775, together with the members of the court, found Bourne guilty and ordered him to be cashiered.[16] A more serious offence had also been committed by another member, Lieutenant (Fireworker) Richard Alexander Powell who was '. . . brought before the Court[17] and accused of shooting a slave being the property of Ensign William Bridson . . .' The court was held on 26 December 1775 and Powell was found guilty and was sentenced to be '. . . Cashiered and to pay Mr Bridson for the deceased slave boy, which he valued at fifty dollars . . .'

Although the killing of the slave boy amounted to murder Powell was, as can be seen, given a light sentence but in July 1776 another Court Martial, which dealt with another murder, brought in a very different verdict. Sepoy Punnah was accused of the '. . . murder of coffree woman named Rose . . .' The Court was held on 6 July 1776 under the Presidency of Captain Thomas Oldham and on this occasion the accused was found to be 'Guilty' and was '. . . Condemned to death by Hanging . . .' and the sentence was confirmed by The Governor and carried out at the gallows.[18]

During the previous year, 1775, the Company mounted a small expedition to Balambangan but the general move to the occupation of the various Dutch settlements in the East Indies did not in fact take place with any serious endeavour for a few years, however William Broff, who was to become Governor in 1776 was appointed, on 21 August 1775, to be:

'. . . Chief of the Expedition to Balambangan, the Islands of Balambangan, Labuan and all other islands, Territories, Factories and Settlements . . . Eastward of the Straits of Sunda and [was to have command of] all Forces employed on the expedition . . .'[19]

In November 1776 Lieutenant Colonel James Kirkpatrick was appointed '. . . Commander of Our Military Forces at Fort Marlborough on the West Coast of the Island of Sumatra . . .'[20] replacing Major Mussenden Johnstone who had been appointed as Commander by a commission dated 12 April 1774.[21]

The next three years were uneventful from a military point of view with the garrison doing normal duties and with the maintenance of the main fortification being the main occupation. Discipline continued to be swift and hard and in many cases the sentence was used as a means to warn the soldiery of the folly of breaking the rules and regulations. Although all the Europeans serving in the Fort Marlborough Local Corps were volunteers many soon realised that once in the Settlement there was little chance of escape. A few tried and when caught were treated as deserters and given severe sentences. One case occurred in October 1779 when '. . . Sergeant John Pavy of the Vacant Company[22] [was] found on board ship about half-a-mile from Rat Island . . .' Pavy was accused of 'Desertion' and on 5 October 1779 appeared before a General Court Martial, President Captain Thomas Oldham, and was found guilty. He was sentenced as follows:

'. . . that he suffer a Thousand Lashes on his Bare back with a Cat-o-Nine Tails by the Drummers of this Garrison being first reduced to their rank . . .'[23]

In 1781 an expedition was mounted at Fort Marlborough consisting of five ships belonging to the Honorable East India Company and a detachment of military from the Garrison. The small force took possession of Padang and other Dutch factories thus securing them for the British.[24]

It was during that year that yet another officer was court martialled at Fort Marlborough. The court was convened on 21 April with Captain William Phillips being appointed as President to consider the charge brought against Captain Frederick Mandiville who was accused of '. . . disposing of a quantity of the Hon'ble Company's Copper Hoops . . .' The Court found Mandiville guilty and sentenced him to be 'Cashiered', however the President of the Council reduced the sentence to '. . . six months suspension [and] should any officer be granted a Captain's commission in that time they will take the rank [seniority], and he pay for the hoops . . .'[25] In the following year there are, on record, details of five general courts martial that were held at Fort Marlborough and it is worth considering the variety of offences and sentences that were given within the garrison. These were:

Lieutenant John Young on 11 February 1782 accused of '. . . un-officer-like and ungentlemanlike behaviour . . .' who was 'Cashiered'.[26]

On 15 April 1782 Nicholas Moey, John Andrew Goslip, Sybrandt Jacobs and William Wistenbourg 'soldiers of this Garrison accused of an intention, making preparations for and attempting to desert between 3rd and 11th instant . . .' All were found guilty and were sentenced to '. . . 500 lashes on their bare backs with a Cat-o-Nine Tails by the Drummers of this Garrison . . .'[27]

In September 1782 Captain Richard Vaughan for 'behaving in an unofficer like manner . . .' Vaughan was found guilty and sentenced to be cashiered but this was commuted to being '. . . Suspended from all military service . . .'[28]

On 16 October 1782 Private Hugh Kelly was accused of '. . . Robbing Mr Henry Lewis of a Bag of Money amounting to 130 dollars . . .' Kelly was found guilty and was sentenced to receive '. . . 1,000 lashes on his bare Back from Drummers of this Garrison and to be drummed with a halter around his neck and afterwards transported . . .'[29]

Finally during 1782 the Court Martial held on 22 October 1782 when '. . . John Marlock and Thomas Symonds, Matrosses . . .' were accused of robbery, but were found 'Not Guilty'.[30]

It was during that year that the magazine at Fort Marlborough was struck by lightning which fired some four hundred barrels of powder causing them to blow up and do considerable damage to the area around the magazine. Marsden in his book *The History of Sumatra* briefly refers to the incident and comments that '. . . providentially few lives were lost . . .'[31] Unfortunately, once again, there are no records in the burial register for the year 1782 so that it is impossible to give the actual number of casualties.[32]

Little has been recorded dealing with the military activities of the forces in Fort Marlborough and the outlying garrisons during the next few years. There was the constant need for vigilance against an uprising and to this end the garrison continued to carry out regular drills and perform the endless guard duties both at Fort Marlborough and in the small forts that had been erected in the Residencies.

In April 1786 Isaac Gardiner, a Mattross, was tried by general court martial having been accused of '. . . Stealing the locks of the Companies arms and offering them for sale to the Malays . . .' Gardiner was found guilty and was sentenced as follows: '. . . To receive 1000 lashes in the usual manner and also that he never be allowed to rise higher than the rank of Private Soldier . . .'[33]

During September of the same year a Sepoy, Shaik Shamshew, who was serving at Natal was found guilty on 14 September of desertion and sentenced to '. . . 600 lashes, and to be drummed out of the Garrison with a halter around his neck . . .'[34] With the change-over in 1785, from being a Presidency to becoming a Residency under Calcutta, the documents relating to the military administration and discipline within Fort Marlborough terminate in the Sumatra Records at the end of 1787. Thereafter, is is assumed that such records will have been deposited in India, and to date these have not been traced. There are, however, two final courts martial where the accused was found to be guilty and sentenced where the crime and the relating sentence are of interest. The first was on 9 November 1786 when '. . . Richard Roach, Private Infantry . . .' was found guilty of 'assault and robbery' and was sentenced to death. The finding was commuted by the President and Council who '. . . upon a perusal and due consideration of the sentence pronounced against the Prisoner . . . mitigate it and order that he be kept a prisoner in the Fort in Irons, and to Draw Water for the troops for the space of five years . . .'[35] The final court martial in the records is dated 13 January 1787 when '. . . John Middleton, Private Infantry, John Coburn Pensioner, Jemidar Saich Hussien . . .' were all accused of forgery. Only Middleton was found guilty and he was sentenced to be '. . . Flogged publicly at a cart-tail with a Cat-o-Nine Tails to the number of 200 lashes . . .'[36]

The very existence of Fort Marlborough and the need for the garrison force of the Fort Marlborough Local Corps was of course dependant upon the produce being exported through the Settlement from the numerous spice plantations. The details of the civil administration and the trading figures are adequately and very thoroughly covered in Doctor John Bastin's book *The British in West Sumatra (1685–1825)* and so long as the trading continued there was the need for the fort and the garrison.

A plan of Fort Marlborough and its environs was drawn by Captain John Macdonald in 1798 and shows the Fort, the Sepoy Lines, the hospital compound and the 'Burying

Bencoolen Bay, the small road and wharf as seen from Fort Marlborough. Sketch by Andrews, 1794–1798. Courtesy of the Director, National Museum, Singapore.

Government House and the Council House, with the entrance to Fort Marlborough, at Bencoolen. Sketch by Andrews, 1794–1798. Courtesy of the Director, National Museum, Singapore.

Fort Marlborough from the south-east (Old Bencoolen). Sketch by Andrews, 1794–1798. Courtesy of the Director, National Museum, Singapore.

Fort Marlborough, the south front. Sketch by Andrews, 1794–1798. Courtesy of the Director, National Museum, Singapore.

Ground' which was in fact adjacent to the road leading from Bencoolen to Fort Marlborough.[37] During the following year a number of lithographs were prepared from a series of sketches by Samuel Andrews who visited Fort Marlborough sometime during 1794–98. The drawings were engraved by Joseph C. Stadler in London during 1799. The sketches show Fort Marlborough as having a moat, with earth ramparts and with a walled area adjacent to the main southern entrance. Stone buildings are shown immediately inside the entrance and in addition there was a three-storeyed tower, which had an open-top battlement, with a flagpole from which the Company's flag was flown.[38] The post of Deputy Governor of the Benkulen Residency was changed, in title, to that of Commissioner and on 29 January 1800 Walter Ewer assumed the appointment and title of Commissioner and assumed control of the Residency.[39] On 6 August of that year Ewer wrote to the President of the Board of Control, Henry Dundas, and reported that the condition of Fort Marlborough was far from satisfactory. He stated that during the five year period from January 1795 to December 1799 the cost of maintenance for the fort had been '... 110,289 piasters ...'[40] He further commented that there were '... Orders from Bengal to finish it [the Fort], I shall let them go on. It will require, I suppose, ps 20,000 more. It will then be complete & look regular; but here your expenses shall stop. I will not suffer another dollar to be laid out, except to keep out the weather ...'[41] It would therefore appear, from his statement, that there was uncompleted work to be carried out at the Fort. The Fort had, by 1800, been in existence for eighty-four years and it is therefore probable that the work referred to related to work on improvements which were required although Ewer does not specify exactly what the work was that was such a drain on his budget. Ewer continued his report by making an attack on the need to have the Fort at all, and says:

> '... When finished it will be useless against a European enemy. In order to make it tenable, the greatest part of the settlement must be destroyed; the most inveterate enemy cou'd not do greater damage than the Commandant, who attempts to defend this fortress. If I may judge from what happen'd on the Thanksgiving Day, when some of the old masonry gave way, it will not stand the firing of its own guns. If, then, it will not bear a salute, how will it be possible to keep it together during a seige, which it will have to maintain against an enemy without, & its own artillery within. A trifle will keep out the Malays, for the most insignificant work, with Sepoys to defend it, would defy all the inhabitants of Sumatra ...'[42]

and later under the title 'Military Expenses' he again complains, writing:

> '... The Military Expenses. These are enormous, particularly [those relating to] the Europeans, which are quite useless, most of them fitter for the hospital than the field. I have sent his Lordship a private report on this subject, as many of my particular friends are concerned in it ...'[43]

Ewer, always quick to criticise, commented in the same correspondence that '... I hear Sir George Leith [Lieutenant-Governor of Penang, 1800–1804] is going to build a fort, a church, & a palace. As for the first, I wish he had mine at a quarter of what it cost, & I wou'd do without it ...'[44] The fort he was referring to was, of course, the fort that was built on Prince of Wales Island (Penang) and was later known as Fort Cornwallis.

The military officers were gradually being replaced by officers of the Bengal Army as the local commissions expired, although four local commissions were approved in 1800 they were all in respect of locally enlisted officers (see Annexure A). The officers serving on the Fort Marlborough establishment at the turn of the century were:

Artillery

Captain H. Douglas	– Acting Commandant
Lieutenant G. F. Parker	
Lieutenant D. Maclean	
Lieutenant F. Beagham	
Lieutenant Charles Hutchinson	– Adjutant to the Garrison
Lieutenant R. Best	– Quartermaster to the Garrison

Infantry

Lieutenants	Thomas Brown,	Hiram Cox,	George Barclay,
	Ambrose Brenan,	Robert Graham,	
	Daniel Gallies		

Bengal Engineers

Captain F. H. Pearson.

Medical Establishment

Surgeons	John Lumsdain,	Batholemew Hartley,
	John Spratt,	Thomas Trail,
	James Powell,	Charles Campbell,
	John Bond,	John Kennedy.

Ordnance Department

Conductor J. R. Fulford.[45]

Ewer was obviously not a military man and his correspondence with Dundas shows his extreme dislike at having to spend any of the money allocated to him on defence. In a further letter to Dundas, dated 20 January 1801, he remarks on the cost of the Garrison at Fort Marlborough as follows:

'... Charges of garrison Marlbro' [*sic*] amount to ps 202,123. I have deducted the European infantry & artillery, which are a great expense, & no use. I shall, however, want a few artillery men of some sort or other. The GollAndazes [*sic*] wou'd do very well. There is no danger from them here, as is apprehended in Bengal, for they will always be consider'd as aliens by the Malays & detested as such.

I have deducted one half of the [cost of the] Sepoys. This subject, however, requires some consideration. I do not believe any foreign enemy will attack the place, especially as they wou'd get more profit by taking an Indiaman in the roads, full loaded for Europe, & this wou'd be a much easier business ...

When the fort is finished, which will be in two or three months, I don't see the use of an engineer. I am told that the tower of the fort, which is old, is cracked, & the rest will never stand. Shou'd it be overturned by an earthquake, or fall of itself, I shall not build it up again, for it is useless. A common bricklayer will answer the purpose, but, shou'd it be thought proper to have an engineer, one half of the salary will be sufficient for the person I have appointed, who is a young officer. He knows but little of the business, 'tis true, but his predecessor did not know much more ...'[46]

The barracks at this time were outside the fort at Mount Argyle and had been built at considerable expense. A Colonel's house was also built in the area as it was reckoned to be healthier than the area near to the fort. Ewer's correspondence to Dundas also contained criticism of the military and in his usual carping manner he commented '... no air will mend such a miserable drunken crew & the sooner the climate carries them off the better. The Sepoys live in huts of Bamboo. I see no need of a Barrack Master for them ...'[47] To have such a man acting as the 'Commander-in-Chief' must have made the position of the

Commandant almost impossible. The Commandant of the Fort Marlborough Garrison at that time was Lieutenant Colonel Thomas William Clayton (18th Regiment, Bengal Native Infantry).[48]

In 1802 an Act of Parliament was passed which enabled the Honorable East India Company to make their settlement at Fort Marlborough '. . . a factory subordinate to the presidency of Fort William in Bengal, and to transfer the servants who on the reduction of that establishment shall be supernumerary, to the presidency of Fort St George . . .'[49] As a result of this change the military establishment of the garrison was reduced and Lieutenant Colonel T. W. Clayton and his detachment of sepoys returned to Bengal.

During the latter part of 1804 the brother of the Sultan of Moco Moco, and two other chiefs raised a small force and attacked the district of Ipu, which was under the control of the Company. The small detachment of the Fort Marlborough Corps that was on duty were driven off and as a result a small force under the command of Lieutenant Hastings Dare, 7th Regiment, Bengal Native Infantry,[50] comprising eighty-three sepoy officers and men, five lascars, twenty-two Bengal convicts and eighteen soldiers of the Bugis-guard, set out from Fort Marlborough on 22 November.[51] The force reached Ipu on 3 December and found that the enemy had retired to a kampong, Koto Tuggoh, which was located near to the coast. The Company's force set out in pursuit and on 17 December a number of the soldiers were wounded as the path had been thickly spread with ranjaus.[52]

The force proceeded at a slow pace due to adverse weather conditions and Dare records on 22 December:

> '. . . We had much difficulty in getting the mortar and its bed down, being obliged to make use of long thick rattans, tied to them, and successively to several trees [the party were, at this time attempting to descend a steep hill]. It was realy [*sic*] admirable to observe the patience of the sepoys and Bengal convicts on this occasion. On mustering the coolies, found that nearly one half had run during the night, which obliged us to fling away twenty bags of rice besides salt and other articles. Our course lay N crossing the river several times. My poor faithful dog Gruff was carried away by the violence of the steam and lost. We were obliged to make bridges, by cutting down tall trees, laying them across the stream, and interlacing them with rattans . . .'[53]

On 23 December the report noted that the sepoy officer and many of the sepoys were ill of '. . . fluxes and fevers, and lame with swelled and sore feet . . .' During a severe storm on the following day much of the powder was damaged and on the 25th, Christmas Day, Marsden records: '. . . Our wine and liquors nearly expended, and we have but one miserable half-starved chicken left, although we have been on short allowance the whole way . . .'

However things improved, as on the 26th the party learned that the enemy were about two to three hundred strong and were armed with muskets, blunderbusses and '. . . wall-pieces . . .' The weather was still bad and on the 29th another of the coolies died and at this stage it was necessary to abandon some of the shells for the mortar.

Eventually on 3 January 1805 the small force met up with the enemy near Koto Tuggoh and came under fire from the enemy who had taken up position on the heights above the track. The force attacked the enemy and drove them back but in doing so lost one sepoy killed and a number of others injured by ranjaus. The force continued to pursue the enemy and it was recorded that:

> '. . . the roads shockingly bad, and we were obliged to leave behind several coolies, and two sepoys, who were unable to accompany us. 4th (January) Obliged to fling away the bullets of the cartridges, three-fourths of which were damaged, and other articles. Most

Fort Marlborough, main gate. Photograph by E. Edwards McKinnon, 1982.

Fort Marlborough, inner gate. Photograph by E. Edwards McKinnon, 1982.

**Fort Marlborough: approach to the south-west bastion.
Photograph by E. Edwards McKinnon, 1982.**

Fort Marlborough, north gate leading to Padri's Battery. The name 'Padri' is from a community of religious fanatics, Wahabite puritans, who had been converted during a pilgrimage to Mecca and who returned to Pedir in Sumatra. The name Pediri was corrupted to 'Padri' although the connection between the battery and the religious community is not known. (*Journal of Malayan Branch, Royal Asiatic Society,* Vol XIX, Part I, 1941. Article 'More on Bencoolen' by R.J. Wilkinson.

Above:
Fort Marlborough: three East India Company graves. Photograph by E. Edwards McKinnon, 1982

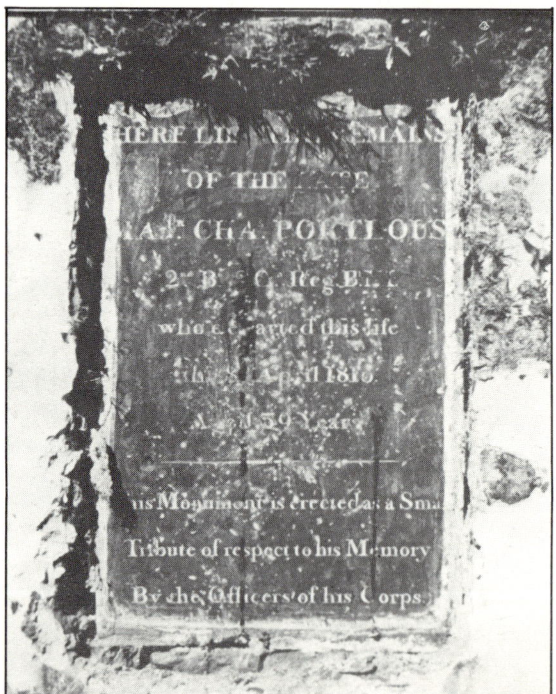

Right:
Memorial to Major Charles Porteous, 20th Regiment of Bengal Native Infantry, in the Bencoolen Cemetery.

Two views of the fort showing the main gate and the moat.
Photographs by E. Edwards McKinnon, 1982.

Fort Marlborough: looking west. Rampart has been restored but not to its original state, 1982.

Fort Marlborough: approach to the west bastion, 1982.
Photographs by E. Edwards McKinnon, 1982.

of the detachment sick with fluxes and fevers, or wounded in the feet . . .'[54]
The enemy were again encountered, this time near a lake, where they had built four small earthworks. The Company's force, attacked and drove the enemy off, killing a number and suffering four sepoys wounded and more received minor wounds to the feet due to the enemy continuing to use ranjaus. Due to the large number of the force suffering from wounds and sickness it was necessary to withdraw to Rantau Kramas and in order to do so the mortar bed and some shells were abandoned. The camp was reached on 6 January but it was then decided that the force would need to proceed to Moco Moco so as to obtain stores and reinforcements. The party eventually arrived at Moco Moco on 22 January and were able to rest in huts that were prepared under the orders of the Resident, Mr Russell. At Moco Moco the military had established a small fort named Fort Ann (also known as Fort Anna). The description given by Marsden is as follows:

> '. . . Fort Ann lies on the southern and the settlement on the northern side of the Si Luggan river, which name belongs properly to the place also, and that of Moco-moco [*sic*] to a small village higher up. The bazar consists of about one hundred houses, all full of children. All the northern end is the sultan's which has nothing particular to distinguish it, but only its being larger than other Malay houses . . .'[55]

The detachment rested until 9 February and then a party consisting of Lieutenant Hastings Dare, Mr Alexander, surgeon, seventy sepoys, including officers, twenty-seven lascars and eleven soldiers of the Bugis-guard and a number of Bengal convicts, set out to make a show of force to the Malay kampongs that had helped the enemy. The small force met up with some of the enemy on 26 February and although the enemy had felled a number of trees to block the path the sepoys attacked and were driven back by the enemy and also by the large number of ranjaus that had been planted along the only way forward. The Company's force split, with a subedar and thirty sepoys, together with the Bugis-guard, moving around a swamp to the right, the whole force on a given signal attacked and captured the enemy positions which were held, it was thought, by some three to four hundred men. Two sepoys were killed and seven were wounded by weapons other than ranjaus. On 27 February the force advanced to Koto Tuggoh and after firing one shot from the mortar (on this occasion a smaller mortar had been brought with the force) and a few musket shots the kampong inhabitants fled. The kampong was then destroyed and the force proceeded to Koto Bharu where negotiations were held and '. . . The chiefs met as desired, and after a short conversation agreed to all that we proposed. Papers were thereupon drawn up, and signed and sworn to under the British colours . . .'[56] The force then withdrew to Moco Moco and on to Bencoolen where the party arrived on 30 March. It was recorded that it was:

> '. . . one of the most fatiguing and harassing expeditions any detachment of troops ever served upon; attended with the sickness of the whole of the party, and the death of many, particularly of Mr Alexander, the surgeon . . .'[57]

The high rate of casualties was occasioned by the fact that the expedition had been sent into the jungle during the height of the rainy season. After the return of the expedition Lieutenant Hastings Dare was commended for his conduct during the small campaign and Governor Ewer could have had little to complain about on this occasion when the military had acquitted themselves admirably.

In July 1805 a small military force had to be despatched to the region of Tapanuli where one chief had attacked a kampong under Company protection and had killed some of the inhabitants and carried others off into slavery. The force was a mixed Sepoy and 'locally recruited' Malay party. The party was accompanied by Mr W. H. Hayes. When the party were in the area one section that was under command of a sergeant was attacked and one sepoy was killed. Because of the difficult circumstances the section had to abandon the body of the sepoy. The report of the succeeding action has been recorded as follows:

'... The village of Singapollam being immediately carried by storm, and the enemy retreating by one gate, as our people entered at the opposite, the accoutrements of the sepoy, who had been killed the day before, were seen hanging as trophies in the front of the houses, and in the town-hall. Mr Hayes saw the head entirely scalped, and one of the fingers fixed upon a fork or scewer, still warm from the fire. On proceeding to the village of Labusucom, situated little more than two hundred yards from the former, he found a large plantain leaf full of human flesh, mixed with lime-juice and chili pepper, from which he inferred that they had been surprised in the very act of feasting on the sepoy, whose body had been divided between the two kampongs ...'[58]

Peace negotiations were concluded and the party returned to Fort Malborough without being able to effectively punish the local Malay chiefs who had, after the peace treaty had been concluded, admitted that cannibalism was their custom.

Ewer handed over the Residency in 1805 to Thomas Parr who was appointed Resident on 27 September of that year. He remained in office until the evening of 23 December 1807 when he was stabbed to death by the local Malays whilst at his garden house at Mount Felix, three miles along the seashore from Fort Marlborough.[59] Parr had caused dissention in Bencoolen as he had:

'... offended Indonesian opinion at Benkulen by enforcing on the people the cultivation of coffee, by attempting to disband the Bugis Corps, and by interfering in long standing judicial practices at Fort Marlborough ...'[60]

It was generally believed that some of the local chiefs had been implicated in the crime but it was thought best to leave them alone and in order to show that the incident could not pass without some retribution '... the dipati of Sukarami and several other minor chiefs who were under suspicion were blown from the guns, and a few villages burned by the British in vengeance ...'[61]

About 150 yards from Fort Marlborough stands the Parr Mausoleum which was erected by '... the Supreme government ...' and M. A. Jaspan, writing in 1974 gives the details of the inscription as follows:

'... Here are deposited the Remains of Thomas Parr Esquire
Resident and Representative of the Honourable E.I. Cy.
at Fort Marlborough in Bencoolen who although a benevo-
lent Father to the Malay Inhabitants and solicitous to
improve their Freedom and Prosperity in the prudent and
gradual Introduction of Spontaneous ...
fell with ... ence of the ... and ... ivalled
atrocity under the misguided and barbarous Fury of a Band
of Assassins in the Night of the 27th of December A.D.
1807 in the ... th Year of his Age ...'[62]

and the inscription of course belies Parr's true temperament.

Shortly after Parr's death W. B. Martin, who assumed the duties of Resident at Fort Marlborough, recorded that he considered that the 'Bugis Corps' should be maintained. He is recorded as writing:

'... Important advantages might be derived from the services of a small regulated and efficient Corps of this description on emergencies when it would be neither practicable nor expedient to detail a party of regular Sepoys into the Country. The Officers of this Corps are men of the first consideration and influence in Bencoolen whose families for a long series of years have been distinguished by their attachment and fidelity to the Company's Government . . .'[63]

The Bugis Corps had been recruited from the local Bugis and had been used to supplement the garrison forces stationed on the west coast of Sumatra. The Bugis under Daing Marupa, had first settled at Indrapura, on the west coast, where they established an important influence in that area and then a party of one hundred and twelve Bugis, under Daing Mabela, the son of Daing Marupa, settled in Benkulen where they formed into a military corps under the orders of the Governor. The Resident, W. B. Martin, also considered it would be advantageous

'. . . to raise a Corps of 50 Bugguesses, 25 of whom are stationed near the former village of Sookooramee and the remainder at Sillebar. But in order to obviate the inconvenience which resulted from the mode in which the Corps was formerly maintained and paid and to ensure a greater degree of future regularity and discipline, the Resident has considered it to be expedient to place it on the same establishment with the Sebundy Corps and to direct that all reports . . . be made through the Adjutant to Government. Under this arrangement the Bugguesses will receive the pay of P4 p.m. instead of P3.2 of which one dollar is to be deducted untill the expense of cloathing [*sic*] which will be provided by the Company shall have been defrayed, and they will receive their rice from the Stores on the same terms as the Sepoys, viz: 12 Bamboes per dollar. The Adjutant will also draw the pay and allowances of the Corps agreeably to the Abstract is to be recorded . . .

The difficulty in enlisting Bugguesses which was formerly experienced and which necessitated the Commandant of the Corps to engage the services of his own private debtors will be obviated by this arrangement which renders their pay adequate to the provision of a comfortable subsistence and will consequently give them an interest in the service, which was never before felt. A weekly report of this Corps is to be made to the Adjutant in order to enable him to make his report to the Resident and a monthly Survey Report is to be made of the Arms and Accoutrements & is to be taken either by the Adjutant or by a careful European non-commissioned officer . . .'[64]

The situation changed little over the next eleven years with the small garrison being supported by the Fort Marlborough Bugis Corps and with the fortress being the main garrison on the west coast, although the fort at Moco Moco, Fort Anna, was still maintained. The next major change was in 1818 when Sir Thomas Stamford Raffles was appointed Lieutenant Governor of the West Coast of Sumatra on 20 March of that year. In April, Raffles wrote that he was making a number of changes in order to save unnecessary expenditure and as part of his cuts he intended to:

'. . . disband the bodyguard of the Resident from the date of my arrival. This corps was maintained on the footing of the Company's cavalry in Bengal, at an annual expense to the Company of not less than 10,000 rupees . . .'[65]

Later in the year he made further cuts. The European establishment at Laye was discontinued from 15 June and the Residents at Saloomah, Manna and Cawoor were told to close their accounts and report to Fort Marlborough by 1 August. The Residency at Moco Moco was also abolished from the same date and with the closing of these residencies Raffles was able to '. . . reduce the whole of the local establishment of irregular Bengal and Bugguesse troops maintained at each station, and all the duties that may hereafter be required from the latter description of troops will be performed by the Marlbro Bugguese corps . . .'[66]

Thus the Fort Marlborough establishment was reduced further and was by this time little more than a force capable of policing duties rather than a defence force. In September 1820 Raffles wrote to Bengal stating that he was intending to carry out a survey of the actual numbers and conditions of the people living in and around Bencoolen and on his staff, acting as a second assistant, was Captain William Gordon Mackenzie of the 2nd Bengal European Regiment.[67]

The situation in the garrison changed little between 1820 and 1825 when the Company surrendered its interests on the west coast of Sumatra and in exchange received the Dutch port of Malacca on the west coast of Malaya as part of the Anglo-Dutch Agreement of 1824.[68]

There were few major changes in the garrison over the last ten years of British rule. The garrison force had been increased in 1815 when the establishment was increased to include a detachment of troops from the 20th Regiment, Bengal Native Infantry. The senior officer of the 20th Regiment, which was a detachment from the main regiment based on Prince of Wales Island, was Captain Charles Porteous. The other officers serving at that time with the Fort Marlborough garrison were:

Lieutenant Charles R. Skardon
Lieutenant William L. Truman
Lieutenant Charles L. Walker
Lieutenant Thomas Haslam.

During the following year, 1816, the detachment was increased by one additional officer, Lieutenant John Gordon, and Porteous had been promoted to the rank of major. Unfortunately Major Charles Porteous was taken ill and died on 8 April 1816 and was buried in the Christian cemetery at Bencoolen.[69]

Although at the turn of the century (1799–1800) the fort was reported to be in a bad state of repair the fort continued to be used by the Dutch after the hand-over in 1825 and in more recent years it has been used by the Indonesian Army. The fort is now known locally as 'Benteng Raffles' and as can been seen from the illustrations is still a formidable building although the three-storeyed tower has disappeared. The other extant memorial relating to the 111 years of the Company's rule on the west coast of Sumatra is the 'Hamilton Obelisk' which is depicted on the sketch by Andrews dated *c*1794–1798 and engraved by Joseph C. Stadler. The obelisk was constructed under the direction of Captain John Macdonald the Military Engineer at Fort Marlborough (1794) and is about twelve feet high and six feet wide. The photograph in the article by Brooks in *The Journal of the Straits Branch of the Royal Asiatic Society*, of 1918,[70] shows the memorial without pillars and chains, but a photograph taken by M. A. Jaspan, which was published in 1974 shows small pillars and chains. Captain Robert Hamilton who was Captain Commandant of the Garrison and second Member of government died on 15 December 1793 as a result of an abcess of the liver.[71]

An examination of the burial records of the settlement (which are held in the India Office Library and Records, and are not complete) gives a good indication of the high number of fatalities among those European soldiers who served on the west coast of Sumatra. It has been possible to trace 497 military deaths during the period 1712 and 1825.

Away from Bencoolen and Fort Marlborough there is little remaining of the old fortifications. A recent visit to Moco Moco by Mr E. Edwards McKinnon revealed the following.

Rat Island and King's Cliff showing the Hamilton Monument. Sketch by Andrews, 1794–1798. Courtesy of the Director, National Museum, Singapore.

Two views of the memorial to Captain Robert Hamilton, Bencoolen. Photograph by E. Edwards McKinnon, 1982.

Fort Anna, Sumatra.

Above: Views of the wall on the river side of the fortification.

Right: Remains of the fort covered with vegetation.

Photographs by E. Edwards McKinnon, 1982.

Fort Anna, Sumatra, cannon found near the school at Moko-moko.
Photograph by E. Edwards McKinnon, 1982.

The English Fortification at Linau, Bengkulu Selatan.

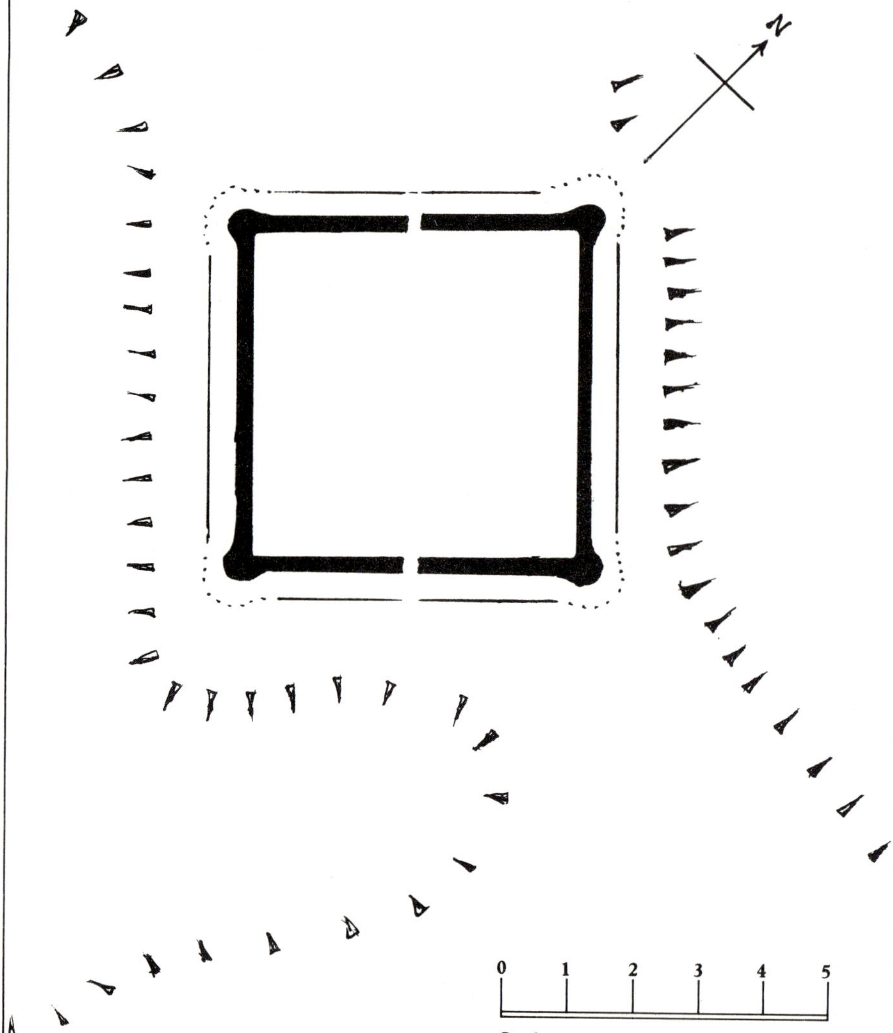

The remaining height of the earthwork is about 1.5 m. The site is situated north of Air Linau, some 200 metres south-east of the Linau, Bintuan Road.

'. . . Fort Anna is, alas no more, The site on the river bank immediately south of Muko Muko [present day spelling of the area] is reminiscent of the aftermath of Verdun. Hardly one brick stands upon another. Rubble litters the trenches that were once the alignments of the walls. Apart from three small sections of the riverside wall (one of which stands to its original height) and a small section of a casemate on the opposite side . . .'[72]

Nearby, at the school, are four iron cannon. The markings on one of these iron cannon show a crown above a capital 'P' plus a date '24.1.?8' and with the intials 'MR' and an emblem. The one other ruin that has been found by Mr Edwards McKinnon, was located at Linau, south of Bintuan. The ruin is of a small square fortification and when visited in August 1982 the remaining height of the earthwork was about 1.5 metres and was situated north of Air Linau, some 200 metres south-east of the Linau–Bintuan Road.

The following tabulation shows the year and appointments, etc of the Europeans of the military establishment who died mainly at Bencoolen and who were buried in the Bencoolen Christian Cemetery, known locally as the Kuburan Inggeris.

																			Dependants		
	Major	Captain	Lieutenant	Lieutenant Fireworker	Ensign	Surgeon	Assistant Surgeon	Sergeant Major	Conductor	Sergeant	Corporal	Private/Centinell	Gunner	Matross	Storekeeper	Drum Major	Drummer	Fifer	Wife	Children	Total
1712													1[1]								1
1713					3								2					2			7
No record for period 1714–1753																					—
1754		1									1	3	3								8
1755												3									3
No record for period 1756–1758																					—
1759												16									16
No record for period 1760–1764																					—
1765			1	1	1					1	2	37		13							56
1766												2									2
No record for period 1767–1778																					—
1779				1							1	10		3			2				17
1780				1		1				6	3	13		8							32
1781		1						1			4	5		5				2			18
1782	No recorded deaths																				—
1783												2									2
1784		1								2	2	4[2]	6	4							19
1785			1							1	3	5	1	4		1					16
1786				1	1	1				2	1		1	3	1				1		12
1787		1	1			1					1	3	1	9[3]				1[4]		4	22
1788							1			1				2[5]			1	1		1	7
1789					1	1		1				4	2	9						1	19
1790		1	2			1							1	3							8
1791										2	2	13	3	4							24
1792						1				1	3	1		2			2				10

	Major	Captain	Lieutenant	Lieutenant Fireworker	Ensign	Surgeon	Assistant Surgeon	Sergeant Major	Conductor	Sergeant	Corporal	Private/Centinell	Gunner	Matross	Storekeeper	Drum Major	Drummer	Fifer	Dependants: Wife	Dependants: Children	Total
1793		1				1						1	2	4							9
1794					1					3	1	6	5								16
1795				1		1		1		1	3	19	9	7			1				43
1796												4	1						1	1	7
1797									1	2	2	8		3							16
1798						1				1											2
1799				1			1	1			1	6								1	11
1800		1								2		9	2							2	16
1801									3	2	1	1									7
1802										1	1										2
1803		1									1										2
1804						1															1
1805						1				1		2								2	6
1806	No recorded deaths																				—
1807			1							1											2
1808			1		1[6]					1			1								4
1809													2								2
1810												1	2								3
1811										1		1[2]	1	1							4
1812														5							5
1813									1												1
1814		1																			1
1815										3	1			6					1		11
1816	1													1							2
1817								1						2							3
1818	No recorded deaths																				—
1819	No recorded deaths																				—
1820	No recorded deaths																				—
1821							1	1			1		1				1				5
1822						1	1		1				2	3						1	9
1823																			2	2	4
1824			1						1		1										3
1825																			1		1
Totals	1	9	8	6	8	12	4	6	7	35	36	179	49	101	1	1	6	7	6	15	497

NOTES

1. Listed as a Gunner's Mate.
2. Includes 1 invalid Soldier.
3. Includes 2 invalid Matross.
4. Includes 1 invalid Fifer.
5. Includes 1 invalid Matross.
6. Cornet of the Native Infantry, Madras Establishment.

There is no record of any deaths between the years 1760–1764 although there are accounts of a number of deaths of 'Europeans' during that period. Wright and Read in their work *Malay Peninsula* give a short account of the French occupation which occurred at this time. The details given are:

'. . . Disease appeared amongst the Frenchmen, and worked such frightful havoc in their ranks that after a lapse of a few months D'Estaing was glad to re-embark the remnants of his force and allow the Settlement to revert to the British. An expedition under the command of Captain Vincent, sent in 1761 to restore British authority, found the place a veritable charnel-house owing to the half-buried remains of the invaders which were poisoning the air in every direction. The terrible conditions caused excessive sickness amongst the occupying [British] force. Of 183 men landed there were, four months later, only 57 effectives . . .'

There were of course an equal number of deaths amongst the civil servants but unfortunately due to a lack of records it is impossible to ascertain the ranks and trades of the servicemen that died.

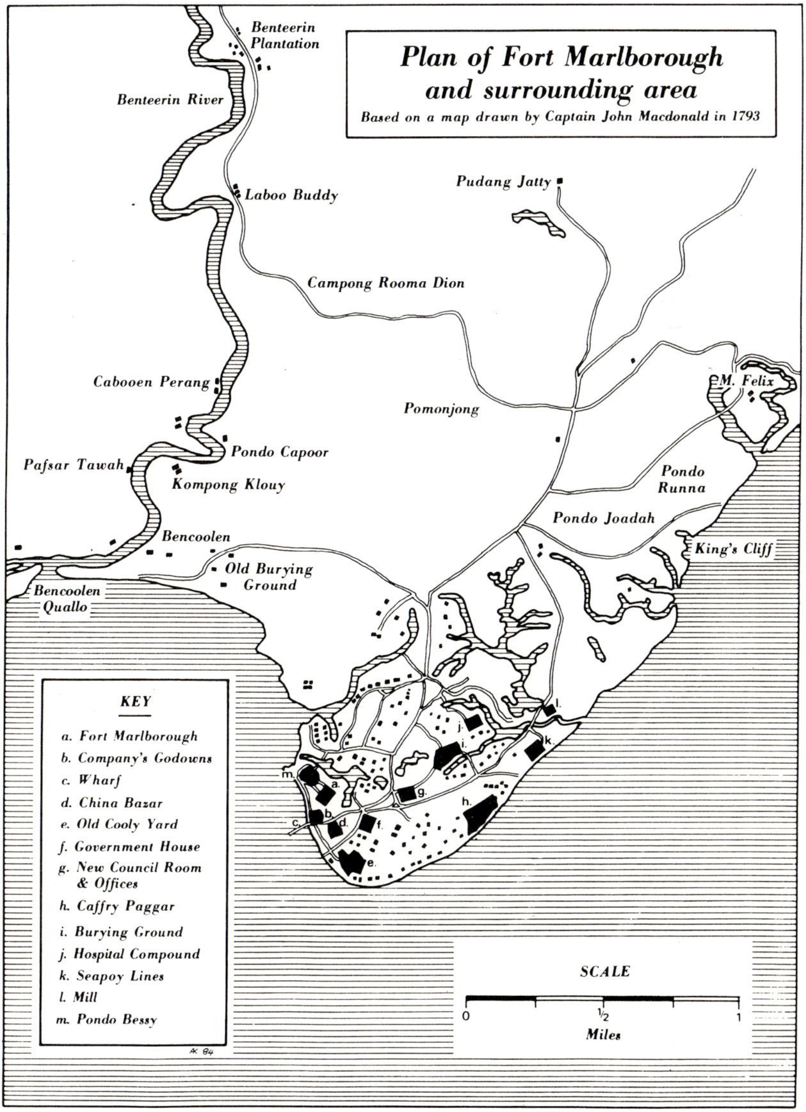

ANNEXURE A

List of Commissions granted for the Fort Marlborough Local Corps 1769 to 1800

Appointment	*Granted to*	*Date*
Ensign of Infantry	John Williamson	28 Jan 1769
Capt of the Train of Arty	John Allison	7 Nov 1769
Capt/Lieut of the Train of Artillery	James Murray	7 Nov 1769
Lieut Fireworker	Francis Isoot	9 Nov 1769
2 Lieut of the Train of Artillery	William Lyndsay	30 Nov 1769
Lieutenant of Infantry	John Bourne	26 Nov 1769
Adjutant for all the Forces	Thomas Baddaley	30 Nov 1769
Ensign of Infantry	John Williamson	27 Nov 1769
(this was confirmation of the commission granted on 28 Jan 1769)		
Ensign of Infantry	Trevor Blunden	10 Dec 1770
Ensign of Infantry	Richard Whitfield Chinnery	20 Dec 1770
Lieut Fireworker	George King	21 Dec 1770
Ensign of Infantry	Thomas Baddaley	22 Dec 1770
Lieut Fireworker	Williame [*sic*] Maitland	13 Mar 1771
Ensign of Foot	William Phillips	5 Aug 1771
Capt/Lieut of the Train of Artillery	Thomas Oldham	19 Oct 1772
Lieutenant	George King	9 Oct 1772
Lieutenant	William Maitland	9 Oct 1772
Lieutenant	Trevor Blunden	3 Mar 1773
Captain/Lieutenant	Sam Reynolds	5 Jan 1774
Lieutenant	Thomas Baddaley	5 Jan 1774
Lieutenant	William Phillips	5 Jan 1774
Lieut Fireworker	Philip Ogilvie	9 Feb 1774
Lieut Fireworker	William Richardson	2 Sep 1774
Ensign	Andrew Dingwell	3 Sep 1774
Lieut Fireworker	Richard Alexander Powell	5 Sep 1774
Ensign	John Swann	6 Sep 1774
Captain of Bugguesses	Datoo Nodeen	31 Oct 1774
Lieut of Bugguesses	Raddeen Temoongoong	31 Oct 1774
Ensign of Foot	Thomas Vetch	23 Dec 1775
Lieutenant of Infantry	Charles Cooke	30 Mar 1776
Lieutenant of Foot	William Cave Dickson	4 Jul 1776
Ensign of Foot	Soloman Smithson	30 Dec 1776
Lieutenant of Foot	Medhop Blunden	– Feb 1777
Captain/Lieut of Infantry	Thomas Baddaley	1 Jun 1777
Lieut of Artillery	Brevet Capt John Townsend	1 Aug 1777
Ensign of Infantry	Matthew Jenour	1 Aug 1777
Adjutant of Infantry	John Townsend	1 Oct 1777
Ensign of Foot	George Greenwallers	8 Dec 1777
Ensign of Foot	William Rogers	6 Jan 1778
Ensign of Foot	Luke Rogers	2 Mar 1778
Ensign of Foot	William Sanderson	3 Mar 1778
Lieut Fireworker	William Harris	1 Nov 1778
Captain/Lieut of Infantry	Thomas Chandler	20 Jan 1779*
Lieutenant of Infantry	Soloman Smithson	20 Jan 1779*
Lieutenant of Infantry	Mathew Jenour [*sic*]	20 Jan 1779*
(Note – *appointments were effective from 1 June 1778)		
Ensign (from 1 Jan 1779)	Richard Vaughan	20 Jan 1779
Major of Infantry	John Hutcheon	3 Jan 1779
Ensign	Thomas Henry Dunning	5 Jun 1779
Ensign	William Robson	5 Jun 1779
Lieut Fireworker	John Campbell	5 Jun 1779

Appointment	*Granted to*	*Date*
Ensign of Arabs	Ishmail	1 Aug 1779
Lieut Fireworker	John Young	1 Aug 1779
Captain of Bugguesses	Dion Maroopah	1 Jan 1779*
Lieutenant of Bugguesses	Raddeen Temongong [*sic*]	1 Jan 1779*
Ensign of Bugguesses	Raddeen Pawong Illum	1 Jan 1779*

(Note – *the three commissions granted to the three Bugguesse officers were granted during 1779 and were backdated to 1 January of that year)

Jemidar (from 1 Dec 1779)	Pilligram Pandy	10 Jan 1780
Captain of Infantry	James Vetch	1 Aug 1780
Captain/Lieutenant of Infantry	Medhop Blunden	1 Aug 1780
Lieutenant Adjutant	Matthew Jenour	1 Aug 1780
Lieutenant of Infantry	William Robinson (Ensign)	1 Aug 1780
2 Lieut of Artillery	Lieut (Fireworker) John Young	1 Aug 1780
Lieut Fireworker	James Rondaldson	1 Aug 1780
Quartermaster of Infantry	Isaac Byers	1 Aug 1780
Captain of Infantry	William Sibthorpe	21 Aug 1780
Captain of Artillery	Thomas Pickering	21 Aug 1780
Captain of Artillery	Frederick Mandeville	21 Aug 1780
Subadah [*sic*]	Marza Mahomet Beg	28 Aug 1780
Jemidar	Mahomet Saib	28 Aug 1780
Ensign of Arabs	See Seine	6 Sep 1780
Captain of Infantry	Frederick Marsden	30 Nov 1780
Lieutenant of Infantry	Isaac Byers	30 Nov 1780
Adjutant of Infantry	Richard Vaughan	1 Dec 1780
First Lieut of Artillery	John Young	28 Mar 1781
2 Lieut of Artillery	John Campbell	28 Mar 1781
Lieutenant of Infantry	Joseph Jefferson	21 Apr 1781
Lieutenant of Infantry	Henry Dunning	28 Apr 1781
Lieut Fireworker	William West	24 Jul 1781
Ensign	William Counsell	24 Jul 1781
Ensign of Infantry	George Carpenter	1 Nov 1781
2 Lieut of Artillery	William Westlake	1 Nov 1781
Lieutenant of Infantry	William Counsel	1 Nov 1781
Ensign of Infantry	Robert Best	1 Nov 1781
Ensign of Infantry	John Leach	1 Nov 1781
Ensign of Infantry	William Beckett	1 Nov 1781
Lieut Fireworker	Samuel Dickens	1 Nov 1781
Lieut Fireworker	George Prosar	1 Nov 1781
Subedar [*sic*]	Ismail	7 Nov 1781
Captain Engineer	Elisha Trapaud	17 Dec 1781
Subaltern – Engineer	Francis Pearson	17 Dec 1781
Lieut Fireworker	Henry Lewis	17 Dec 1781
Lieutenant of Artillery	John Campbell	6 Mar 1782
2 Lieut of Artillery	William West	6 Mar 1782
Ensign	John Veitch	6 Mar 1782
Major in the Corps of Militia of Fort Marlborough	John Marsden	4 Apr 1782
Captain in the Corps of Militia	Lawrence Bogle	4 Apr 1782
Captain in the Corps of Militia	Robert Nichols	25 Apr 1782
Lieutenant, Corps of Militia	Robert Lockhart	24 Apr 1782
Lieutenant, Corps of Militia	Henry Dun	25 Apr 1782
Captain of Infantry	Medhop Blunden	21 May 1782
Lieutenant of Infantry	Richard Vaugh	21 May 1782
Lieutenant of Infantry	George Carpenter	21 May 1782
Lieut Fireworker	Henry Blyth Mann Simpson	20 Jul 1782
Surgeon of the Troops	Charles Wasmus	1 Feb 1783

Appointment	*Granted to*	*Date*
Ensign	James McKillup	13 Jan 1783
Subedah [*sic*]	Mahomed Sahib	17 Apr 1783
Lieutenant of Infantry	John Leach	12 May 1783
Major of Artillery	Thomas Oldham	2 May 1783
Lieutenant of Infantry	William Becket	23 Jul 1783
Capt/Lieut of Artillery	John Campbell	1 Jun 1783
First Lieut of Artillery	William West	1 Jun 1783
2 Lieut Artillery	Samuel Dickin	2 Jun 1783
Lieut Fireworker	Charles Hamilton	8 Dec 1783
1st Lieut of Artillery	Samuel Dickin	10 Dec 1783
2 Lieut Artillery	George Rose	6 Jan 1784
2 Lieut Artillery	Henry Lewis	1 Feb 1784
Lieut Fireworker	Daniel McLane	1 Feb 1784
Lieutenant of Infantry	John Veitch	5 Apr 1784
Adjutant	George Carpenter	5 Apr 1784
Quartermaster	William Beckett	5 Apr 1784
Ensign of Infantry	Charles Hamilton	5 Apr 1784
Ensign of Infantry	John Moore	5 Sep 1784
Ensign of Infantry	John Daly	6 Sep 1784
Lieut Fireworker	Campbell Blyth	6 Sep 1784
Major by Brevet	William Phillips	6 Sep 1784
Captain of Infantry	Isaac Byers	14 Jun 1784
Captain by Brevet	George Carpenter	15 Jun 1784
Captain Engineers	F. H. Pearson	16 Jun 1784
Capt of Infantry by Brevet	William Beckett	17 Jun 1784
First Lieut of Artillery	Henry Lewis	4 Feb 1785
2 Lieut of Artillery	Daniel MacLean	4 Feb 1785
Subedah [*sic*]	Abdul Kadur	4 Feb 1785
Exchanged into Infantry	Samuel Dickin	4 Feb 1785
Lieutenant of Infantry	John Baines	4 Feb 1785
Lieutenant by Brevet	Charles Hamilton	4 Feb 1785
Lieutenant Fireworker	Robert Lloyd O'Brien	19 Mar 1785
Lieutenant of Infantry	Charles Hamilton	3 Feb 1785
Ensign of Infantry	John Chandler	19 Aug 1786
Ensign of Infantry	Samuel Alexander	20 Aug 1786
Lieut Fireworker	Robert Best	21 Aug 1786
Temporary Captain of Infantry	Charles Wasmus	8 Nov 1786
Captain of Infantry	Charles Wasmus	10 Jan 1787
Brevet Captain of Infantry	Thomas Brown	31 Dec 1793
Ensign of Infantry	William Stephenson	14 Feb 1794
Lieutenant of Bugguesses	Radin Naka	1 Jan 1796
Ensign of Bugguesses	Radin Mahomet	1 Jan 1796
Ensign of Bugguesses	Dion Usoph [or Usolph]	1 Jan 1796
Lieutenant of Bugguesses	Dion Hindra	1 Aug 1800
Lieutenant of Bugguesses	Diang Usolph	2 Aug 1800
Ensign of the Arab Guard	See Adjam	1 Aug 1800
Lieutenant of Bugguesses	See Iammee	1 Aug 1800

From an examination of the commissions it is apparent that when an officer was moved from one post to another he was given a new commission, the commissions were, therefore, tied to the appointment rather than to the officer.

Appointments shown are as given on the document.

ANNEXURE B

Record of Military Burials on the West Coast of Sumatra

The following list gives details of the burials of military personnel who died during their service on the west coast of Sumatra. The list has been compiled from various documents originating in Sumatra contemporary with the occurrence, and from the copies of the burial register which, unfortunately, has a number of gaps due to the loss of some of the 'returns' that were despatched from Bencoolen.

Name	*Rating*	*Date of burial*	*Cause of death*
Bantal			
William Mathews	Gunners Mate	22 Aug 1712	
Augistine Derosario	Topass Centinell	6 Feb 1713	
Alexander Gamble	Ensign	13 Mar 1713	
Stephen Olliver	Ensign	2 May 1713	
John Sone	Gunner	6 May 1713	
James Horsly	Gunners Mate	21 May 1713	
William Newton	Europe Centinell	7 Aug 1713	
Charles Wickstead	Ensigne [*sic*]	13 Sep 1713	
Fort Marlborough			
William Jackson	Centinel	8 Aug 1754	Dropsy
Peter Farrell	Gunner in the Train	12 Aug 1754	Fever
William Bullmore	Soldier	28 Aug 1754	Fever
James Paul Morrissett	Capt of the Military	5 Oct 1754	Gout
William Hampton	Gunner in the Train	9 Oct 1754	Fever
James Johnson	Gunner's Servant	9 Oct 1754	
Joseph Mullins	Corpl	19 Nov 1754	Fever
John George	Soldier	17 Dec 1754	Flux
Mersa Merleerah	Soldier	12 Jan 1755	Dropsy
Francis D. Mount	Topass Centy	27 Jan 1755	Fever
Peter Brew	Soldier	27 Jan 1755	Flux
John Berry	Soldier	23 Apr 1759	
John Porter	Soldier	27 May 1759	
John Kidd	Soldier	14 Jun 1759	
Robert Baxter	Soldier	24 Jun 1759	
George Nash	Soldier	10 Jul 1759	
Willm Thomson	Soldier	1 Aug 1759	
Edwd Campkin	Soldier	3 Aug 1759	
Richd Dalton	Soldier	10 Sep 1759	
Saml Spooner	Soldier	11 Sep 1759	
Richd Correll	Soldier	25 Sep 1759	
Thos Hailey	Soldier	30 Sep 1759	
Richd Palmer	Soldier	22 Oct 1759	
Richd Alliston	Soldier	27 Oct 1759	
George Griffiths	Soldier	3 Nov 1759	
Thos Philistine	Soldier	15 Nov 1759	
Richard Phelpps	Soldier	27 Nov 1759	

The next entry from the burial register commences in January 1765 and gives details of the cause of death and the age at time of demise.

Date of burial	*Name*	*Employment or rating*	*Cause of death*	*Age*
Fort Marlborough				
1765				
Jan 25	Mr Day	Matross	Billious fever	45
Jan 30	Mr Mattson	Private	Flux	22
Feb 2	Joseph Cader	Private	Flux	50
Feb 21	Lawrence Shengon	Private	Nervous fever	32
Feb 22	Mr Brown	Private	Flux	53
Feb 23	Thos Griffiths	Private	Flux	34
Mar 12	Simon Hallsett	Matross	Flux	30
Mar 20	Mr Blakey	Lieutenant	Consumption	30
Apr 12	Mr Dawson	Private	Flux	34
Apr 18	Danl Lowney	Private	Nervous fever	20
Apr 28	Mr Morgan	Matross	Nervous fever	30
Apr 29	Mr Wilkinson	Private	Billious fever	58
May 5	Mr Ryan	Sergeant	Flux	30
May 5	James Johnson	Corporal	Flux	32
May 5	Mr Hardest	Private	Flux	22
May 17	Wm McKnight	Corporal	Flux	40
May 20	Thos Dods	Private	Billious fever	50
May 21	James Agan	Private	Billious fever	26
May 21	Andrew Griffiths	Private	Billious fever	50
May 21	James Ellis	Matross	Flux	36
May 23	Mr Simpson	Matross	Flux	22
May 23	Thos Dean	Private	Flux	30
Jun 7	James Matthews	Private	Billious fever	37
Jun 7	Edwd Butler	Private	Flux	20
Jun 3	Wm Sattler	Matross	Flux	30
Jun 7	Mr Bryant	Matross	Flux	25
Jun 9	Samuel Williams	Matross	Flux	31
Jun 10	Thomas Lockhart	Private	Dropsy	23
Jun -	Mr Mason	Ensign	Consumption	22
Jun 18	James Harrison	Private	Flux	24
Jun 20	Bernard Sumiere	Private	Flux	25
Jun 24	Edwd Barker	Private	Billious fever	40
Jun 23	- Burton	Private	Billious fever	28
Jun 23	James Jones	Private	Billious fever	24
—	Mr D Prevara	Private	Billious fever	42
—	Thos Dale	Private	Billious fever	37
—	Nathl Shirley	Lieut Fireworker	Billious fever	28
—	Davd Roberts	Matross	Flux	30
—	Mr Gardiner	Private	Flux	18
—	Xv Carbus	Matross	Flux	22
—	Mr Daniel	Matross	Flux	35
—	Robt King	Matross	Billious fever	26
—	Wm Clary	Matross	Dropsy	35
—	Mr Lincoln	Private	Flux	27
—	— Merridith	Private	Putrid fever	30
—	Wm Powers	Private	Flux	35
—	Henry Stone	Private	Flux	23
—	Fracs Goodick	Private	Flux	30
—	Richd Smith	Private	Billious in the spleen	26

Date of burial	*Name*	*Employment or rating*	*Cause of death*	*Age*
—	Frans Price	Private	Billious fever	24
—	Thos Norris	Private	Billious in the spleen	50
—	Captain McTatam	of ye Infantry	Consumption	40
—	Mr Shoebrick	of ye Infantry	Flux	38
—	Mr Davis	of ye Infantry	Flux	37
—	Wm Maddos	of ye Infantry	Fever	34
—	James Green	of ye Infantry	Flux	34
1766				
—	Saml Bell	Private	Nervous fever	28
—	Thomas Flat	Private	Flux	26

No burial records available for the years 1767 to 1778.

Date of burial	*Name*	*Employment or rating*	*Cause of death*	*Age*
1779				
May 21	Thomas Larkins	Soldier	—	—
May 24	William Dalby	Soldier	—	—
May 27	Henry Foss	Soldier	—	—
Jun 8	Peter Flood	Soldier	—	—
Jun 16	Paul Duizar	Drummer	—	—
Jun 27	Benjamin Westfield	Soldier	—	—
Jul 1	Lewis Feignham	Soldier	—	—
Jul 9	Michael Fleming	Soldier	Flux	28
Jul 14	William Skeldon	Matross	Fever	30
Jul 17	William Hugh	Soldier Infantry	Flux	34
Jul 27	Robert Arnold	Corporal	Nervous fever	26
Jul 27	John Smith	Soldier	Nervous fever	32
Jul 28	John Wilson	Drummer	Billious fever	24
Aug 1	Sutton J. Leiche	Lieut Fireworker of Artillery	Billious fever	23
Nov 15	James Sangwell	Soldier Centinel	Consumption	26
Nov 17	John Burgess	Matross	Putrid fever	30
Dec 27	William Taylor	Matross	Putrid fever	30
1780				
Jan 25	Richard Hynes	Soldier Centinel	Flux	25
Mar 3	William Perry	Private - Matross	Billious fever	29
Mar 12	James Heard	Private - Matross	Billious fever	24
Mar 31	John Young	Private - Matross	Billious fever	40
Apr 2	John Jones	Serjeant	Putrid fever	40
Apr 6	William Harvey	Serjeant	Flux	40
Apr 12	Rich Lewis	Private - Centinel	Billious fever	25
May 7	Nicholas Brown	Corporal of Infantry	Billious fever	27
May 13	Richard Nicholas	Private - Centinal	Putrid fever	29
May 21	Francis Scarbrush	Private - Centinal	Flux	26
Jun 23	John Ross	Serjeant of Infantry	Billious fever	36
Aug 1	Edw Merryman	Private Centinel	Putrid fever	26
Aug 22	Edw Silverlocks	Private - Centinal	Billious fever	30
Aug 22	James Cornwall	Private - Matross	Billious fever	40
Aug 25	James Peacock	Corporal of Infantry	Billious fever	27
Aug 27	James Pool	Sergeant	Flux	26
Sep 5	James Benjamin	Lieut Fireworker	Putrid fever	18
Sep 9	Robert Jiacock	Private Centinel	Putrid fever	36
Sep 16	Derby Conner	Private Centinel	Bloody flux	33
Sep 17	John Bone	Private Centinel	Consumption	28
Sep 20	Thomas Hall	Private - Matross	Apoplexy	32
Oct 2	Fargas Norton	Private - Matross	Billious fever	24

Date of burial	*Name*	*Employment or rating*	*Cause of death*	*Age*
Oct 13	John Hawkins	Private - Matross	Dropsy	45
Oct 17	William Trooby	Serjeant of Infantry	Bloody flux	29
Oct 20	William Hariman	Corporal of Infantry	Billious fever	25
Oct 21	Felix Rode	Private - Centinel	Bloody flux	19
Nov 2	John Watts	Private - Centinel	Consumption	40
Nov 7	George Linch	Serjeant	Putrid fever	30
Nov 17	William Ball	Private Matross	Flux	32
Nov 20	Isac Hanaker	Private Centinel	Inflamation of the stomach	46
Dec 7	Duncan Campbell	Senior Surgeon	Inflamation of the lung	40
Dec 14	Joseph Shepard	Private Centinal	Bloody flux	32
1781				
Jan 1	William Pickering	Captain of Artillery	Bloody flux	28
Jan 4	Edward Perry	Private Matross	Bloody flux	45
Feb 22	Joseph Mills	Private Matross	Billious fever	24
Feb 28	Benjamin Izard	Corporal of Infantry	Malignant fever	38
Mar 11	Michael Cox	Serjeant Major	Putrid fever	45
Mar 31	John Fearson	Fifer	Putrid fever	23
Apr 7	Ralph Wright	Private Centinel	Drowned in the swamp	28
Apr 10	Conrad Chariges	Corporal of Infantry	Gangrine of the face	40
Apr 26	James Bushel	Private Matross	Bloody flux	36
May 3	Peter Scott	Private Matross	Consumption	24
May 4	James Church	Private Centinel	Bloody flux	27
May 22	Henry Youlden	Fifer	Bloody flux	34
Jun 4	Thomas Norton	Corporal Infantry	Inflamation	31
Jun 6	Robert Burman	Corporal Infantry	Malignant fever	36
Jun 16	Denis Skilly	Private Matross	Mortally wounded whilst firing a cannon	34
Jun 23	George Hobler	Private Centinel	Bloody flux	36
Jul 9	John McDonald	Private Centinel	Dropsy	27
Aug 5	John Cork	Private Centinel	Bloody flux	27

Burial register for the period October 1781 until October 1783 missing.

1783				
Nov 6	Arnold Spence	Private soldier	Malignant fever	34
Nov 14	Joseph Pickering	Private soldier	Malignant fever	32
1784				
Mar 3	Thomas Kinns	Gunner	Dysentry	30
Mar 14	Tybrant Jacobs	Matross	Apoliptic [*sic*]	26
Mar 16	Thomas Woodstock	Gunner	Fever	27
Apr 7	John Hend'k Boodin	Invalid soldier	Dysentry	42
Apr 29	Mullys Riber	Private soldier	Putrid fever	28
May 29	Chrip Routbook	Private soldier	Peupnaimo [*sic*]	28
Jun 25	Jacobus Uniack	Gunner	Fever	25
Aug 13	John Nicholls	Gunner	Dysentry	22
Aug 25	John Di Lanti	Gunner	Putrid fever	24
Sep 19	George Campbell	Captain, Military	Gunshot wound - found on board vessel at Madras	40
Oct 2	Anthony Wiggatt	Matross	Dysentry	42
Oct 5	Benj Calloway	Serjeant of Infantry	Dysentry	30
Oct 11	John Lisne	Matross	Dysentry	50
Oct 22	John Hastings	Corporal	Dysentry	44
Nov 12	John Rokeham	Matross	Dysentry	26
Dec 2	Lewis Johnson	Gunner	Dysentry	27
Dec 6	Job Watkins	Corporal of Artillery	Dysentry	28

Date of burial	*Name*	*Employment or rating*	*Cause of death*	*Age*
Dec 13	Hak Hunnington	Soldier	Malignant fever	20
Dec 31	John King	Sergeant of Infy	Dysentry	40
1785				
Mar 27	John Poole	Drum Major	Malignant fever	25
May 13	Thomas Smith	Sergeant of Infantry	Fever	40
Jun 10	John Sandell	Bombadier	Dropsy	46
Jun 16	William Adams	Corporal of Infantry	A wound received	27
Jul 31	James Hargrave	Private of the King's 98th Regt	Putrid fever	22
Sep 9	William Meredith	Gunner	Putrid fever	32
Sep 11	Samuel Dickin	Lieut of Infantry	Dysentry	20
Sep 14	Henry Potts	Private of the King's 98th Regt	Dysentry	42
Oct 6	John Cording	Mattross	Dysentry	34
Oct 14	James Steward	Private of the King's 98th Regt	Accidentally strangled	23
Oct 17	James Ward	Private of the King's 100th Regt	Putrid fever	30
Oct 26	Robert Cox	Matross	—	40
Oct 27	John Seabrooke	Bombadier	—	38
Oct 28	James Beventon	Private of the King's 23rd Regt of Lt Dragoons	—	33
Nov 15	James Gordon	Mattross	—	36
Dec 13	Daniel Boyvin	Mattross	—	35
1786				
Jan 22	George Middleton	Sergeant of Infantry	—	30
Mar 5	Conner Shune	Mattross	—	26
May 27	Mr Adden Hurdd	Mily Storekeeper & Reader	—	37
Aug 12	Charles Williamson	Gunner, Arty Company	—	32
1786 at Croee				
—	—— O'Brien	Lieut Fireworker	—	36
1786 at Nattal (also spelt Natal)				
—	John Moore	Ensign	—	34
Fort Marlborough				
Sep 1	Isaac Lamon	Mattross of Arty Comp	—	34
Sep 12	Maclane Yang	Bombadier	—	32
Oct 3	James King	Mattross	—	30
Oct 28	Nancy Scott MacDonald	Wife of Capt MacDonald of Artillery Company	—	25
Dec 26	Mr John Vaux	Surgeon	—	33
Dec 29	Joseph Morris	Sergeant of Infantry	—	30
1787				
Jan 10	Joseph Morton	Surgeon	Putrid fever	26
Jan 13	William Hale	Mattross	Small pox	21
Jan 27	Thos Foreman	Gunner	Fever	26
Jan 31	Jane Pavy	Daughter of Sgt John Pavy	—	3½
Feb 3	James Ellis	Mattross	Dropsy	28
Feb 13	John Pavy	Son of Sgt J Pavy	—	5½ months
Feb 17	Willm Gray	Matross	—	—
Mar 18	—— Bridges	Daughter of Capt Bridges	—	5
Mar 26	John Stanley	Matross	Small pox	21

Date of burial	*Name*	*Employment or rating*	*Cause of death*	*Age*
May 1	John Trudget	Invalided Matross	Putrid fever	43
May 18	Abrah Harrowell	Invalided Matross	Dropsy & dysentery	50
May 20	Jeremiah Cranwell	Private	Dropsy	30
Jun 2	William Clarke	Invalided Fifer	Dysentery	40
Jun 27	—— MacDonald	Daughter of Capt MacDonald	—	2
Jul 16	Thos Vincent	Capt of Artillery	Accident	20
Jul 21	James Aldridges	Mattross	Dysentery	32
Aug 19	Willm Swan	Bombadier	Consumption	39
Aug 19	Robt Mainwaring	Matross	Putrid fever	27
Oct 4	Wm Pleasant	Private	Dysentery	20
Dec 7	Samuel Spruson	Mattross	Dysentery	22
Dec 10	Mr John Vietch	Lieutenant	Accidental	28
Dec 29	John Shepper	Private	Putrid fever	19
1788				
May 11	Willm Talbot	Mattross	Dysentery	46
Jun —	Danl Beech	Fifer	—	—
Jul 25	Thos Elliot	Assistant Surgeon	Putrid fever	24
Aug 3	James Whatley	Invalided matross	—	30
Aug 18	Thomas Cossen	Son of Thos Cossen, Matross	—	4
Oct 7	John Green	Sergeant of Artillery	—	28
Nov 4	John Ferdinand	Drummer	—	—
1789				
Feb 19	John Shockly	Matross	—	34
Mar 20	Phenis Mackley	Gunner	Dysentery	28
Mar 24	William Smart	Matross	Dysentery	26
Apr 9	Thos Morgan	Surgeon	Putrid fever	24
Apr 12	Wm Evans	Private 36th Regt	Sciritious [*sic*] viscera	42
Apr 13	John Bomain	Matross	Dysentery	41
May 14	John Terry	Son of J. Terry, Matross	Fever	6
May 20	Robt Mason	Pte 36th Regmt	Dysentery	21
May 31	Wm Carter	Matross	Fever	20
Jun 6	John Chandler	Ensign	Putrid fever	25
Jul 19	John Winch	Gunner	Fever	29
Jul 20	Timothy Divan	Sergt Major Sepoy Det	Dysentery	41
Jul 26	Anthony Hovel	Matross	Dysentery	32
Aug 7	Edwd Warth	Matross	Dysentery	31
Aug 13	Alix McKenzie	Private 16th Regt	—	29
Oct 6	Franc Lumar	Matross	—	20
Oct 17	Edwd Mason	Matross	Consumption	40
Dec 6	John Kilroney	Soldier in His Majt Service	Fever	38
Dec 18	Thos Hooper	Matross	Dropsy	35
1790				
Jan 2	Thos Roberts	Gunner Arty	Diseased viscera	41
Apr 13	John Brown	Lieut of Infantry	Hectic fever	25
Apr 14	Edwd Wiseman	Surgeon	Diseased viscera	49
Jun 8	Lawrence White	Matross	—	25
Jun 20	William Watson	Matross	—	26
Jul 12	Wm Harris	Matross	Putrid dysentery	22
Jul 14	Henry Willis	Lieut of Artillery	Putrid fever	22
1790 at Nattal				
Aug 4	James Turner	Capt of —— Infantry	Putrid fever	28

Date of burial	Name	Employment or rating	Cause of death	Age
Fort Marlborough				
1791				
Feb 4	John Saunders	Matross	Putrid fever	30
Mar 13	James Cox	Matross	Putrid dysentery	31
1791 at Nattal				
Mar 21	John Nicholas	Sergeant of Artillery	Shot	30
Mar 25	Thos Penney	Gunner Artillery	Drowned at sea	20
Mar 28	George Mitchell	Corporal Artillery	Consumption	30
Fort Marlborough				
Jun 8	Lawr White	Matross	Consumption	35
Jun 22	John Walding	Matross	Consumption	38
Jul 7	Alexr Smith	Gunner	Consumption	37
Jul 14	Josp Shepherd	Gunner	Putrid dysentery	40
Aug 22	Thos Wheatly	Recruit – Bengal	Putrid fever	38
Aug 23	Thos Chapman	Recruit – Bengal	Malignant fever	19
Aug 24	Michl Francis	Sergeant – Bengal	Malignant fever	35
Aug 25	George Dochril	Recruit – Bengal	Malignant fever	18
Aug 27	Peter Brodrick	Recruit – Bengal	Malignant fever	18
Aug 28	Saml Hamilton	Recruit – Bengal	Malignant fever	30
Aug 31	Hugh Thompson	Recruit – Bengal	Malignant fever	33
Sep 3	David Young	Recruit – Bengal	Malignant fever	34
Sep 4	Wm Enticott	Recruit – Bengal	Malignant fever	22
Sep 7	Thos O'Brien	Recruit – Bengal	Malignant fever	19
Sep 13	Michl Morris	Recruit for Bengal	Malignant fever	29
Sep 15	Gilbt Toler	Recruit for Bengal	Malignant fever	20
Oct 6	John Grace	Recruit for Bengal	Putrid fever	20
Oct 28	Thos Brown	Corpl of Artillery	Colliquative diarrhoea	28
Nov 17	Richd Siddel	Recruit for Bengal	Malignant fever	18
1792				
Feb 20	Wm Hay	Recruit for Bengal	Malignant fever	18
May 12	John Hy Haacker	Drummer Hanovern Regt	Fever	40
May 18	Thomas Clewer	Corporal Hanovern Regt	Fever	30
1792 at Natal				
May 26	Henry Francis	Sergt of Artillery	Fever	30
Fort Marlborough				
May 31	John Moore	Corpl of Artillery	Dropsy	20
Jul 3	William Bell	Surgeon	Putrid fever	33
Jul 5	Thos Symour	Drummer of Artillery	Fractured tibia	32
Jul 12	John Healey	Matross	Suicide	30
Aug 4	Robt Chapman	Matross	Dysentery	28
Aug 31	John Evans	Corpl of Artillery	Insane	30
1793				
Mar 7	Michl Carney	Kings soldier	Dysentery	32
Mar 11	Willm Ross	Gunner of Artillery	Suicide	30
Mar 29	George Delfruman	Mattross	Dysentery	25
Apr 28	Robert Field	Mattross	Drowned	20
May 20	Nichs Houison	Surgeon	Dropsy	30
Aug 3	William Stokes	Gunner Artillery	Fever	27
Aug 15	Nichl Brown	Mattross	Fever	21
Aug 30	George King	Mattross	Fever	30
Dec 15	Robert Hamilton	Captain Commandant	Abcess of the liver	40

Date of burial	Name	Employment or rating	Cause of death	Age
1794				
Jan 10	James Burgess	Private Artillery	Malignant fever	40
Jan 10	William Brooke	Private Artillery	Fever	23
Jan 22	James Hoshen	Sergeant Infantry	Consumption	22
Feb 9	Robert Norton	Sergeant Artillery	Obstruction	30
Feb 11	John Campbell	Private Infantry	Dysentery	24
Mar 11	Francis Morgan	Private Artillery	Consumption	24
Mar 29	John Neale	Corporal Infantry	Fever & dysentery	29
Apr 17	Thomas Thompson	Private Infantry	Liver obstruction	24
Apr 19	Richard Hill	Private Infantry	Flux	22
Apr —	Samuel Waugh	Private Artillery	Malignant fever	30
Jun 12	Thomas Cossar	Sergt Artillery	Dropsy	45
Aug 7	Strickland Pearson	Private Infantry	Fever	22
Nov 13	Samuel Alexander	Ensign	Obstruction	25
Nov 13	Thomas Tate	Private Artillery	Consumption	35
Nov 16	Samuel Taylor	Private Infantry	Obstruction	36
Nov 19	Christopher Grogan	Private Infantry	Fever and jaundice	24
1795				
Feb 17	John Finn	Private Artillery	Accident – fracture of the head	22
May 12	Richard Robinson	Private Artillery	From this date the burial registers ceased to record cause of death.	
May 23	James Ferris	Private Infantry		
May 24	William Taylor	Private Artillery		
May 26	John Cook	Corporal Artillery		
May 31	Hugh Calon	Private Artillery		
Jun 4	George Chaldner	Private Artillery		
Jun 26	George Simpson	Sergt Artillery		
Jul 3	John White	Private Artillery		
Jul 8	John Cockran	Private Artillery		
Jul 25	Mr Thomas Traill	Surgeon		
Jul 30	Michael Smith	Matross		
Aug 22	Lachland Dandy	Private Infantry		
Aug 25	John Fulmack	Private Infantry		
Aug 30	Thomas Loveday	Private Infantry		
Sep 1	John Smith	Matross		
Sep 1	Robert Murray	Matross		
Sep 4	John Fitzgibbons	Private Infantry		
Sep 4	John Taylor	Private Infantry		
Sep 16	David Coggan	Private Artillery		
Sep 19	John Martin	Private Infantry		
Sep 26	Francis Allen	Private Infantry		
Sep 27	Alex Sutherland	Act. Sergt Major		
Sep 27	John Wilkinson	Private Infantry		
Sep 29	John Riely	Private Infantry		
Sep 29	John Clark	Matross		
Oct 4	James Morton	Private Infantry		
Oct 8	John Roney	Private Infantry		
Oct 9	Thos Hawthorne	Mattross		
Oct 22	John Jenkins	Corporal of Infantry		
Oct 27	P. Macmillan	Mattross		
Oct 27	Thos Jones	Private Infantry		
Nov 4	John Ewing	Private Infantry		
Nov 15	W. Simmons	Gunner		
Nov 18	Thomas Randall	Corporal of Infantry		

Date of burial	*Name*	*Employment or rating*
Nov 26	Wm Hughes	Private Infantry
Nov 28	Wm Peck	Matross
Dec 1	John Glub	Private Infantry
Dec 12	John Steptons	Drummer
Dec 20	John Corsey	Private Infantry
Dec 21	James Sexton	Private Infantry
Dec 22	John Newland	Private Infantry
Dec —	Henry Lewis	Lieutenant Fireworker, Bengal Artillery
1796		
Jan 18	Richd Harper	Private Infantry
Feb 3	Ann Sewell	Wife to I. Sewell Private Infy.
Feb 10	Miles Kirney	Son of F. Kirney Corpl of Arty
Jun 6	Ralph Bloomer	Private Infantry
Oct 3	William Rourke	Gunner Artillery
Nov 13	Philip Rust	Private of Infy (Drowned)
Nov 13	Robert Hogan	Private of Infy (Drowned)
1797		
Jan 8	Edward Drayton	Corporal Artillery
Jan 20	John Kennedy	Corporal Infantry
Jan 29	Mr John Eversley	Conductor of Ordnance
Mar 10	John Kidner	Matross Artillery
Mar 15	John Gregory	Matross Artillery
Apr 26	Samuel Beckford	Private Infy
Jul 5	Thomas Stapleton	Private Infy
Aug 1	Thomas Newton	Private of Infantry
Aug 31	William Quick	Private Infy
Sep 28	Thomas Tucker	Private Infy (Age 28)
Oct 1	Wm Stanley	Sergt Arty (Age 40)
Oct 14	Arch Maclane	Private Infy (Age 21)
Nov 2	John Hennesy	Sgt of Infy
Nov 23	Peter Flynn	Pte of Infy (Age 24)
Nov 27	John Cogan	Matross
Dec 13	Wm McCullock	Pte Infy (Age 26)
1798		
Apr 9	Saml Woolaston	Sgt of Arty (Age 28)
Jul 9	Alexander Monteath	Surgeon
1799		
Jan 3	Mr Harrison	Assistant Surgeon
Jun 10	John Ross	Private
Jul 17	Joseph Hook	Son of Sergeant Major Hook
Jul 21	John Love	Private
Oct 14	Joseph Hart	Sergt Major
Nov 16	C. A. Hana	Private
Nov 19	W. Wanders	Private
Nov 20	R. Mousell	Lieut Fireworker
Nov 27	C. Plummer	Private
Dec 1	—— Brown	Private
Dec 4	—— son	Corporal [document damaged]
1800		
Jan 10	William Jones	Private
Jan 28	Robert Duncan	Private
Mar 19	John McNeal	Private

Date of burial	*Name*	*Employment or rating*
Apr 5	Thomas Lyes	Private
Apr 11	Thomas Tyrell	Gunner
Apr 15	Patrick Connelly	Private
May 29	Eliza Hartley	Daughter of Surgeon B. Hartley
Jul 24	J. Roberts	Private
Aug 23	Eliza Clayton	Daughter of Colonel Clayton (age 2 yrs 1 month)
Oct 2	John Christian	Sergeant
Oct 20	S. Whelps	Gunner
Nov 15	Thos May	Private
Nov 17	John Burgess	Private
Nov 19	James Rafter	Private
Nov 21	J. Mitchell	Sergeant
Nov 28	H. H. Pearson	Captain of Engineers
1801		
Jan 9	J. Evatt	Sergeant
Mar 21	Joseph Sangster	Private
Apr 1	—— Wimpole	Sergeant
Apr 30	—— Eliott	Corporal
Aug 12	—— Fitzgerald	Conductor of Ordnance
Aug 29	Willm Harding	Conductor of Ordnance
Nov 1	J. Smith	Sergeant Major
1802		
Jan 23	Wm Burrows	Corporal
Jun 18	W Davis	Sergeant
1803		
Mar 1	Wm Price	Corporal
Oct 22	Captain Henry Douglas	Bengal Artillery
1804 at Tappanooley		
Jul 17	John Spratt	Surgeon
1805		
May 26	Andrew Hunter	Serjeant
Jun 21	John Dykes	Private
Jun 22	William Cox	(Sons of Lieut W. B. Cox)
	Philip Cox	
Aug 14	John Alexander	Surgeon
Sep 20	S. Rose	Private
1807		
Feb 7	John Newland	Sgt of Artillery
Oct 19	James McNaughton	Lieut 20th Regt NI, Bengal Est.
1808		
Feb 24	Serjt Johnstone	of the Artillery
May —	Henry Martin	Cornet, N.I. Madras Est
Aug 10	Thomas Pearce	Lieut 20th Regt N.I. Bengal Est.
Nov 12	Edward Ogden	Gunner Artillery
1809		
Oct 6	James Shea	Private in the Artillery
Nov 16	William Par	Private in the Artillery
1810		
Feb 2	Edward Slatery	Private in the Artillery
Apr 5	J. Cox	Private HM 14th Regt
Jul 2	Robert Berries	Private in the Artillery

Date of burial	*Name*	*Employment or rating*
1811		
Jan 6	J. Faggot	Private of Artillery
Jul 3	J. Scott	Pensioner of the Marine Regt Barracks
Aug 30	William McKannah	Matross of the Bengal Artillery
Oct 4	James Robertson	Acting Sergeant, Magazine Establishment, Matross of the Bengal Artillery
1812		
Jun 19	John Murray	Matross of Artillery
Jul 18	Dennis M'Guire	Matross of Artillery
Jul 20	John O'Brien	Matross of Artillery
Oct 17	John Kelly	Matross of Artillery
Dec 14	John Burroughs	Matross of Arty.
1813		
Jun 3	Robert Leicester	Conductor of Ordnance (Age 54)
1814		
Feb 8	William Beckford Cox	Captain of Bengal Invalids
1815		
Jan 11	John Wood	Matross Bengal Artillery
Mar 27	Joan Morgan	Wife of Mr Morgan, Bengal Ordnce (Age 35)
Apr 14	Peter Ontzidar	Matross of Artillery (Age 32)
May 8	Thomas Fox	Matross of Artillery (Age 22)
May 8	Joseph Cogran	Matross of Artillery (Age 32)
Jul 1	Hugh Stewart	Matross Bengal Arty
Sep 14	John Dunn	Sergeant Bengal Artillery
Oct 26	Owen Doyle	Corporal of Marines, HMS *Revolutionaire* (Age 36)
Oct 31	Cornelius Crovans	Serjeant of Marines, HMS *Revolutionaire*
Nov 5	John McLaughlin	Matross, Bengal Artillery (Age 24)
Nov 12	John Brewin	Serjeant, Bengal Artillery (Age 32)
1816		
Apr 9	Charles Porteous	Major in the 20th Regiment Native Infantry
Nov 20	John Brickford	Matross of Artillery (Age 36 years, 5 months)
1817		
Feb 23	John Colling	Matross of Arty (Age 25 years, 8 months)
Sep 10	Thos Skinner	Sergt Major, 1st Batt of the 20th Regt NI (Age 42)
Dec 20	Walter Clyde	Matross of Artillery (Age about 24)
1821		
Aug 8	Thomas Matthew	Corporal of Artillery (Age 29)
Sep 8	James Walsh	Gunner of Artillery (Age 22)
Sep 16	Mr Jn Lancaster	Assistant Surgeon (Age about 30)
Dec 6	Charles Boyland	Fifer of Fort Marlbro Local Corps (Age 19)
Dec 7	John Manning	Sergeant Major of Fort Marlbro Local Corps (Age about 30)
1822		
Jan 17	Edward McMahon	Gunner of the 3d Compy 2d Battn Artillery detachment (Age 24)
Jan 19	Alexander Allan	Matross of 7th Compy 3d Batt. Artillery detachment (Age 26)
Feb 7	Albert Blist Mailien	Matross of 3d Compy 2d Batt. Artillery detachment (Age 28)
Mar 5	John Seldon	Matross of 7th Compy 3d Batt Artillery detachment (Age 23)

Date of burial	*Name*	*Employment or rating*
Jun 23	William Rush	Gunner of the European Artillery detachment (Age 27)
Jul 9	Mr Acting Assistant Surgeon Duncan McCalman	Assistant Surgeon (Age 27)
Jul 23	Isabella Margaret Hurst	Daughter of Thos Hurst Corporal of the European Artillery detachment (Age 8 months and 20 days)
Sep 15	H. Jack Esq	Private Surgeon to the Lieut Governor (on the Bengal Medical Establishment)
Nov 16	Joseph Boardman	Conductor of Ordnance Bengal Establishment (Age 48)
1823		
Feb [illegible]	Ann Hurst	Wife of Thos Hurst, Gunner of the European Artillery detachment (Age 34)
Mar 27	Mrs McInnes	Wife of Major McInnes, Commandant of Fort Marlbro Local Corps (Age 22)
Oct 17	Mary Ann Riley	Daughter of the late Sergeant Riley (Age about 7 yrs)
Dec 17	Frederick Patrick Devine	Son of Serjt Devine (Age 3 years and 4 months)
1824		
Feb 2	—— Watson	Lieutenant of Fort Marlbro Local Corps
Feb 4	Thomas Hurst	Corporal of the European Artillery detachment (Age about 33)
Oct 26	Robert Eaton	Conductor of Ordnance (Age 39)
1825		
Mar 9	Jane	Wife of Sub Conductor P Devine of the Ordnance at Fort Marlbro (Age 33)

NOTES TO CHAPTER TWO

1. William Marsden, *The History of Sumatra,* (Oxford University Press, Oxford in Asia Historical Reprints, 1966), p452.
2. Ibid.
3. Ibid.
4. John Bastin, *The British in West Sumatra (1685–1825),* (University of Malaya Press, Kuala Lumpur, 1965), p68.
5. The Governor at Bencoolen continued to be in charge of the military garrison and was therefore directly concerned with the administration and discipline of the military force that formed the garrison. The actual order giving him the necessary authority reads:

> '. . . [to] rule and Govern all and every [one of] our Factors and Servants under the said Presidency and all Soldiers under the said Presidency and all Soldiers and inhabitants of our said Fort Marlborough and Town of Bencoolen . . .'

India Office Library and Records, Reference H/Misc/24.

6. IOLR, Sumatra Factory Records, 31. Fort St George was the headquarters of the Madras Presidency and the title was frequently used to denote the 'Madras Presidency'. East-India Register and Directory for 1803, pages 137 to 222, and subsequent editions of the Register and Directory, use this title for all entries referring to Madras.
7. Marsden, op cit, p452.
8. Bastin, op cit, p71.
9. IOLR, H/Misc/24.
10. IOLR, G/35/152.
11. Ibid.
12. Ibid.
13. Bastin, op cit, p70, note 252.

14. IOLR G/35/152.
15. Ibid.
16. Ibid.
17. The composition of the Court was:

President – Captain John Hutcheon

Members – Captain/Lieutenant Thomas Oldham
Lieutenant William Maitland
Lieutenant Trevor Blunden
Lieutenant Thomas Baddeley
Lieut (Fireworker) James Richardson
Lieut (Fireworker) Charles Disfountain
Ensign Charles Cook.

Reference – IOLR G/35/152.
18. Ibid.
19. Ibid.
20. IOLR H/Misc/24.
21. Ibid.
22. The 'Vacant Company' was the term used for the Company that had a 'vacancy' on its establishment for the Company Commander, i.e. there was no Company officer 'in post'.
23. IOLR G/35/153.
24. Marsden, op cit, p452.
25. IOLR G/35/153.
26. Ibid, pp 75 to 85.
27. Ibid, pp 86 to 91.
28. Ibid, pp 98 to 106.
29. Ibid, pp 106 to 113.
30. Ibid.
31. Marsden, op cit, p452.
32. The Burial Register was maintained at Bencoolen and in accordance with Company policy a copy of the register was despatched periodically to the East India Company office in London. The relevant section was signed and sent by hand of the Captain of a ship leaving Bencoolen for England. The registers held in the present-day India Office Library and Records are the copies that arrived in England. There are gaps which generally could be accounted for by the loss of vessels between the west coast of Sumatra and England. There were many such vessels lost during this long and difficult passage. The whereabouts of the 'Bencoolen' copy of the register is not known. It was in all probability transferred to India in 1825 when Sumatra was relinquished to the Dutch. A complete copy of the register is, possibly, held in some Go-down in India.
33. IOLR G/35/153, p166.
34. Ibid.
35. Ibid, pp 204 to 215.
36. Ibid, p 217.
37. IOLR Map Misc 1/326.
38. John Bastin and Pauline Rohatgi, *Prints of South East Asia in the India Office Library*, (HMSO, 1976), pp 68 to 73.
39. Bastin, op cit, pp xxvii and xxviii.
40. The piaster was also known as the Spanish dollar. The accepted rate for the Spanish dollar at the beginning of the nineteenth century was five shillings sterling.
41. Bastin, op cit, p106.
42. Sumatra Factory Records, letter Ewer to Dundas dated 6 August 1800.
43. Bastin, op cit, p107.
44. Ibid, p108.
45. East India Kalendar, 1800.
46. Bastin, op cit, pp 109 and 110.
47. Ibid, p110.
48. The Hodson Index, (National Army Museum, London). Thomas William Clayton was born 1763/64 and died at Barrackpore on 22 September 1804. He was a Cadet in 1771; arrived in India on 10 December 1771; his appointments and promotions were: Ensign 30 March 1773, Lieutenant 27 May 1778, Captain 21 January 1784, Major in 1797 and Lieutenant Colonel 29 May 1800. He served in the 3rd Bengal European Regiment, became a Major in 2/1st Native Infantry and transferred as Lieutenant Colonel to 2/18th Regiment Bengal Native Infantry and then transferred to the 20th (Marine) Regiment Bengal Native Infantry when it was formed in 1803.

49. Marsden, op cit, p452.
50. East-India Register and Directory, 1804, p50.
51. Marsden, op cit, pp 308 and 309.
52. Ranjaus are slips of bamboo sharpened at each end, the part that is stuck into the ground being thicker than the upper end which is sharpened to a fine thin point. The thin end was hardened by dipping it in oil and applying it to the smoke of a lamp near to the flame. Ranjaus were planted on footpaths, sometimes erect and sometimes sloping so that when trodden upon they would pierce the foot and cause painful and crippling wounds. Ranjaus were also widely used during the Naning campaigns of 1831 and 1832.
53. Marsden, op cit, p312.
54. Ibid, p315.
55. Ibid, p319.
56. Ibid, p320.
57. Ibid, p324.
58. Ibid, p463.
59. Bastin, op cit, p128. Bastin quotes the date of Parr's death as being 27 December and gives the details of his death as follows: '... he was stabbed to death and his head removed ...' In his article 'More on Bencoolen' R. J. Wilkinson states that in addition to Parr being murdered '... one of his writers, Mr Charles Murray ...' was also killed whilst defending Mrs Parr.

The West Coast of Sumatra Burial Register (IOLR N/7/1) shows the following entries:

'1807

Dec 24. Thomas Parr Esq. Resident at Fort Marlbro [*sic*]. Murdered by the Malays at Mount Felix on the night of 23rd ...'

It is therefore certain that the correct date of Parr's death was 23 December and that he was buried on 24 December 1807. Murray was not killed outright but died of his injuries a few days later. He was buried on 7 January 1808 and this date is confirmed by an entry in the Burial Register.
60. Bastin, op cit, p 129.
61. Ibid.
62. Sumatra Research Bulletin 4, 1 of 1974. Article 'Bencoolen (Bengkulu) 136 Years After' by M. A. Jaspan.
63. Sumatra Factory Records, 112. Minute of 25 February 1808.
64. Bastin, op cit, p 103.
65. Ibid, p 157.
66. Ibid, p 168.
67. Captain William Gordon MacKenzie was born at Edinburgh on 9 May 1785. He became a Cadet in 1801 and arrived in India on 6 July 1802. He served as Third Assistant to the Resident at Amboina between 1812 and 1816. He was promoted Captain from 31 May 1816 (East-India Register and Directory for 1817, p40 (Bengal)) and was appointed Second Assistant to the Resident at Fort Marlborough in October 1817. He later acted as Resident of Malacca for a very short period during 1825–6. He retired from Company service on 19 September 1840 and died on 20 July 1842. (East-India Register and Directories and Hodson Index, National Army Museum).
68. Nicholas Tarling, *Anglo-Dutch Rivalry in the Malay World 1780–1824,* (Cambridge University Press, 1962).
69. Charles Porteous was born at Perth, Scotland on 8 June 1776 the son of Alexander Porteous, a merchant of Perth, and Helen Bell his wife. He became a Company cadet in 1795 and an ensign on 2 November 1796. He arrived in India on 3 March 1797 and spent the next nineteen years, until his death, in the East in the service of the Company. He became a Lieutenant on 30 October 1797 and married Elizabeth, the daughter of Lawrence Rawstone at Berhampore on 24 April 1802. He was appointed as Adjutant, 1st Battalion, Marine Regiment in 1803 and promoted to Captain on 30 November 1804. In 1806 he moved with his unit, which, in 1803 had been renamed 20th (Marine) Regiment Bengal Native Infantry, to Prince of Wales Island (Pulo Pinang). He then served, on detachment, at Malacca from 1800 and was finally sent, as the Senior Captain of the Regiment, to Fort Marlborough, Benkulen. He visited India prior to moving to Bencoolen and had a will made at

Barrackpore on 20 January 1814. He was promoted to Major on 1 October 1815 and died six months later in April 1816. At the time of his death his wife and three children were resident at Barrackpore. The children were Sophia Mary born 15 December 1804; Henry Alexander born 10 April 1810 and Frederick William born on 3 May 1813. (Hodson Index, National Army Museum).

70. C. J. Brooks, 'English Tombs and Monuments in Bencoolen', published in *The Journal of the Straits Branch, Royal Asiatic Society*, No 78 (1918), plate V.

Hodson Index, National Army Museum. Captain Robert Hamilton joined the Company as a Cadet in 1772. He became a Fireworker on 22 February 1773, Lieutenant 7 March 1778 and Captain/Lieutenant 1 October 1781. He was promoted to Captain on 2 February 1785 and died at Bencoolen on 15 December 1793. The plaque on the obelisk reads:

'Underneath this Obelisk are interred
the Remains of
CAPTAIN ROBERT HAMILTON
who died on the 15th of Decr 1793
at the Age of 38 Years.
in the Command of the Troops,
and
Second Member of the Government'

Details of the inscription on the obelisk kindly provided by Mr E. Edwards McKinnon.

71. IOLR N/7/1. Burial Register of Fort Marlborough (Bencoolen).

72. Correspondence from Mr E. Edwards McKinnon dated 18 September 1982, and 2 November 1982.

CHAPTER THREE

The Penang Garrison, 1786–1830

Prior to the Honourable East India Company's establishment of a settlement on Penang Island in 1786 an attempt had been made to form a trading post and factory in Kedah, on the Malayan Peninsular, but this ended in failure. In 1772 The Honourable Edward Monckton[1] was appointed commander of the mission to Kedah but this mission was doomed to failure from the start. Monckton, a young man, had no trading experience but was selected by Madras because of his connections, which was hardly a recommendation for the qualification of leadership of the important mission to establish trading with the Sultan of Kedah. He had with him a former British Navy midshipman Francis Light.[2] Monckton was to obtain from the Sultan the port customs of Kedah, or the right to collect customs. This was to help offset the cost of any troops that might be garrisoned at the port. He was also to negotiate the trading rights for tin, wax and pepper as well as 'elephants teeth' all for the China market. In addition there was the need to obtain a site for the establishment of the factory.

Prior to this time Francis Light had established a trading post in Kedah at the mouth of the river and had been given trading rights for his employers who were, at that time, Messrs Jourdaine, Sulivan and De Souza of Madras. In return for the trading concessions he agreed to help the Sultan should his territory be invaded by the Bugis who dominated the southern state of Selangor. In a letter written in 1771 to De Souza, Light comments on Penang as follows:

> '. . . Withinside of Pulo Pinang is a fine clear channel of 7 and 14 fathoms through which a ship may work any time. I remember it was once your opinion that a House upon Pulo Pinang would be very useful; it would be extremely so because the Europe ships[3] can easily stop there. There is plenty of wood, water and provisions there, they can be supplied with tin, pepper, beetlenut, rattans and birdsnests, and the Macao ships will be glad to stop there, and all other vessels passing through the Straits may be as easily supplied as at Malacca. Whether this would not suit the Company[4] better than our Association I leave you to judge . . .'[5]

Monckton departed from Madras leading an expedition of two ships and a military force, of Sepoys and artillery, under command of Lieutenant D. McClintock. Light met Monckton and his force at the mouth of the river. Light had by this time become friendly with the Sultan's naval commander (Admiral), the Laxsamana, who was also related to the Sultan by marriage. Without waiting for an invitation, or taking advice from Light, Monckton set off with a guard of eighteen Sepoys and a dozen artillerymen, together with his interpreter, to meet the Sultan. Light accompanied the party, who proceeded up river in a long boat and two prahus, and throughout the negotiations with the Sultan on the following day Monckton appears to have ignored any advice from Light. Needless to say the negotiations broke down and Monckton returned to the coast. Light remained with the Sultan and obtained a concession that the Company could build a factory on a stretch of land at the mouth of the Kedah river. He also gave the Company the option on all the tin that was to be exported and the right to purchase black pepper and elephants' teeth. Customs duties on foreign vessels were also granted to the Company on condition that one half of the amount collected was to be paid to the Sultan. In return he wished the Company to provide defence in the form of a vessel to be sent should Kedah be attacked from Selangor. Even these concessions were badly handled by Monckton and later as a result of Monckton's report, Madras considered that Kedah was of little use as a trading post. Monckton was recalled to Madras and left on 13 December 1772, having embarked his military force on the *Tankerville*; he arrived back in Madras on 12 January 1773.[6] The records however do not show from which elements of the army the detachment under Lieutenant McClintock had been raised, nor are there any details of casualties given.

Francis Light continued to trade in the area of the Malayan Peninsular and in 1782 he was permitted to settle on P'huket, or Junk Ceylon, an island which is a mile off of the Kra Peninsula. There had at one time been a large Portuguese settlement there and it has been described as follows '... a fine market street, composed of large brick buildings, among which rose the spacious houses belonging to the Europeans that used to reside here while their ships lay at anchor in the harbour ...'[7] Junk Ceylon was attacked by a war fleet from Siam and many of the residents were killed and the remainder of the European and Eurasian community were made homeless and dispersed. Fortunately Francis Light was in Calcutta at the time of the attack and never returned to the island. At the time it had been a possibility that the Company may have set up a trading port on the island but now with its loss there was a need to find an alternative site. In the following year, 1786, the Dutch attacked and took Rhio Lingga, and Junk Ceylon was attacked by the Burmese. There now seemed to be little choice for the Company for a new trading post and settlement and it was at this stage that Francis Light, who was in Calcutta, wrote to the Company:

> '... The Dutch now possess both sides of the Straits of Malacca; nothing is left but the small kingdoms of Queda and Acheen. As I understand from you that it is the wish of the Government to secure some useful and convenient port for the protection of the merchants who trade with China ... I made use of the influence and friendship I had with the King of Queda (Kedah) and his Minister to obtain a grant of the island of Penang ...'[8]

The Company considered Acheen and decided against this as it would have been necessary to occupy the area with a military force in order to keep the Sultan under control so that the Company's servants could go about their business. This obviously was not a satisfactory solution on the finance aspect alone and it was therefore agreed that Pinang would be the choice. The Company agreed on 2 March 1786 '... to accept the King of Queda's offer to the Company of the Harbour and Island of Pinang ...'[9] and it was also agreed that '... in consideration of the Board's favourable opinion of Captain Francis Light, his knowledge of the Malay language and the high esteem he stands with the King of Queda and other Malay chiefs, that he be vested with the charge and superintendence of the island of Pinang on the part of the Company until their pleasure be known or until further orders ...'[10]

As a result of these decisions Captain Francis Light, who was at this time a sea captain of the Honourable East India Company, arrived off Pinang Island with his three ships on 17 July 1786 and proceeded to take possession of the island on behalf of the Company from his friend Sultan Abdullah of Kedah (Queda). The Company's new Settlement was named 'Prince of Wales Island', a name that was to be used in official records and files for the next fifty years. The island, also known as Pulo Pinang, later to be known as Penang Island, was described by the nineteenth century author John Crawfurd as being '. . . situated towards the western end of the Straits of Malacca, separated from the mainland . . . by a channel about two miles broad, and distant from the nearest point of Sumatra about 150 miles . . . The island is about 15 miles long and from 7 to 8 miles broad and is computed to contain an area of . . . 160 statute square miles, so that it is by size 30 square miles less than that of the Isle of Wight . . .'[11] Prior to the arrival of Francis Light the island had been visited by '. . . only a handful of Tamils who had . . . come to trade for betel nuts (pinang) . . .'[12] It is from this trade that the island was given its name. The Malay translation of the name, 'Pulo Pinang' is 'Areca palm island'.

On setting his party on shore, Light instructed his marine detachment to set about clearing wood and undergrowth from the shoreline in order to make room for the tented lines. Cameron, another historian of the nineteenth century, recorded the landing as follows '. . . early on the morning of the 17th Mr Light disembarked with the Marines and Lascars and the small body of European officers who had accompanied him. On landing they found extending down to the strip of sand on which the boats had grated nothing but dense jungle with an impenetrable undergrowth of shrubs and creepers . . .'[13] The sandy shoreline where the landing had taken place was where Light later built the fort that was to become the main defence post of the Settlement. The three ships, which formed the expedition, had anchored opposite the landing point so as to give cover to the operation. The ships were the *Eliza*, the *Speedwell* and the *Prince Henry* which continued to house the Marine detachment until 18 July when sufficient tropical foliage had been cleared to allow the tented camp to be set-up.

By mid-August the Settlement was well established and the local Malays had been recruited by Light to help clear further areas of tropical undergrowth. The tents proved to be quite inadequate in the tropical rain and were replaced by wooden huts. On 10 August the first two East Indiamen arrived off Prince of Wales Island, *en route* for China. They were the *Vansitart* and the *Valentine*.[14] The following day the Captains of the two ships landed to take part in the ceremony of raising the flag on the recently-erected mast. The Marine detachment formed up, the Malays and Chinese gathered about the area and Captain Francis Light and the thirteen Englishmen assembled for the ceremony. Light wrote, at that time, in his diary '. . . all the gentlemen unitedly hoisted the flag taking possession of the Island in the name of His Britannic Majesty, George III, and for the Honourable East India Company. The artillery and the ships fired a royal salute, and the Marines three volleys . . .'[15] The officers who were recorded as being present at the ceremony were:

Lieutenant James Gray – Commandant of the Marine detachment
Captain Elisha Trapaud – Captain of Engineers
Captain George Howell – Captain of Artillery.[16]

Once the Settlement had been established Light ordered the start of the building of a fort. The fort was mainly a timbered stockade with a ditch '. . . bastions solid enough to hold six-pounders landed from the ship . . .'[17]

Light diplomatically named his fort after the new Governor General in Calcutta, Lord Cornwallis. In many Company records, Penang is often referred to under the title of the Fort – ie Fort Cornwallis. Lord Cornwallis was not impressed and commented upon the building of the fort critically saying '. . . I will not go so far as to say that in the present embarrassed state of the company's finances I would have recommended the undertaking . . .' he concluded by saying '. . . my objections go rather to the time in which this measure has been resolved upon, than to the measure itself . . .'[18]

The size of the Settlement continued to grow and the town that had grown up adjacent to the fort had been named Georgetown in honour of His Majesty King George III. The Company had staff, servants and 'followers' which soon amounted to 1,000 persons. Light had recorded on his landing at Pulo Pinang that the total Malay population of the island was '. . . 50-odd . . .'[19]

With the enlargement of the Settlement, so the military garrison increased. The main duties of the garrison troops were to act as guards and a number of the Sepoys were used in the role of a Settlement Police force.

In the East India Kalender for the year 1788 three officers are shown as serving on Prince of Wales Island – these were:

Lieutenant John Glass – Artillery,

Lieutenant Fireworkers George Rahen and William Murray

However, it is considered that at least one officer of the Marines would have been on station in addition to the Artillery officers but he is not identified. In that year the fort was reinforced under Light's orders when the stockade was doubled. He later had the stockade trebled and mounted some cannon to the seaward side. The fort was however not a success and in 1790 he wrote complaining of its condition saying '. . . the fort is decayed . . . the fort has rotted and whole faces are falling down . . .'[20] The situation on Prince of Wales Island gave cause for concern and as a result of an attempt to enter Penang harbour by a fleet of eighteen long prahus armed with guns belonging to the Lanuns, which was fortunately stopped by the three ships under Light's command, Calcutta decided to increase the defence of the Settlement by sending two companies of native infantry and ten eighteen-pound guns for the defence of the port. In the meantime, on the island, Captain Glass, who was acting as the Commandant, had under his command '. . . two weak companies of Sepoys . . .' In February 1791 Commodore William Cornwallis arrived with his squadron. In March, Light discovered that the Sultan of Kedah had appointed the Tungku Muda his brother-in-law, Rading Mahomed his new son-in-law, and the Laxsamana to command his forces and authorised them to either negotiate or attack as they considered necessary. Captain Scott[21] was sent to Kedah with terms for a treaty and he met the Laxsamana at Qualla (Kuala) Muda, the mouth of the river, and a treaty was agreed with the Laxsamana agreeing that the Lanun Pirates would be sent away. With Scott having secured the peace treaty, Cornwallis sailed away with his fleet heading for Madras. He sailed on 15 March and on the 19th the Lanuns moved down the Kedah rivers and assembled at Prai, immediately opposite Georgetown. The following day the Sultan despatched his own force to join the pirates and an armada of some ninety prahus assembled and sailed past Fort Cornwallis in an act of defiance. The Malay-Lanun force commenced erecting stockades and forts on the mainland opposite Georgetown.

The merchants in Georgetown panicking over this threat of an attack, ensured that all the merchant ships had quit the harbour by 23 March. One East Indiaman remained and that was the *Bombay Castle*. With the enemy force still building up on the shore near Prai, Light decided to strike first. Captain Glass embarked three companies of Sepoys and concentrated his whole force in one combined operation so that in the early morning darkness of 12 April the expedition set off from Prince of Wales Island and headed towards Prai Point. It was essential for the success of the operation that the unwieldy ships' boats that were used kept together despite the strong currents and tides and that the naval support force kept with the assault force. Fortunately all went well, the embarkation, the journey across the water and having once formed up, the troops, under Captain Glass, dispersed the large force that had been collected together for the defence of the Malay fort. The first fort was overrun and the second fort made an attempt at resistance but once the Sepoys had gained the ramparts they soon put the defenders to flight. The forts were then burned and the enemy fled retreating into the jungle. In the meantime the gunboats advanced and attacked the fleet of prahus which was estimated to be about 200 strong. Light had succeeded in defeating an enemy of some 10,000 Malays and Lanuns and 250 armed prahus had been dispersed. The losses suffered by Captain Glass's force were extremely light with four men being killed, three Sepoys and one Company slave.[22]

Captain John Glass was the Commandant of the Garrison and as such had control not only of the artillery but also of the Sepoys. Complaints had been made against some of the Sepoys who were employed as police. Malay traders accused them of harassment and of intimidation but Glass dealt with these problems quickly and efficiently. He and Light worked well together and appear to have been an efficient team for the administration of this new Settlement. It was unfortunate for Light that Glass died on 5 April 1793 and was buried in the Northam Road Cemetery. Glass's monument has the oldest decipherable inscription of all the military graves in this cemetery.[23]

With the war breaking out in Europe in that year there was cause for alarm within the Settlement over the condition of the defences. The commander of the Artillery at Fort Cornwallis wrote a report in which he complained '. . . our Fort is nothing more than an old rotten stockade, with decayed platforms than cannot hold guns . . .'[24] At this time the Garrison strength was listed as being seventeen men of a detachment of the Royal Artillery, thirty non-European infantrymen and a force of 100 Sepoys of the Marine Regiment, Bengal Native Infantry.

Under Glass's successor, Captain Hamilton,[25] the military discipline deteriorated but the situation improved as the size of the garrison force increased and more experienced officers arrived in the Settlement. During 1794, Bengal considered that the whole question of the defences at Georgetown should be reviewed and in that year Major Alexander Kyd, of the Bengal Engineers,[26] was ordered to go to Prince of Wales Island. He was to report on the state of the fortification and also to examine the state of readiness of the troops to see if they could defend the Settlement against a moderate attack and, if not, he was to make recommendations as to the number of reinforcements that would be required to secure the Settlement against an enemy. However by this time Light, again without prior approval of his superiors at Bengal, had rebuilt Fort Cornwallis as a stone structure at the cost of some $67,000.[27]

Francis Light died in October 1794 which caused a serious setback to the control and administration of the Settlement. The date shown on the inscription on his tombstone in the Northam Road Cemetery is 21 October 1794[28] but Tregonning records in his book that an official despatch sent by Philip Mannington, Light's successor, shows Francis Light as having died on 25 October 1794.[29] The burial register for the cemetery for that period no longer exists, and therefore the correct date cannot be verified.

Major Kyd, who had in fact visited Penang in 1787 as a Captain in the Bengal Engineers to report on the condition of the harbour, completed his inspection of the fort, the defences and the harbour and submitted his report to Bengal in August 1795, some ten months after the death of Light. He describes the fort:

> '. . . The fort is a square of masonry of 500ft exterior side with a very inconsiderable ditch, with no covered way or outwork of any kind, nor is there as yet even a parapet put

upon it, and as the Chinese who contracted for the completion of it has failed in his undertaking, no other means has been taken to go on with it, so that the fort is as undefendable as it was at the beginning of the war when it was undertaken and at first carried out with great spirit.

It is exactly on the same spot, and of the same dimensions, as a stockade of trees that was constructed on first taking possession, intended merely as a security against the Malays, for which purpose then it was adequate . . .'[30]

It became clear from Kyd's report that the fort was in need of additional work and alterations if it were to be considered as the main defensive position for the town and Settlement. He also commented in his report that the fort was '. . . surrounded on all sides with buildings, many of them masonry, within a pistol shot of the walls [and] so close to the water's edge that ships can anchor 100–150 yards off, overlook it from the main mast, or even the poop of a large ship . . .'[31] He concluded his report by stating that in his opinion it was hard to imagine a spot '. . . worse chosen, it can defend neither town nor harbour . . .'

During the time that Kyd had been on Prince of Wales Island inspecting and compiling his report, the French had overrun Holland in January 1795 and by establishing the new Republic of Batavia, which concluded an alliance with France, the Dutch in Malacca were then considered to be a threat to the Company's settlement. An expedition was mounted from Madras with four European battalions being embarked under Major Archibald Brown and the harbour at Prince of Wales Island was used as a staging port. The use of the port for this function was seen as a great advantage should future eastward-bound expeditons be mounted from India. There was, however, still uncertainty about the future of the fort as Kyd, although praising the harbour facilities, failed to convince Bengal of the justification for further large expenditure on the fort. The Governor General gave authorisation for local repairs only to be made to the fort. He considered that if Penang was to be defended it would be by the fleet rather than the 'crumbling stone of the fort'.

Although Mannington had replaced Francis Light[32] his appointment was not confirmed and a military man was given the appointment. On 14 May 1795, Major Forbes Ross MacDonald took command of the island and ruled it with the firm hand of a strict military disciplinarian. The population had increased by this time to 20,000, as estimated by Major A. Kyd, whereas only four years prior to Kyd's arrival Light had, in 1790, estimated his island's population as being 10,000. McDonald's tour of duty was not a happy one as he was in frequent conflict with the civilian merchants and his own civil servants. He resigned in 1799 and returned to Madras. At this time the post of Superintendent of the Settlement was abolished and a new post was introduced, that of Lieutenant Governor.

Penang had in the meantime also proved a useful staging post as in 1796 the convict settlement on the Andaman Islands was closed and the convicts moved to Prince of Wales Island. The expedition sent, in 1797, to invade Manila used the harbour as a rendezvous but the expedition was recalled and in the event did not press home the attack on Manila.

By 1800 the Settlement had increased in size and the military establishment was increased so as to include an Engineer officer and Medical officers. The officers posted to Prince of Wales Island were:

Bengal Engineers

Lieutenant Joseph Stokoe.

Bengal Medical Department

Surgeon James Hutton

Assistant Surgeon Michael Cheafe

Assistant Surgeon Henry Waring.

Sir George Leith was appointed Lieutenant Governor of the Settlement and in order to deal with the constant problem of pirates and also to obtain additional land for the production of food he negotiated with the Sultan of Kedah for a tract of territory on the mainland of Malaya. The Sultan was in favour of the idea, and was able to convince those of his ministers who were averse to allowing the Company to get a foothold on the mainland. It was eventually agreed that a coastal area near Prai would be the most suitable for both parties. By promising '. . . $2,000 to the Sultan's ministers and harem, every obstacle was removed . . .' and for an annual payment of $4,000 the Sultan ceded a section of land which was three miles in depth and stretched from the Muda River in the north to Soonghy (Sungai) Crean in the south and this area included Prye (Prai) town and the mouth of the Prye river. Formal possession was taken of this area on 7 July 1800 and was named Province Wellesley after the governor-general in India, the elder brother of Arthur Wesley, (who later changed his name to Arthur Wellesley) who was to become the Duke of Wellington. Arthur Wesley visited Prince of Wales Island during the autumn of 1797 as part of the expedition intended to attack Manila. His own unit although required to live on the troop transports in the harbour were brought ashore to carry out training.

With the additional area to defend, it was again necessary to increase the establishment of the Settlement. In 1801 the establishment was increased and the following officers were stationed on the island – filling the appointments as shown:

Commandant	
Major Thomas Polhill	7th Regt, Bengal NI
Adjutant and Quartermaster	
Lieutenant Norman MacAlister	Bengal Artillery
Commissary of Ordnance	
Lieutenant Charles Sealy	Bengal Artillery
Engineer	
Brevet Captain Joseph Stokoe	Bengal Engineers
Medical	
Surgeon James Hutton	Bengal Medical Department
Assistant Surgeon Henry Waring	Bengal Medical Department
Paymaster	
Mr P. Mannington	
Infantry	
Lieutenant John Lindsay	10th Regt, Bengal NI
Lieutenant John Leslie	5th Regt, Bengal NI
Lieutenant Conell Simpson	14th Regt, Bengal NI.[33]

Thomas Polhill was promoted to the rank of Lieutenant Colonel on 4 January 1801 but remained in command of the garrison at Prince of Wales Island.[34]

The importance of Pulo Pinang increased during the early years of the 1800s as the harbour was used as a staging post both for naval vessels in the area and as a port of call for ships plying the China trade routes. In 1805 it was decided by the East India Company that the status of the Settlement should be upgraded to that of a Presidency. One of the objectives of this move was to improve the harbour facilities and construct a dockyard and to have the ability to build and repair His Majesty's ships. In addition the Government of the newly-created Presidency were tasked to do '. . . whatever may be necessary to put the island in a tolerable state of defence . . .' and that '. . . a regular fortification for the permanently securing against a European force should be constructed . . .'[35]

During the changeover period of the administration in 1805 it was actively considered that a naval base should be constructed at Pulo Jeraja, an island some six or seven miles along the eastern coast from Georgetown. Bengal required that a new fort should be constructed to safeguard the new base and Georgetown. No single fort could be expected to cover both the naval base and Georgetown, which apart from any other considerations were miles apart. Fort Cornwallis was conveniently located to the town and could, if developed properly, cover both the town and the harbour. Whilst these considerations were discussed the policy was to keep the existing fort in a reasonably serviceable condition and this continued until 1811. In the meantime the proposal of the construction of the new fort at Pulo Jeraja was debated and it was finally agreed that the cost of its construction would be too high and the plan was abandoned.

Captain Joseph Stokoe, the Settlement military engineer, died at Penang on 12 January 1801 and was replaced by Lieutenant John Gordon, also of the Bengal Engineers. He, in turn, died whilst serving at Penang on 12 November 1804.[36] He was succeeded by Captain Thomas Robinson who took up the appointment which was now graded as 'Engineer and Surveyor, Prince of Wales Island' in June 1806. Other changes in the military staff of the island included Lieutenant Charles Morse, of the Bengal Artillery, being appointed on 3 December 1805 as Deputy Commissary of Ordnance in place of Captain Charles Sealy.[37] Unfortunately Morse's tour of duty in the Settlement was cut short by his early death in February 1806 at the age of twenty-six. The burial register for Prince of Wales Island shows Lieutenant Morse as being interred on 16 February 1806.[38] Although there was a continual increase in population of the Settlement during the early 1800s it is probable that a series of epidemics hit the Settlement during 1806 as the Burial Register records 101 burials for that year, whereas the three previous years' average was only 25 persons buried each year. Of the 101 buried in 1806, 72 were military or naval personnel with the remaining 29 being civilians which included some members of military families. Unfortunately the Register does not give the cause of death and a careful examination of the memorials and gravestones in the Northam Road Cemetery has failed to reveal any inscriptions giving the causes of death. It is unfortunate that during the bombing of Georgetown during 1941 bombs fell on both ends of the cemetery destroying graves and memorials at the eastern and western ends. It was at the eastern end that the majority of the graves dating from the 1806 period would have been located.

During the second half of the year 1806, Captain Thomas Robinson, the Settlement's new engineer and surveyor, was instructed to submit plans for the defence of the town against a European naval attack and to find a suitable site for a second fort, which was to be located near to the proposed naval base. He submitted an elaborate scheme which included reinforcing the Garrison to a strength of 2,000 men and purchasing and demolishing a large area of Georgetown so as to clear an area around Fort Cornwallis to give a clear field of fire. His estimate for the repair and modernisation of the fort was $20,000.[39] These very costly plans and the proposed destruction of parts of the town were, as was to be expected, rejected by the Penang Council, which in its new capacity as a Presidency could make such a decision, and thus the fort remained in its unsatisfactory condition.

With the development of the Settlement and the increase of the garrison military strength, the need for additional accommodation grew and the Sepoy lines were built eastward of the town. With Fort Cornwallis being on the shoreline the various buildings developed eastwards and to the south of the fort. The Church of St George the Martyr is located east of the fort and was probably built about 1798–9 as the first recorded Anglican service in Penang is given as being in 1799.[40] From 1800 regular services were held in the church which still stands looking very much as it did in 1819 when it was consecrated by the Bishop of Calcutta the Rt Rev T. F. Middleton, who at the same time consecrated the Northam Road Cemetery which itself is just east of the church. Moving eastward again, the site of the old lines or barracks can be traced today through the name 'Barrack Road'.

During 1805 an order was issued by the Court of Directors in London which placed the Governor, Philip Dundas, in command of the whole of the Settlement with jurisdiction over the military garrison. The order reads:

'... Whereas we, the said United Company, some time since obtained by cession [*sic*] from a Native Prince in the East Indies an Island heretofore called Pulo Penang and now called Prince of Wales Island situate in the Straights [*sic*] of Malacca in the East Indies, and also a tract of Country on the Peninsula of Malacca opposite the said Island and we have caused a town and citadel or Chief Fort to be erected on the said Island. And whereas we have resolved to make a Settlement and establish a Factory there, and to provide for the Government thereof as far as we lawfully may or can ...

Philip Dundas Esq., to be President and Governor of our said Factory of Prince of Wales Island ...

[He should] exercise all such other Power and authorities as are incidental and necessary for the Government of the said Factory and Places, and our Servants and Soldiers therein ...'

In addition the following authority was given to Philip Dundas:

'... To Philip Dundas,

[we] appoint you to our Governor and Commander-in-Chief of our Citadel or Chief Fort at Prince of Wales Island in the Straights [*sic*] of Malacca in the East Indies, and of the Garrison of the said Citadel and Fort, and also Governor and Command-in-Chief of the Town which is or shall be ... [to] command all Commission[ed] Officers, Non Commission officers, Soldiers and others belonging to our Military Forces, and all the People and inhabitants employed or residing in our Citadel, Fort and Town ...'

This order was dated 10 April 1805, and on the same day an order was given to the Commander of the garrison which reads:

'... To Colonel Norman Macalister,

Commander of all our Military Forces at our Factory of Prince of Wales Island in the Straights of Malacca in the East Indies and all places subordinate or annexed thereto, save and except our Forces which are or shall be employed in our Citadel or Chief Fort at Prince of Wales Island, or in the Town which shall be the seat of Government of the said Factory ...'

The order authorised Colonel Macalister to take command of the troops that were allocated to the defence of the Settlement.

The life in the garrison for the military remained routine with the normal round of guard duties and drills, and the occasional range practice. Ammunition was expensive and although the Sepoy regiments followed the British army pattern of training there would have been a need for economy in a garrison such as Penang. An examination of the orders of the Settlement gives some indication of the types of problems and movement within the garrison. Although, in general correspondence the Settlement was referred to as 'Prince of Wales Island' the internal title used was 'Penang'. The following gives extracts from 'Penang General Orders and Notices' for the years 1805 to 1807:

Plan of Prince of Wales Island, *c*1820. Courtesy of The British Library (Ref 266548).

Fort Cornwallis: seaward defences. Photograph by Mrs J. Harfield, 1977.

Memorial of Captain John Glass, Northam Road Cemetery, Pulo Pinang. Photograph by Mrs J. Harfield.

Memorial of Lieutenant Charles Morse, Northam Road Cemetery, Pulo Pinang. Photograph by Mrs J. Harfield, 1977.

'*General Orders*
1805

Nov 13 The 73rd Regt of Foot to embark on board the HC Ship 'Dwaynes' returning to Europe. (The 73rd Regiment was merely in transit in the Settlement at this time).[41]

Nov 19 Captain James Dewar, 20 Regt NI and Lieut Vincent of the same Regt to proceed to Bengal to stand trial before a Court Martial on charges given in against them by Major Dick . . .'

James Dewar returned to the Settlement as a Major and as the senior officer in command of the detachment of the 20th (Marine) Regiment, Bengal NI and this is confirmed by the entry in the East-India Register and Directory for 1805.

'Nov 22 Lieut Charles Morse of the Bombay Artillery to be Deputy Commissary of Ordnance . . .'

This was qualified by a further order dated 3 December 1805 which was published as 'General Order by Government'

'Fort Cornwallis 3rd December 1805.

The Hon'ble the Governor and Council have been pleased to appoint Lieutenant Charles Morse, of the Bombay Artillery doing duty with the detachment on board the 'East Camden' to be Deputy Commissary of Ordnance in the room of Captain Charles Sealy who has resigned. Lieutenant Charles Morse to take charge accordingly. Captain Sealy has permission of the Hon'ble the Governor and Council so soon as he may have delivered over charge of his office, to proceed to Calcutta on his private affairs.

By order of the Hon'ble the Governor and Council
(Signed) H. S. Pearson.
Secy to Government'[42]

Morse, of course, died after only just over three months in his new appointment.

Other orders that were issued in November of 1805 concern movement of officers for various reasons but illustrate the type of order issued at this time:

'Nov 24 Capt William Gill of the Marine Regt to proceed to Fort William . . .'[43]

'Fort Cornwallis. 29th November 1805.

Captain Dewar and Lieutenant Vincent are directed to send on to the Adjutants Office a list of Evidence they mean to call on their approaching trial, that they may be directed to hold themselves in readiness to proceed to Fort William on the Transport daily expected from Balambangan . . .

By order of the Hon'ble the Governor and Council.
(Signed) H. S. Pearson.
Secy to Government'[44]

In addition to movements being notified in General Orders changes of appointments and miscellaneous military instructions were also promulgated through these orders. The entries for the year 1806 give a good example of the types of orders that were published.

'*1806*

Mar 10 Lieut Phillips Actg Adjutant to the Troops having requested permission to resign his appointment, Captain Drummond is directed to resume charge of that office.[45]

May 1 Lt Leonard Whitesmith of the 1st B.V. Brigade to be adjutant of the Militia in the room of Lt Oake resigned.[46]

Jun 19 [This order commented on] the zeal and readiness displayed by the Militia of PW Island and the unit is now disbanded.

Fort Cornwallis. 29th August 1806.

The Hon'ble the Governor and Council are pleased to direct that from the last day of the present month Acting Ensigns Barrett and Moser, shall be struck off the strength of this Garrison. On this occasion, the Hon'ble the Governor and Council have much pleasure in testifying their entire approbation of the Conduct of these officers, and regret that the general usage of the Service has rendered it necessary to dispense with their Services, however sensible they may be of the great utility which might otherwise be derived therefrom.

By order of the Hon'ble the Governor and Council

(Signed) Thomas Raffles.

Actg Secy to Government'[47]

A General Order was issued on 17 October 1806 ordering the assembling of the troops on the Esplanade to hear the Proclamation read announcing Colonel N. MacAlister's succession to the Government of Prince of Wales Island. The number of officers stationed in the garrison had slightly increased in 1806 and the following officers are shown as serving on the establishment. These are annotated in the East-India Register as being on 'P. W. Is', or 'P. W. Island', etc.

6th Regiment, Bengal NI

Captain/Lieut George Herbert.

10th Regiment, Bengal NI

Captain/Lieut John Lindsay

20th Regiment, Bengal NI

Major James Dewar

Captain/Lieut Henry Oake

24th Regiment, Bengal NI

Lieut R. H. Buchanan. Lieut Richard James.

Regiment of Artillery (Bengal)

Captain/Lieut Norman Macalister

Medical Department

Assistant Surgeon Henry Waring. Assistant Surgeon James Heriot.

By early 1807 the arrival and departure of many high ranking personages, including numerous Admirals, necessitating the firing of a large number of ceremonial salutes caused part of the one of the walls of the fort to be damaged and to partially collapse. The guns of the fort were, it was said, its worst enemy. A large section of the rampart was then dismantled and rebuilt. Prince of Wales Island continued to act as a convenient staging post and during 1807 the island was once again used as a staging area, on this occasion it was for the '2nd Battalion of His Majesty's 1st or Royal Regiment'. Prior to this troops who had been assembled for the Moluccas campaign assembled at Penang and after serving as an occupying force at Balambangan from 29 September 1803[48] until 1805 a force of some 800 troops staged through the Settlement and the Company requested the Governor to prepare the Sepoy Lines so as to accept '. . . the troops expected from Balambangan . . .' The troops arrived on 16 December 1805 under command of Lieutenant Colonel Gales.[49] The 2nd Battalion 1st or Royal Regiment arrived and disembarked. An order, published at Fort

Cornwallis, shows the following administrative details '... [the] Regiment having arrived at this Presidency they are to receive the same allowances with the Hon'ble Company's Troops stationed here ... the Paymaster of that Corps [is] to draw on account on the Military Paymaster here a sum not exceeding one month's pay and allowances according to the present strength of that Regiment, and issue the same, deducting the amount due to the Military Storekeeper and the Contractor which he will cause to be paid weekly by his Quarter Master [*sic*] to them respectively on account of Liquor, Meat and Bread furnished ...'[50] This order was signed by the Acting Secretary to the Government, Thomas Raffles.[51] The 1st Regiment were to remain on the island until late April 1808 and during this time they suffered a number of casualties due to fever. During the year 1807 twenty-eight men died and in the following year, up until the date of their departure, they lost a further fourteen soldiers. All were recorded as being buried in the Northam Road Cemetery but despite a careful search no trace could be found of any of the graves of the men of the 1st Regiment.[52]

The Garrison Engineer and Surveyor, Captain Thomas Robinson, again submitted a proposal for the improvement of the defence of the Settlement. This second plan involved the abandonment of Fort Cornwallis and the construction of a new fortification '... eight times the size of the existing fort, [to] be erected 120 yards further inland ...'[53] It would have again meant the destruction of the larger part of the town and would have achieved the very result that the fort was supposed to guard against. This plan was also rejected by the Presidency Council, as had been the first plan, and so Robertson set to work on a third plan. With the assistance of Colonel Norman Macalister[54] a plan was drawn up and submitted in July 1808 which was then approved. Under this latest scheme stone batteries 600 feet in length, were added. The new batteries were twenty feet thick and were capable of mounting seventeen heavy guns. The work was undertaken and completed by the end of January 1809 at a cost of just over $4,500. During 1809 an additional battery was constructed extending out over the north-west bastion of the fort and was designed to mount seven 32-pounder cannon. This new section reached down to the beach and gave artillery fire to cover the harbour. With the completion of this work to improve the fortifications Georgetown became one of the best-defended ports in the Far East.

The East-India Register for 1808 shows an increase in the number of infantry officers and men garrisoned on the island. There were in that year nine officers of the 20th (Marine) Regiment, Bengal Native Infantry and four Bengal Artillery Officers –

20th (Marine) Native Infantry

Major James Dewar	Captain John B. Sealy
Lieut John M'Innes	Lieut Nicholas Manley
Lieut Thomas Travers	Lieut John Gordon
Lieut John Sappings	Lieut Thomas Black
Lieut Thomas Griffiths	

Bengal Regiment of Artillery

Captain J. P. Drummond	appointed Town Major
Captain Horatio Greene	
Captain/Lieut William Richards	
Captain/Lieut Allan Graham[55]	

Lieutenant John M'Innes was appointed to act as the Brigade Major under Penang General Order dated 19 June 1806.

By this time an Aide de camp had also been appointed to the Governor and this appoint-

Fort Cornwallis, Pulo Pinang, north wall. Photograph by Mrs J. Harfield, 1977.

Fort Cornwallis, Pulo Pinang, east wall. Photograph by Mrs J. Harfield, 1977.

Fort Cornwallis: Arms Cote and Ammunition Store. Photograph by Mrs J. Harfield, 1977.

Fort Cornwallis: Arms Cote. Photograph by Mrs J. Harfield, 1977.

ment was filled by Lieutenant John Veatch of the 4th Regiment, Bombay Native Infantry.[56]

The Governor Philip Dundas, and his staff, contributed to the social life of the Settlement which had previously been regarded as a rather primitive seaport, but was now well provided with social events such as balls, picnics, etc, many of which have been recorded in the editions of *The Prince of Wales Gazette*. The attap huts and sheds that abounded along the streets of Georgetown were replaced by brick and stone buildings and by 1811 the suburban hillsides were thickly studded with 'handsome villas and picturesque bungalows and intersected with pleasant carriage roads all of which gave a favourable impression to the visitor to the Settlement.'[57]

Once again the peaceful routine of Settlement was destroyed when, in 1811, naval vessels and troop transports *en route* to the invasion of Java, called at the port prior to their main assembly area which for this campaign was to be Malacca. During the period of the capture and occupation of Java the harbour of Pulo Pinang was frequently used as a staging post, and the fort was maintained in a state of readiness against an anticipated attack by French privateers.

The engineer post of the Settlement was regraded and became the post of 'Superintending Engineer, Prince of Wales Island'. This post was filled on 11 November 1814 by Lieutenant Robert Smith. Robert Smith was born in 1787 at Bideford, Devon the third son of James and Mary Smith. James Smith had been private secretary to the Marquis of Hastings and was living in Bengal in the early 1780s. Robert Smith was admitted as a Cadet in 1803 and became an Ensign in the Infantry on 29 April 1805. He transferred to the Bengal Engineers and was stationed at Calcutta and was fortunate enough to be there at the same time as the well-known artist George Chinnery. He had already shown promise as an artist whilst at Bideford but it is quite likely that meeting such artists as Chinnery and Thomas Hickey and Robert Home, who were all resident in Calcutta during Richard Smith's period of duty there, his interest and expertise in painting grew. He was promoted to the rank of Lieutenant on 17 September 1807 and was appointed to the post of Superintendent of Works at the Bengal Gun-carriage Agency. In 1808 he was responsible for the construction of the lighthouse at Kedgeree. In 1809 he was appointed Adjutant of the Corps of Engineers and in 1812 he was appointed Aide de camp to the Commander-in-Chief, Sir George Nugent. Smith's expertise with brush and pencil was noted by Lady Nugent in her journal commenting that she had '. . . received a present of drawings from Mr Smith, an engineer ADC. He draws beautifully and his sketches are so correct that I know every place immediately . . .'[58] In 1815 Smith briefly left the island to take up the appointment of Assistant Field Engineer, 2nd Division during the Nepal War and was awarded the India Medal. He returned to the island and remained there probably until 24 July 1819 when he was granted furlough on a sick certificate from that date until 30 October 1822. During his service on the island he was promoted to the rank of Captain, this was on 17 September 1817. It was during this second period of his service in the Settlement that he completed his set of views of the island. These now form part of the collection in the India Office Library and also part of the collection of the Penang State Museum. His subject matters were all dated 1817–1818 and were given titles. The following list of his 'Penang' paintings gives an indication of his range of work.

View of the Chinese Mills, Penang.

View of the Cascade, Prince of Wales Island.
Inscribed to George Chinnery Esq. The cascade or waterfall is located in what is now the Botanical or Waterfall Gardens, although it is less dramatic than when Smith painted the scene. The Gardens are abour four miles from Georgetown.

View of the Great Tree, Prince of Wales Island.
Inscribed to Captain Thomas Maddock (Maddock was of the 7th Regiment Bengal NI and was employed as Assistant Secretary to the Military Board). The Great Tree was situated in the Relau valley and was a favourite spot for picnics.

View from the Convalescent Bungalow, Prince of Wales Island.
Inscribed to Major J. MacInnes (MacInnes, spelt as M'Innes, of the 20th (Marine) Regt Bengal NI was serving in the appointment of Brigade Major in the Settlement). The bungalow was erected by the East India Company near to the top of Penang Hill for accommodation of sick officers and their families. It was also used by visitors who wished to spend a night on the hill but permission to do so was required to be obtained from the Government. In the early days the ascent to the bungalow was made on horseback, and ladies were transported in a sedan chair borne on the shoulders of Malay bearers. The path at this time was a steep ten-foot-wide track. Adjacent to the bungalow was the residence of the medical officer. Later the bungalow was generally used for the accommodation of visitors to the Settlement.

View of the North Beach from the Council House.
Inscribed to Lieutenant General Alexander Kyd. The inscription is to the same officer who, as a Major, had been sent by Bengal to report on the defences of the Settlement. The dedication may well have been added later as Smith proceeded on furlough on 24 July 1819 but Kyd was not promoted to Lieutenant General until 12 August 1819.[59] The painting shows the reservoir which stood on the ground where the Eastern & Oriental Hotel now stands. In the distance can be seen the house 'Runneymead' which was purchased in 1807 by Thomas Stamford Raffles. The peak named Mount Erskine can also be seen and it was on this peak that the signal station had been built.

View of Glugor House and Spice Plantations, Prince of Wales Island.
Inscribed to David Browne Esqre, the owner. Glugor House was situated about five miles south of Georgetown and was built about 1812 for David Browne.

View of Halliburton's Hill, Prince of Wales Island.
Inscribed to Sir Ralph Rice. The bungalow was situated on the eastern side of Penang Hill, below the level of the Convalescent Bungalow. It was the home of Thomas Halyburton (as his name was normally spelt) who was a merchant and at one time Sheriff of Prince of Wales Island.

View from Strawberry Hill, Prince of Wales Island.
Inscribed to Sir Charles D'Oyly, Bart. The view from Strawberry Hill was spectacular and from this location (which is today where the Penang Hill Railway terminus is built) the town and the fort can be seen, the fort is some four miles from the foot of the hill. The name Strawberry Hill was derived from a garden on the hill at this point which had strawberries and roses growing.

View of Suffolk House, Prince of Wales Island.
Inscribed to the Hon'ble the Governor W. E. Phillips Esq. The house was built by W. E. Phillips, probably about 1809. Phillips later became Governor of Prince of Wales Island.

View of Mount Erskine and Pulo Ticoose Bay, Prince of Wales Island. Inscribed to John James Erskine Esq., The view is from Mount Erskine looking across Pulo Ticoose (Tikus) Bay and Georgetown with the hills of Kedah in the background. Erskine was, at one time, Assistant Superintendent of Marine of the Island.

All the paintings were engraved by William Daniell and published by him in London on 1 January 1821.[60]

With the ending of the war in Europe the need to maintain Fort Cornwallis in a state of readiness for the defence of Georgetown ceased to exist and the Presidency Council were reluctant to spend any further funds on the fortifications of the Settlement. The Settlement continued to expand and at this time the military establishment was beginning to develop so that troops could be provided for Malacca and the newly-formed settlement at Singapore at the eastern end of the Straits of Malacca.

Another officer who had been posted to Penang and who was to have an influence on the development of the Settlement was Lieutenant James Low.[61] Low moved to the Settlement and came under the orders of the Government of Prince of Wales Island on 1 January 1818. He settled into the routine well and in June of that year he is mentioned as appearing as a witness at the double wedding of the elder daughter and niece of Colonel Alexander Bannerman, the Governor of the Settlement.[62]

In 1820 a local unit was again raised and on this occasion it was raised to save the expense of having a Regiment of the Madras army permanently stationed in the garrison. Lieutenant Low was given charge of the unit. The unit was mainly recruited from the Upper Provinces of Bengal and it was to be used as a defence force for Province Wellesley and to act against the pirate menace which was still very much a problem. Low remained in command of this unit until mid-1827 when it was disbanded[63] and the majority of the soldiers returned to their homes in Bengal.[64] Details of James Low's long career in the Settlements is recorded in Chapter 8.

During 1824, due to his expertise with the Thai language,[65] Low was seconded from the Penang Garrison and was employed on a political mission to the Raja of Ligor in an attempt to obtain river craft for an expedition to Ava.[66] In September of 1824 he was again loaned for similar work and this time accompanied Lieutenant Colonel Snow on a mission to Tennasserim.

In September 1826 he was given command of a detachment of forty sepoys and a cruiser and the expedition set off from Prince of Wales Island to Perak. His mission was to warn the Siamese who had invaded Perak that they must leave the state. The Siamese had, in fact, withdrawn even before Low and his detachment had arrived. Low and his troops stopped at the capital of the state for a month and on 18 October 1826 he grossly exceeded his authority by signing a Treaty with the Sultan of Perak which gave the Sultan the right to expell any Siamese he did not wish to have in Perak. In addition Low's agreement committed the Company, in other words the Commander of the Penang Garrison, to render military aid to the Sultan should he request such help. The Penang Government were not empowered to approve such a Treaty. The matter was referred to the Governor General but the Treaty appears never to have been formally agreed.[67] After a lengthy battle by correspondence the Supreme Council agreed to the conditions of the Treaty although it was never formally ratified. Low's treaty then became an accepted obligation and in later years when the Sultan of Perak appealed to the Government of the Straits Settlements for aid the terms were accepted as binding and aid was provided during the years 1844 and 1853.

At this time, 1826, the three Settlements had developed and Penang, Malacca and Singapore were brought together under one Presidency and from that time became known as the Straits Settlements Presidency.

It was during 1826 that a series of piratical raids were carried out on the shipping in Penang harbour. Captain Low was ordered to deal with the problem and set out with half a Company of his Local Corps to attack and destroy the stronghold of a local chief Nakhoda (Captain) Udin who, it was said, was responsible for the raids. The stockade was located, attacked and then destroyed with Udin and some of his followers captured and brought to Penang by Captain Low. The Court had no jurisdiction over this minor chief and he was sent to the Raja of Ligor to be punished. The Raja refused to deal with the case declaring that the Chief was one of his own officials and complained to the Penang Government about Low's handling of the situation. The complaints were received and were passed on by the Secretary - Captain Henry Burney (Low had been an official at Burney's wedding to Janet Bannerman the niece of the late Governor) - to the Supreme Government. The outcome was that the Governor Robert Fullerton, and Captain Low were both censured. Low was suspended from all political employment but Fullerton had prepared a good defence which convinced the Supreme Government that the situation had been correctly handled by Low and the suspension was lifted.[68]

Life in the military garrison on the island continued to be a routine of guard duties and training. An examination of the Penang Orders and Notices for the period 1828 to 1830 reveal only routine matters such as:

'9 July Captain Bunbury to be acting Mily Secy and aide de camp vice Lieut Hay proceeding to Europe.[69]

21 July Lieutenant Hicks to be Acting Fort Adjutant vice Lieutenant Snow proceeding to Malacca and Singapore . . .

21 July Lieutenant Boulderson to be acting Executive officer during Lieut Hodge's absence.[70]

10 Sep Captain James Low appointed Suptd External Police and Superintendent Lands on a salary of 500 Rupees per month . . .'

Under the heading of buildings the following expenditure was noted for the year 1828:

'1 April Artillery Barracks and Officers Guard Room for repairing the same assessed to Rs 220.

21 July Signal Sergeants bungalow on the hill repaired for Rs 55-13-0.

20 Oct Sepoy Hospital, Officers Guard Room, Artillery Barracks, (and) Armoury. Proposed estimate for repairing same amounting to Rs 150 for Hospital and 50 Rs for the others.'[71]

Leave from the Garrison was also promulgated in the orders and entries appeared in the following format:

'21 July Lieut Colonel Sale and Lieut Hodge granted leave to proceed to Madras.[72]

1 Oct Lieut Trotter of the 35th Regt MNI granted leave to proceed to Europe on Sick certificate.[73]

3 Nov Lieut Chauvel NI to go to Europe on Sick certificate.[74]

22 Dec Lieut Boulderson permitted to proceed to Madras on private affairs.'[75]

An examination of the General Order book for Penang shows not only entries covering expenditure, leave, appointments, etc but also other administrative details such as movement of companies, and some of the more unusual administrative problems that had to be dealt with by the Garrison staff. Some of these entries are reproduced in order that a picture of the day-to-day problems can be seen:

1828

'Military

3 Jan 2 Companies of the 25th Regt MNI to embark on the 'Speke' to relieve the 4 companies of the 35th Regt in Singapore.

21 Apr Captain Bond appointed to the command of the Artillery.[76]

14 Aug Arsenal Establishment reduced and the surplus men returned to Madras on the 'Hastings'.

1 Oct The services of Lieutenants McLellan and Coldingham and Sergeant Smith placed at the disposal of this Government.[77]

20 Oct [The] Commanding Officer replies [to the Penang Government] respecting the Province Wellesley Guard – the stationing of a European Officer there deemed unnecessary.

— Nov Town Major requested to post a Guard at the Collector and Accountants Office.

3 Dec Town Major to take charge of Gunner Lascar Barracks on the North Beach.

5 Dec [The] Commanding Officer reports the inadequacy of the Forces at this station for Guards, Reliefs, etc. [The Town Major was requested to submit a like statement giving grounds of his opinion].'[78]

Entries in the Order book confirm the disbandment of the Penang Local Corps and one order dated 1 July gives a request by the Superintendent of Family Money, Bengal which asks for '. . . descriptive rolls of the Commissioned, non-commissioned officers and Sepoys of the late PWI Local Corps . . .' The entry is annotated '. . . sent to the Military Paymaster . . .'[79] Another entry on the same date shows that Subadar Sheik Dobie of the late local Corps is '. . . permitted by the Supreme Government to reside and draw his pension at Penang . . .'[80]

Two entries of this period also show that life was not always harmonious within the Garrison. An entry dated 9 July reads '. . . Ensign Goodenough having been shot in a duel by Ensign Mayhew . . . are ordered to be detained . . .'[81] The next entry on 21 July reads '[The] Commanding Officer submits an affidavit of the Medical Officer [showing] Ensign Goodenough to be out of danger and recommends that Ensigns Mayhew and Nicholls join their Corps at Malacca and Singapore . . .' The entry is endorsed 'Ordered. . .Suptd of Police advised . . .'[82] Ensign Goodenough obviously made good his recovery as on 12 December 1828 he was granted absence from duty until 31 December 1828.

Early in 1829 Lord Bentinck, the Governor General, arrived in Penang and as a result of his tour he came to the conclusion that to retain Prince of Wales Island Presidency was no longer viable. When the Presidency was first formed it was expected that Penang would be developed into a Marine Station, for the Royal Navy in the Eastern seas, as well as being a Naval Arsenal for ships of the Royal Navy. Neither of these projects ever came to fruition and it was therefore considered that the administration of the three Straits Settlements could be achieved by a reduced staff. This was agreed, the Presidency was abolished, and the three Settlements were constituted as a single Residency on 1 May 1830 and placed under the Presidency of Fort William. The headquarters of the residency remained at Penang until 1832.[83]

During the final months of the Presidency a tragedy struck the military community of Prince of Wales Island. Lieutenant William Edward Brooshooft, who was serving with the 35th Regiment, Madras Native Infantry, was accidentally drowned whilst bathing at the Waterfall at Battu Feringee (now spelt Batu Feringi), leaving his young wife Charlotte to mourn his loss.[84]

NOTES TO CHAPTER THREE

1. The Honourable Edward Monckton, the fifth son of the first Viscount Galway, he married the daughter of Lord Paget, the Governor of Madras in 1776. He was appointed Sheriff of Madras in 1770 at the early age of twenty-six. H. D. Love, *Vestiges of Old Madras,* (London, 1913).
2. Francis Light was born, illegitimately, in 1740 and joined the Royal Navy in 1759 at the age of nineteen. He left the service in 1763 and then moved to the East although not initially as a Company servant. Whilst in the Royal Navy he saw service afloat as a midshipman which stood him in good stead later when he arrived in the East. He was able to obtain a position as an agent working for Jourdain, Sulivan and De Souza. His first appointment was to assist the agent in Acheh where he soon became aware of the trade being carried out between Sumatra and Kedah in pepper, tin and other items. In 1771 he moved to Kedah and arranged a trading concession between his employers and the Sultan.
3. In this instance Light is referring to the East Indiamen trading between Europe and China.
4. The Honourable East India Company.
5. Letter from Francis Light to De Souza dated 25 November 1771.
6. K. C. Tregonning, *The British in Malaya, The First Forty Years,* (The Association for Asian Studies, 1965), p20.
7. G. E. Gerini, Article translated by, 'Historical Retrospect of Junk Ceylon Island', *Journal of the Siam Society,* vol 2, 1905.
8. Straits Settlements Factory Records, vol 1, in the Library, University of Singapore. Letter from Captain Light to Bengal, 23 January 1786.
9. Straits Settlements Factory Records, vol2, Bengal Consultations, 2 March 1786.
10. Ibid.
11. John Crawfurd, *A Descriptive Dictionary of the Indian Islands and Adjacent Countries,* (London, 1856), p331.
12. *The Penang Information Guide,* 1976, p3, gives the translation of Pulo Pinang as 'Areca palm island'.
13. John Cameron, *Our Tropical Possessions in Malayan India,* (London, 1865), pages 307 and 308.
14. Tregonning, op cit, p44.
15. Captain E. Trapaud, *A Short Account of the Prince of Wales Island, or Pulo Pinang, in the East Indies, given to Captain Light by the King of Quedah,* (London, 1788).
16. Straits Settlements Factory Records, vol1 (1786) in the Library, University of Singapore. Additional notes from the Hodson Index, National Army Museum, London.

Captain Elisha Trapaud, Madras Engineers

Born London 1750/1751. Ensign 14 April 1779, Captain 16 May 1784, At the time of the raising of the flag on Prince of Wales Island Captain Trapaud was on furlough (3 years from 15 July 1876). He continued to serve and was promoted to Lieutenant General on 4 June 1814 and died at Cheltenham on 24 March 1828.

Captain George Howell, Bengal Artillery

Cadet 1772, Fireworker from 17 July 1776, Lieutenant 15 July 1778, Captain/Lieutenant 29 January 1784. Captain 26 May 1786. Captain Howell was on furlough from 13 March 1786. Howell was lost at sea during August 1794 on a voyage to Bencoolen, Sumatra.

Lieutenant James Gray

An officer of the Royal Navy who was commissioned on 12 February 1781. *The Commissioned Sea Officers of the Royal Navy, 1660–1815,* (1954), vol2.

17. Tregonning, op cit, p45.
18. Straits Settlements Factory Records, vol2, in the Library, University of Singapore. Minute by Lord Cornwallis dated 13 December 1786.

19. The Reverend Keppel Garnier,'Early Days in Penang', *Journal of Malayan Branch, Royal Asiatic Society,* (1923).
20. Straits Settlements Factory Records, vol2, Penang to Bengal.
21. James Scott travelled to India as a Purser on the ship *Anson* in 1774. He obtained his discharge and became a 'country trader'. By 1779 he was on Junk Ceylon Island and became a firm friend of Francis Light. They agreed that once Scott was established on Pulo Pinang that Light would administer the island and accept the inevitable loss that the Company salary involved, and Scott would trade and make enough for both. Tregonning records that he did this and continued to prosper even after the death of Francis Light. He controlled his own company and had the virtual monopoly of the import and export trade. He built a large house which he named 'Kelso'. He died on 19 September 1808 and was buried in the Northam Road Cemetery, Penang.
22. Straits Settlements Factory Records, vol4, in the Library, University of Singapore, Light to Bengal, 19 April 1791.
23. The Hodson Index. The record shows John Glass as being a Cadet, 1778, Fireworker from 5 October 1778. Lieutenant from 6 July 1782. Moved to Penang in 1786 and given local rank of Captain from 23 December 1786. Appointed Commandant of Prince of Wales Island garrison 1786. Died at Penang 6 April 1793, however an examination of his memorial inscription at Northam Road Cemetery shows the date as being 5 April 1793.
24. Straits Settlements Factory Records, vol3.
25. Hamilton, Robert. Hamilton became a Cadet in 1772, and promoted to Capt/Lieut on 1 October 1781.
26. Major V. C. P. Hodson, *List of Officers of the Bengal Army, 1758–1834,* (Constable, 1928). *Kyd, Alexander.* The son of Captain James Kyd, RN, born in Scotland on 14 March 1754. Cadet 1775, Ensign 11 Dec 1775, Lieutenant 9 October 1778, Captain 13 February 1781, Major 23 February 1793. After his period of duty in Penang he was promoted Lieut Colonel 25 April 1797 and married Elizabeth Hay (née Wagstaff) the widow of Edward Hay, BCS, at Clifton, Somerset on 13 November 1804. He was then promoted Colonel 1 January 1805, Major General 25 July 1810 and Lieut General 12 August 1819. He died at Albemarle Street, London on 25 November 1826.
27. Tregonning, op cit, p130.
28. Manuscript list of inscriptions taken from the Northam Road Cemetery monuments, Georgetown, during July 1976 shows the date as being 21 October 1794.
29. Tregonning, op cit, p49.
30. Straits Settlements Factory Records, vol7, Major Kyd to Bengal dated 2 August 1795.
31. Ibid.
32. A statue of Francis Light can be seen in the Penang State Museum. Light had five children by Martina Rosels. The eldest, a son, born in 1784, was sent back to England for his education in 1790 and was commissioned into the British Army and served as ADC to the Duke of Wellington.
33. *The New East India Kalendar for 1801.* (Printed by J. Debrett, Piccadilly, 1801).
34. The Hodson Index. Lieutenant Colonel Thomas Polhill, born 1746/1747. Cadet 1772, Lieutenant 3 July 1778, Captain 30 December 1792, Major 31 August 1798 and promoted to Lieutenant Colonel on 4 January 1801 on transfer to the 1st Regiment Bengal NI. Died at Kalpi, United Provinces 9 May 1804.

East-India Register and Directory for 1803 shows under 'First Regiment Native Infantry' Lt. Col. Thomas Polhill as being 'Com. at Prince of Wales's Island'.
35. Tregonning, op cit, p136 and Straits Settlements Factory Records.
36. The 'Georgetown, Prince of Wale's Island [*sic*] Register of Funerals 1800–1818' was kindly made available to me by The Reverend D. Rees of Penang. The register starts at 1800 and shows all the funerals at the Northam Road Cemetery up until 22 December 1818. There is no trace of the burials registered between 1819 and 1926 and it is considered that these were lost during the Japanese occupation of Penang Island during the Second World War. The entry in the register shows Captain Joseph Stokoe, Bengal Engineers as being buried on 12 January 1801. The East-India Register and Directory, 1806 shows an entry under Bengal Casualties – 'Engineers. Lieutenant John Gordon. 12 Nov 1804. Prince of Wales Island'. However the entry in the Register of Funerals shows 'Sir John Gordon Bart. Lt of Engineers. 1804 November 12 Bengal Establishment'

The Hodson Index card at the National Army Museum reads: 'John Gordon. – born Cork 2 July 1776. Cadet 1792. Ensign 8 Nov 1793. Lieut 10 Dec 1800 died, unmarried Penang 12 Nov 1804'.

Burke's Peerage, 1859, gives the following additional detail – 'Sir William (Gordon) d. 7 Jan 1804 and was succeeded by his eldest son (5th and eldest surviving son) Sir John (8th Baronet), a Lieut of engineers in the Hon East India Company's Bengal Army, on whose decease unm at Prince of Wales's Island, 12 Nov 1804, the title devolved upon his only surviving brother'.

37. Penang General Orders and Notices, November 1805 – September 1807, in the National Library, Singapore, General Order dated 3 December 1805.

38. Note from author's manuscript list of inscriptions from the Northam Road Cemetery shows the Morse memorial inscription as reading

'Died
on the February
1806
Lieut Charles Mor(se) [stone damaged]
of the
Bombay Artillery
Aged 26.'

The inscription does not give a day nor has sufficient room been left to include the date '16' which is the date given in the register.

39. Tregonning, op cit, p 137.

40. Anthony C. Dumper, *A Short History and Guide to The Church of St George the Martyr, Penang,* (Penang, 1964).

41. Richard Cannon, *Historical Record of the 73rd Foot,* (Parker, Furnivall & Parker, London, 1851). The 73rd Regiment embarked at Fort St George Madras on 8 September 1805 and arrived at Greenwich in the beginning of July 1806. They obviously travelled via Prince of Wales Island as the General Order by Government dated 20 November 1805 at Pinang read: '... The Detachment of His Majesty's 73rd Regiment to embark this afternoon at 2 o'clock on board the Hon'ble Companys Extraship "Dwaynes". Boats are ready at their Barracks for this purpose ...'

42. Penang General Orders and Notices Nov 1805 – Sep 1807, in the National Library, Singapore.

43. Although the order states that Captain William Gill was of the 'Marine Regt' The East-India Register and Directory for 1805 shows him as being in the 6th Regiment, Bengal Native Infantry.

44. Penang General Orders and Notices.

45. This was Captain J. P. Drummond of the Bengal Artillery.

46. The unit referred to in this order is the 1st Bengal Volunteer Brigade. However at this time a Local Militia had been formed within the Settlement. Lieutenant Leonard Whitesmith, of the 7th Regiment, Bengal NI was, at that time, serving as a Volunteer officer with the Battalion.

47. During the reorganisation of the Settlement's defences in 1805 a local militia was formed and raised. It was generally known as the Prince of Wales Island Local Corps. Its existence was, however, short-lived and it was disbanded in July–August 1806. The two Acting Ensigns Barrett and Moser are not listed in the East-India Register and it is therefore assumed that they were recruited for service with the Local Corps and with the disbandment of the corps there was no longer any need for their services. No other reference to service by these two junior officers can be found.

48. Tregonning, op cit, p 134.

49. Tregonning, op cit, p 135.

50. Penang General Orders and Notices. General Order dated 23 September 1807 at Fort Cornwallis.

51. Thomas Raffles arrived at the Prince of Wales Island Settlement in September 1805. In a few years he was to be promoted to be Lieutenant Governor of Java after the campaign of 1811, where he remained until 1816. On his return to London he was knighted in May 1817 and became Sir Stamford Raffles. In March 1818 he was granted the appointment of Lieutenant Governor of the Settlements on the west coast of Sumatra. Whilst in that post he became involved in the search for an additional site for a trading post and port and on 29 January 1819 landed on the island of Singapura and founded the Settlement that was to become known as Singapore.

52. Georgetown Register of Funerals, 1800–1818.

53. Tregonning, op cit, p 138.

54. Norman MacAlister (also shown as Macalister) died during September 1810. He is recorded in the 'Bengal Military Casualty' section of the East-India Register of 1812 as follows: 'Artillery. Colonel Norman MacAlister. Sep 1810. Lost in ship "Ocean" in the China Sea.' MacAlister, his wife and children, all perished in the ship *Ocean* along with some 200 other passengers and crew. It was not unusual during the first half of the nineteenth century for ships to be lost without trace in far eastern waters.

55. Other detachments of the 20th (Marine) Regiment, Bengal NI arrived in 1809. Their arrival was recorded in the General Order dated 8 June 1809 which read: '... The 2nd Battn of the 20th Regt Native Infantry having arrived at this Presidency the Hon'ble the Governor and Council direct that they be landed from the Transports with all practical expedition. The Comdr-in-Chief is requested to issue such further orders as he may deem necessary for carrying the same into effect ...'
56. Penang General Orders and Notices.
'Fort Cornwallis. 9th September 1806. The Hon'ble the Governor and Council have been pleased to appoint Lieut John Veatch, of the Bombay Establishment, Aide de camp to the Hon'ble the Governor. His allowances to take place from the date of his leaving his station with the Bombay Army.
(Signed) Thomas Raffles ...'
57. George Woodcock, 'Penang: Britain's First Settlement in Malaya', *History Today*, vol 6, no 12, December 1969.
58. Raymond Head, 'From Obsession to Obscurity - Colonel Robert Smith, Artist, Architect and Engineer' *Country Life*, 21 May 1981.
59. Hodson, op cit, p 613.
60. John Bastin and Pauline Rohatgi, *Prints of Southeast Asia in the India Office Library*, (HMSO, London, 1979).
61. Hodson Index, National Army Museum. James Low was born at Kettle, Co Fife on 4 April 1791 and was educated at Edinburgh College. He was commissioned as an Ensign in the 25th Regt, Madras Native Infantry on 11 June 1812 and was appointed Adjutant of the regiment in May 1817. Promoted to Lieutenant in August 1817 and joined 23rd Regiment, Madras Native Infantry.
62. *Prince of Wales Island Gazette*, vol 4, no 164, 4 July 1818.
63. James Low, 'An account of the origin and progress of the British Colonies in the Straits of Malacca', *The Journal of the Indian Archipelago*, vol 4, 1850.
64. Penang Orders and Notices 1828, Military. Entry for 20 February reads 'The party of the late Local Corps sent to Bengal on the ship Flora ...'
65. L. A. Mills, Mills states '... In some cases a much larger bonus was granted (for language qualifications) Lieutenant Low for example was given £450 in 1824 for proficiency in Siamese ...' *British Malaya 1824–1867.*
66. Ibid, p 137.
67. Ibid, p 160.
68. Ibid, p 161.
69. Captain M. A. Bunbury, 40th Regiment, Bengal Native Infantry.
70. Lieutenant J. C. Boulderson, 35th Regiment, Madras Native Infantry. Lieutenant Peter P. Hodge, 1st Regiment, Madras Native Infantry, holding appointment of Executive Officer of Public Works, Prince of Wales Island. (East-India Register, 1828).
71. Penang Orders and Notices 1828–1830, in the National Library, Singapore.
72. Lieutenant Colonel Henry W. Sale, 35th Regiment, Madras Native Infantry.
73. Lieutenant Arthur Trotter, 35th Regiment, Madras Native Infantry.
74. Lieutenant Charles G. Chauvel, 35th Regiment, Madras Native Infantry.
75. Lieutenant J. C. Boulderson, 35th Regiment, Madras Native Infantry.
76. Captain Frederick Bond, Madras Artillery.
77. Not listed in East-India Register and Directory.
78. Penang Orders and Notices 1828–1830.
79. Ibid.
80. Ibid.
81. Ensign Edmund Goodenough, 25th Regiment, Madras Native Infantry and Ensign Alfred Mayhew, 35th Regiment, Madras Native Infantry.
82. Ensign Henry James Nicholls, 25th Regiment, Madras Native Infantry.
83. Tregonning, op cit, pp 72 and 73.
84. Hodson Index. Lieutenant William Edward Brooshooft was born on 16 October 1803 the son of Bernard Emanual Brooshooft. Appointed Lieutenant 1 May 1824 and served with the 35th Regiment Madras Native Infantry. From 27 August 1824 he was appointed 'Quartermaster & interpreter and paymaster'. He married Miss Charlotte Toller, the daughter of Sir Samuel Toller, at Bangalore on 6 December 1824. Although deleted from his regimental roll in the East-India Register his death on 1 March 1830 was not recorded in the casualty section.

CHAPTER FOUR

The Spice Islands and Java Campaigns

On completion of the occupation of Malacca by the Honorable East India Company's forces in 1795 it was considered necessary that other Dutch possessions should be occupied in order to exclude them from French control. The fort at Malacca had surrendered to the force under command of Major Archibald Brown, Madras Army, on 18 August 1795.[1] At the close of the year an expedition was ordered to be mounted to proceed to Amboyna and the Spice Islands that were occupied by the Dutch in the East Indies. The original intention was that the force should be found from Bengal Native regiments but the sepoys refused to embark for service outside India and the Supreme Government, not wishing to face a major confrontation with the native troops, allowed the Bengal sepoys to remain in India. The force was then formed from the Madras troops that were already serving at Malacca. The force sailed from Malacca on 6 January 1796[2] in the naval squadron under the command of Commodore Peter Rainier.

The expedition that embarked was under the command of Major Urban Vigors and consisted of:

'Detachment Madras Artillery.
Two flank companies 3rd European battalion.
Two battalion companies 2nd European battalion.
Seven companies 17th Native battalion, Captain Gordon.
Fifty-one pioneers under Lieutenant Heitland . . .'[3]

Wilson, in his *History of the Madras Army* records that the force that remained at Malacca consisted of:

Madras Artillery	-	37	
European Infantry	-	150	
17th Madras Native Infantry	-	217	
Gun Lascars	-	80	
Pioneers	-	60	(under command of a European Sergeant).

He also notes that 129 Malay soldiers formed part of the garrison.

Amboyna lies in the Molucca Sea and Crawfurd describes the island as '. . . lying off the southern coast of the large island of Ceram . . . Its native name is Ambun, said to be derived from that of its chief town . . .'[4] The island was, in the nineteenth century referred to by its native inhabitants as Hitoe, or Hitu. The island had, at one time, held the trading posts of both the Dutch and the English. During 1623 the Dutch reputedly learned through a locally-recruited Javanese soldier in the employment of the English, that there was a conspiracy to overrun the Dutch fort and kill the garrison. The Dutch acted promptly and seized the English settlers and although a few confessed others were tortured in order to obtain a confession. The Dutch governor then ordered ten of the English to be beheaded, which they were, and the remaining two were spared and sent back to England as a warning to others.[5]

The Dutch at Amboyna surrendered to the Company's force on 17 February 1796 and as a result the arms and armaments were taken over by the Madras force. These consisted of one hundred and sixty-four guns, seven mortars, 51,900 lb of gunpowder and 1,118 muskets as well as a large quantity of stores. A garrison force was formed for Amboyna which consisted of two of the companies of the European battalions and three companies of the 17th Native Regiment. The 17th Native Regiment had voluntarily embarked for the Malacca expedition in October 1795 and there had been no shortage of volunteers from the 16th and 32nd regiments who had been requested to provide men for the vacancies in the ranks of the 17th due to sickness, personnel on leave, etc.

Having secured Amboyna, Commodore Peter Rainier set sail with the remainder of the force to secure Banda. Crawfurd describes this group of islands as follows:

'... The Banda or Nutmeg Islands consists of a group of mere islets, said to be five in number. Banda, correctly Bandan, means in Javanese the thing or things united, or with the word Pulo "united islands". Pulo Nera is the "island of palm-wine" Lontar, written by Europeans Lonthor, is the name of the palm or leaf of which is used for writing on, the word being half Sanscrit and half Javanese; Pulo Ai, properly Pulo Wai, means "Water island"; Pulo Pisang "Banana island"; Pulo Run (Rung), "Chamber island"; Pulo Suwanggi, "Sorcery island" ...'[6]

The main defence position on the islands was the Dutch fortress of Belgica on the island of Nera, which was the island on which the principal settlement had been established: Nera is only a mile and half long and half a mile wide. The largest of the islands is Lontar, which was called Great Banda and the whole of Banda group covered an area of '... no more than 17.6 geographical square miles ...' and is of volcanic formation.[7]

The expedition, under Rainier's command, landed and took possession of Banda Nera (sometimes spelt Neira) and the main Dutch fortress. This brought a further one hundred guns and seven mortars to the mass of arms and ammunition already captured from the Dutch.

The force which took possession of the island of Banda Nera consisted of:

Madras Artillery	–	21
Gun lascars	–	28
Madras European Infantry	–	118
17th Madras Native Infantry	–	129
Pioneers	–	38

In addition Wilson records that there was a part of '... 68 Wirtemberghers ...' and comments that no particulars were found regarding these men but it was probable that they took service with the Honorable East India Company after their capture in Ceylon in 1795.[8] Rainier sent a despatch dated 27 March 1796 in which he praises the 17th Madras Native Infantry as follows:

'... The orderly and dutiful behaviour of the 17th battalion of natives amidst the peculier hardships of a service totally novel to them, will, I hope, be noticed with some particular distinction. Much credit is due to their commander, Captain Thomas Gordon ...'[9]

The garrisons at Amboyna and Banda were reinforced in July 1796 when the following left Madras for the two recently occupied settlements. The force was under command of Major Macneile[10] and consisted of:

Madras Artillery

2 Sergeants, 2 Corporals, 4 Gunners, 16 Matrosses and 1 company of Gun Lascars.

Infantry
The 'Coast Volunteer Battalion' which consisted of
11th Madras Native Infantry - 3 companies
16th Madras Native Infantry - 1 company
32nd Madras Native Infantry - 2 companies

The force was also supplemented by twenty-five volunteers from the 10th Madras Native Infantry, who were to be attached to the 32nd Regiment. Once again the Madras troops had been willing to volunteer for overseas service and this brought forth a commendation from the Commander-in-Chief, Major General Alured Clarke.

Early in 1801 the Resident at Amboyna and the officer commanding the troops in the area, Colonel Daniel Burr, organised an expedition against the Dutch-held spice island Ternate, one of the Molucca or Clove Islands. The force consisted of 348 men and this was considered to be sufficient to take command of the island as the senior naval officer on station, Captain Astle had promised his co-operation and the Dutch force was reputed to be fifty Dutch Europeans and 2,000 locally-recruited native soldiers. The Company's force was made up of artillery and infantry and consisted of:

Madras Artillery
Captain David Ross with 22 European gunners and 28 gun lascars.

Infantry

Madras European regiment	- 83 NCOs and soldiers under command of Captain Walker, of the 4th Madras Native Infantry.
1st Bengal Native Infantry	- 1 Havildar, 8 Sepoys
Wirtemberg Company	- Captain Gaupp and 36 men
Bengal Marine Regiment	- Lieutenant Gill and 60 sepoys
Volunteer Corps	- Captain Thomas Boles and 53 sepoys
Pioneers	
Madras Pioneers	- Lieutenant John Wissett, 9th Madras Native Infantry
Bombay Marine	- Lieutenant John Hayes, Indian Navy 40 European Marines.[11]

The expedition left Amboyna on 22 January with a naval force consisting of HM frigate *La Virginie*, the sloop *Hobart*, and two armed vessels belonging to the Company, these being the *Ternate* and the *Splinter*.[12] The fleet arrived and anchored off Ternate on 10 February and unexpectedly the Dutch refused to surrender and in consequence arrangements were made to attack Fort Orange on the following day. However Captain Astle persuaded Colonel Daniel Burr that a detached defence work should be attacked and taken before any attempt was made to take the main fort.

On 11 February the boats containing the troops departed from the ships at about 11.00am and arrived at the shore at 1.00pm. In the meantime the frigate and the sloop opened fire on the shore installations, but as they kept out of range of the fort much of the shot fell into the sea. The landing place was ill-chosen as it was surrounded on all the land-sides by dense jungle and this necessitated the Company's force making a considerable detour. The troops were led by guides and it was an hour before they came into sight of the fort and were then travelling in single file because of the narrowness of the jungle track. The troops halted for a short while and then advanced on the fort. They were received with fire from two field artillery pieces and racked with grape and musketry fire so that they had to retreat leaving some wounded behind. These were captured by the Dutch.

The troops were evacuated back to the ships but as the sea had become rough and the surf high on the beach the two guns that had been brought on-shore had to be spiked and abandoned.

The senior naval officer refused to co-operate in another attempt to take the fort and the expedition returned to Amboyna having failed in its mission. Colonel Daniel Burr reported to the Madras Government that the expedition had in effect failed due to the lack of co-operation by the senior naval officer, Captain Astle, who had refused to commit any of his men to the attack.[13]

Due to a delay in the repatriation of 150 sepoys to Madras it was decided that a second expedition should be mounted and on 2 April a force of 329 men set out from Amboyna to take the island of Ternate. On this occasion the escort ships were from the Company's squadron and composed of the *Swift*, commanded by Lieutenant John Hayes, and the *Star*, *Splinter* and the brig *Resource*. The fleet arrived off Ternate on 23 April and the troops were landed and the land operation commenced. On 8 May the Dutch redoubt of Koya-meira surrendered and the Company's force proceeded to capture the hill of Talangnamy. The Dutch defended the hill vigorously and it was not until the 22 May that the hill was abandoned by the defending force. In the meantime the naval force had commenced firing on the fort but without much effect. Lieutenant John Hayes lost men killed and wounded on his ship *Swift* as well as receiving a great deal of damage to his masts and rigging.

The small size of the attacking force made the campaign a long one as it was not until 20 June that the garrison opened negotiations for the surrender of the fort and this was merely brought about by the lack of provisions within the fort and the effective blockade that had been established by the combined naval and military force. On the evening of 20 June four Commissioners came into the Company's camp to discuss terms. The capitulation was agreed and the fort surrendered on the morning of 21 June. Fort Orange was then taken over by the Madras Army detachment under command of Captain Joseph Walker.[14] This small force remained on Ternate while the remainder of the expedition returned to Amboyna.

It was decided by the Government that the Bengal Marine Infantry battalion would be sent to the Spice Islands to relieve the Madras troops serving at Malacca and Amboyna. On 11 August 1800 the Bengal Marine Infantry Regiment embarked at Madras under command of Captain Thomas Brougham. The unit was about 747 strong with 289 followers, including women and children. The Madras troops due to be relieved had been delayed and in consequence took part in the first expedition to Ternate but with the failure of this venture the Madras troops embarked for the coast and arrived back at Madras in May 1801. A general order was issued on their return commending their service which read:

> '... 29th May 1801. Major General Braithwaite has much satisfaction in availing himself of this opportunity to express his entire approbation of Lieutenant Colonel McNeile's conduct during a long course of arduous and important command, as well as of the exemplary zeal and spirit which have so eminently distinguished the 2nd battalion, 1st regiment, (this was formerly the 17th Madras Native Infantry) and the Volunteer corps, throughout a very trying service, and an uncommonly tedious passage by sea.
>
> The General derives particular pleasure also in announcing to these corps the resolution of the Governor in Council to testify his sense of their faithful attachment to the British Government during a period of more than five years on foreign service, by

bestowing on them individually a badge to be worn as an honorary mark of distinction; and which the General hopes may be in readiness for delivery previous to the departure of the corps from the Presidency . . .'[15]

The Madras troops had suffered heavily during their foreign tour and the official number of casualties recorded for their service in the 'Eastern Islands' was published in 1802 and gave a total figure of 397. This being made up as follows:

2nd Battalion, 1st Regiment Native Infantry	233
Volunteer battalion	93
Gun lascars	49
Pioneers	22
	397[16]

In 1801 it was decided that a new Volunteer corps would be raised within the Fort St George army and that it would serve in the Moluccas for a period of three years. The establishment of this new unit was fixed at 500 non-commissioned officers, rank and file, with the '. . . ordinary proportion of Native commissioned officers, drummers and puckallies, &c . . .'[17] The European staff consisted of a Captain Commandant, an adjutant, a sergeant-major, and a quartermaster sergeant. The Native staff allowed was a jemadar adjutant, drill havildar, a drill naik, a drum-major and a fife-major. An order was published on 23 September which set out the conditions of service for the Volunteer battalion.

'. . . For the purpose of carrying into effect the resolution of the Right Honorable the Governor in Council as published in the Government General Order of the 22nd instant, the Commander-in-Chief is pleased to direct that such men of the under-mentioned corps as may make a voluntary offer to serve in the Madras Volunteer battalion during a period of three years at the Molucca Islands, shall be permitted to do so on the following terms, viz, jemadars to enter the Madras Volunteer battalion as subadars, havildars as jemadars, naigues as havildars, and the requisite proportion of privates as naigues, according to seniority. During the time the Madras Volunteer battalion may remain embodied, the promotion shall be confined to it: the Native commissioned, non-commissioned, and rank and file, will for the present be struck off the strength of their respective corps, but on their return to the coast, be re-admitted in the several ranks which they may have then attained on the establishment of such corps as may best suit the convenience of the public service, and they will be permitted to leave family certificates.

Corps from which men are permitted to volunteer for the Madras Volunteer battalion viz:

Detachment 1st battalion 4th regiment at Fort St George
The 2nd battalion 14th regiment at Fort St George
The 1st battalion 17th regiment at Fort St George
The 1st battalion 18th regiment at Fort St George
The 1st and 2nd Madras battalions . . .'[18]

On 13 October a further order was issued which allowed an Assistant Surgeon to be posted to the battalion and also three recruit boys to each company. The battalion embarked from Madras on 16 October 1801 under command of Captain Andrew McPherson of the 14th Regiment Madras Native Infantry.

As a result of various political moves both in London and in India it was decided by Lord Wellesley the Governor General, and Robert Farquhar, who later became Lieutenant

Governor of Penang during 1804 and 1805, that the occupation of the Moluccas was serving no useful purpose, neither commercially nor as a military garrison and Lord Wellesley ordered the removal of the administration from the islands, and of course the withdrawal of the military garrison in the Spice Islands. It was recommended that a more usefully strategic settlement should be established on Balambangan. Balambangan, or Blambangan, which in Malay means 'place of palm-planks', is an island situated in the Sulu Sea and is 13 miles from the most northerly point of Borneo. The island is about '. . . 300 miles in circumference . . .',[19] and was ceded to the English in 1762 by the Sultan of Sulu as a reward for liberating him from his captivity at Manila. In 1775 it was occupied by the Honorable East India Company who set up a trading post and garrison on the island. However it was abandoned after the garrison had been attacked and defeated by the subjects of the Sultan of Sulu. Crawfurd describes the island as '. . . sterile, uninhabited, and in the most piratical and barbarous neighbourhood of the whole Archipelago . . .'[20] The occupation of the island of Balambangan, Wellesley hoped, would secure the trade route between Penang (Prince of Wales Island) and Canton. Robert Farquhar evacuated his 800 or so troops from the Moluccas and, using Malacca as a base, formed an expedition with which to occupy the island of Balambangan. Farquhar and his force arrived off the island on 29 September 1803 and commenced setting up his trading settlement and garrison. By February 1804, Robert Farquhar had left Balambangan and had taken over the post of Lieutenant Governor of Penang. In September 1804 both Malacca and Balambangan were placed under the control of the Lieutenant Governor of Penang.[21] When the Honorable East India Company authorities in London became aware of what had transpired and the cost of the expedition and setting up of the trading post, port and garrison at Balambangan they immediately ordered the withdrawal from the island. It was considered at that time that the Company had sufficient expense in dealing with foes in India, and the safeguarding of the coastline. The fact that ships trading between Penang and Canton were rarely attacked made the cost of this new, and to London, unnecessary base unjustified. The troops remained inactive on Balambangan throughout 1804 and much of 1805. In September 1805 the authorities on Prince of Wales Island were requested to '. . . repair the Sepoy lines "for the troops expected from Balambangan". The detachment arrived and took up quarters on 16th December under the command of Lt Col Gales, whom Farquhar had left in charge of the island . . .'[22] This ended the East India Company's temporary occupation and administration of the Spice Islands and the island of Balambangan. The next move towards the eastern islands came in 1809 when it was again considered expedient to occupy the Dutch-held islands and in October 1809 the frigate *Dover* a ship of 38 guns, the *Cornwallis*, 44 guns, and the sloop *Samarang*, 18 guns, sailed for the Spice Islands with a detachment of Madras Artillery, under command of Captain Major H. Court (Major being a forename in this instance), one company of the Madras European Regiment under command of Captain Richard Phillips.[23] Captain Tucker of the *Dover* was in command of the expedition. The force arrived at Amboyna on 13/14 February 1810 and anchored off-shore so as to reconnoitre the defences. The main defence was Fort Victoria which had a battery on each flank. There were also two batteries on the heights above the town with a distance of some 1,500 yards between the two sites. The Dutch garrison consisted of 130 Europeans, 220 officers and seamen and 1,000 locally enlisted soldiers, with an estimated 215 pieces of ordnance mounted on the fort and lower batteries and twenty-one guns on the two batteries above the town.[24] The Company's force left the ships at about 1.00 am on 15 February but missed the landing place in the dark and aroused the enemy who sent up rockets from the fort which resulted in the attacking force withdrawing and returning to the ships. On the following day, at about 2.00 pm the troops landed under cover of fire from the ships. One party consisting of 30 artillerymen, 85 seamen and marines from the *Dover* and a company from the Madras European Regiment, all under command of Captain Phillips advanced against the battery at Wanitoo (one of the batteries on the heights above the town) succeeded in taking the position and turned three of the guns against the other battery, at Battoo-Gantong, with Lieutenant Duncan Stewart, of 1st Battalion, 25th Regiment Madras Native Infantry, (then with the Artillery) in command. Although wounded, Stewart continued to command the guns, meanwhile Captain Phillips led the majority of his small force against the other battery but was unable to reach the site due to the way being barred by a deep ravine.

Captain Court with a party of fifteen artillerymen; 140 seamen and marines from the *Cornwallis* and the *Samarang*, together with a company of the Madras European Regiment under command of Captain Charles Forbes gained the height near to the battery at Battoo-Gantong and was able to drive the enemy from the position.

With both batteries on the heights now able to direct fire onto the town and Fort Victoria, and with the ships also able to direct fire from the seaward side the Commandant of the Garrison had little option but to capitulate, which he did on the following day, 18 February. The capitulation included the whole of the island and the dependencies of Saparona, Hila, Haroeko, Laricque, Bouro and Manipa.[25]

Wilson, in his book *The History of the Madras Army* records that the Dutch and Javanese troops were sent to Java and the locally recruited Amboynese soldiers, who were '. . . well disposed towards us . . .' were recruited into the Company's service. The force, known as the Corps of Amboynese, had the following Establishment:

European					
Captain	–	1	Adjutant (Lieutenant)	–	1
Serjeant Major	–	1	Drill Serjeant	–	1
Natives (Amboynese)					
Adjutant	–	1	Serjeants	–	20
Corporals	–	20	Drummers	–	10
Privates	–	600	Watercarriers	–	7
Carpenters	–	2	Smiths	–	2
Bellows Boy	–	1	Chickledar (a)	–	1
Chuckler (b)	–	1			

(a) A chick is a kind of screen-blind made of bamboo and twine to cover doors or windows of a house or tent. The chickledar would be responsible for these items at the officers quarters, or tents, his duties would equate to those of the punkah-wallah.

(b) A cobbler.[26]

The Amboynese were regarded as reasonably good soldiers and of a good physique. The pay allowed for the whole unit was '. . . 2,994 Spanish Dollers . . .' per month. Each soldier was allowed one dram of arrack and one measure of rice per day and two measures of salt each month.

The losses during the capture of Amboyna were relatively small being given as one seaman, one marine, and two privates of the Madras European Regiment killed. One Lieutenant and ten others were wounded[27] which was considered to be a very light casualty list and so the force, under Captain Tucker, moved on to the Dutch settlement at Manado which was on the north-eastern coast of the island of Celebes. This was considered to be an essential move in order to prevent supplies reaching the Dutch garrisons on Banda-Neira and Ternate.

Reinforcements were despatched on 6 May 1810 from Madras on the frigates *Caroline* (36 guns), *Piedmontaise* (38 guns) which were accompanied by the brig *Barracouta* (18 guns). The reinforcements were under the command of Captain George L. Nixon, and consisted of two companies of the Madras European Regiment and apart from Captain Nixon of that regiment two other officers of the regiment accompanied the troops; these were Lieutenant Patrick Brown and Ensign William Jones Daker (sometimes spelt Dacre). The small convoy called at Penang where scaling ladders and stores were loaded. The *Barracouta* called at Malacca and embarked twenty men of the Madras Artillery and '. . . a couple of field-pieces . . .' The Madras Artillery detachment was under the command of Lieutenant Charles W. Yates.[28] The force arrived off Banda-Neira, the principal island of the Banda group on 8 August 1810 and it was decided that the island should be taken even though the two main defence works, Fort Belgica and Fort Nassau, presented a formidable target. There were in addition several detached batteries and the total number of guns facing the force amounted to 138, of which fifty-two heavy guns were mounted in Fort Belgica. A landing party consisting of some 200 men, including infantry, artillerymen, marines and sailors landed at just before daylight and successfully captured one of the detached batteries near to the beach. This was achieved without the alarm being raised and the majority of the party advanced towards Fort Belgica and with the help of the scaling ladders the attack was made with such speed and determination that the Fort was taken by 5.30 am without any casualties being incurred by the attacking force. The Dutch Colonel Commandant of the Fort was killed as were ten of his soldiers. Four Dutch officers and forty artillerymen were captured but a number of the defenders were able to escape, though not for long. With the main Fort in the hands of the Company's troops the remainder of the garrison surrendered which in turn saw the surrender of the other island troops. The total Dutch garrison had consisted of 700 regular soldiers and 800 militiamen.[29]

The Governor-General issued an order at Fort William on 16 February 1811 commending the force as follows:

> '. . . His Lordship in Council considers the rapid conquest of a place so strongly fortified by nature and by art as the island of Banda, in the face of a superior force, without the loss of a man, as forming a singular event in the annals of British enterprise, and the Governor-General in Council has great pleasure in recording a public testimony of the approbation and applause with which he contemplates the distinguished zeal and gallantry of the Commanding Officer of the troops, Captain Nixon, and the courage, discipline, and firmness of the general body of the officers and troops under his direction, in carrying into effect, with such a signal and rapid success, the plan of operations for the reduction of that Settlement, concerted by Captain Cole of His Majesty's Ship 'Caroline'.[30]

Letters of commendation were also sent to Lieutenant Charles W. Yates (Wilson refers to the officer as Lieutenant Yeates) and to Ensign Richard Allen, who was serving with the 21st Regiment, Madras Native Infantry, and who volunteered to serve with the force whilst he was stationed on Penang Island.

The next move in the conquest of the Moluccas was the taking of the island of Ternate. On 21 August 1810 a force under the command of Captain David Forbes, of the Madras European Regiment, and Lieutenant Charles Forbes and Ensign John Cursham of the same regiment, together with 174 soldiers embarked on HM Ship *Dover* at Amboyna and

Boats of HM sloop *Procris* with detachments of the 14th and 89th Regiments attacking and capturing six French gunboats off the coast of Java on 31 July 1811. Courtesy of Royal Irish Fusilier Regimental Museum.

sailed for Ternate. The force consisted of:

Infantry and artillery	74	
Marines	36	
Seamen	32	
Amboynese	32	(from the Corps of Amboynese).

The island was reached on 25 August and plans were made to capture the fort and the detached works. Fort Orange and the defence positions of Kayo-Meira and Kota-Barro were garrisoned by 500 Dutch troops and 500 locally-recruited soldiers. The attacking force left the ships on the night of 27 August but did not effect a landing until 7.00 am on the following day. Captain Forbes marched inland to take a hill which appeared to overlook the defence position of Kayo-Meira but on reaching the summit it was found that the jungle was too thick and that the Dutch position could not be seen. After a short break Captain Forbes and his party moved off in the direction of the Dutch position and eventually reached the beach at 10.00 pm. They then advanced along this but encountered an outlying picket which raised the alarm and opened fire. Forbes and his party pressed their attack and succeeded in taking the fort with the loss of three men killed and fourteen wounded. They captured one Dutch officer and sixty–eight men.[31] At about 2.00 pm on 29 August the *Dover* opened fire on the battery at Kota-Barro and silenced that position. The guns at Fort Orange opened fire on the *Dover* and whilst Captain Tucker was dealing with that problem the guns at Kota-Barro re-opened fire on his ship. The situation was saved when a land force lead by Ensign Cursham attacked the battery at Kota-Barro from the rear and after taking the position turned the guns onto the town. This permitted the *Dover* to move close to Fort Orange and open direct fire onto the fort. The whole garrison surrendered at about 5.00 pm that day.[32]

A further ninety guns, which were still mounted, and twenty-nine dismounted guns were captured by the Company's force. The colours that had been captured at the fort were sent to India, via Amboyna. Captain David Forbes remained in charge of the garrison at Ternate. This then closed the campaign for the possession of the Spice Islands. The islands remained under the control of the Honorable East India Company and were garrisoned until 1816 when they were restored to the Dutch in accordance with the terms agreed at the convention of August 1814. A number of officers were given administrative appointments during the period that the islands were occupied by the British. Captain David Forbes held the post of Deputy Governor of Banda, Lieutenant Charles Forbes was appointed Assistant to the Resident at Palembang in Sumatra. He was permitted to retain this appointment after the return of his regiment to India. Captain Richard Phillips, also of the Madras European Regiment, became Resident and Commandant at Macassar and died whilst still holding that post on 3 December 1814. A number of officers and men of the Madras European Regiment and other officers attached to the force died during the short occupation of the Spice Islands due to the unhealthy climate. The following deaths were recorded:

1811	
Lieutenant W. H. D'Avenant	Died at Banda
Assistant Surgeon John Milne (serving with 16th Regt MNI)	Died at Banda
1814	
Captain Richard Phillips	Died at Macassar on 3 December.
Lieutenant William Jones Daker	Died at Banda Neira on 13 December.[33]

1815

Captain David Forbes — Died at Banda on 18 April.[34]

Lieutenant William C. Carbery — Died at Ternate.

With the return of the Spice Islands, which included Amboyna to the Dutch in 1816 the Company's local force (the Corps of Amboynese) was disbanded.

Along with the Spice Islands the large island of Java had been incorporated with the French Empire and the island now remained under Dutch control. It was therefore agreed that an expedition would be mounted to take control of the island. The island of Java is 640 miles in length and 80 to 140 miles in width. It has been computed as being '. . . 37,029 geographical square miles (and as having a coastline of) about 1,400 English miles in extent . . .'[35] The expedition to secure Java from the Dutch was mounted at Malacca where the following units assembled:

First, or Advance Brigade

Rifle and Light Companies, 89th Regiment of Foot (200 strong).[36]

20th (Marine) Regiment, Bengal Native Infantry.

Volunteer Battalion, Bengal Native Infantry.

Three squadrons of the 22nd Dragoons.

Detachment of the Madras Horse Artillery.

The left flank battalion was formed by the:

Rifle and Light Companies of the 78th (Highland) Regiment of Foot.

Light Company of the 69th Regiment of Foot.

Grenadier Company of the Bengal Native Infantry, Volunteers.

The battalion was commanded by Major Fraser of the 78th Highlanders.[37]

Second Brigade

This brigade was commanded by Colonel Gibbs of the 59th Regiment of Foot.

14th Regiment of Foot.

59th Regiment of Foot.

Battalion of the Bengal Native Infantry Volunteers.

Third Brigade

This brigade was commanded by Lieutenant Colonel Adams of the 78th (Highland) Regiment of Foot.

69th Regiment of Foot (less the Light Company).

78th (Highland) Regiment of Foot (less the Rifle and Light Companies).

Battalion of the Bengal Native Infantry Volunteers.

Reserve Brigade

Battalions of Bengal Native Infantry Volunteers.

The force was supported by detachments of the Bengal and Madras Artillery and Engineers. The whole expeditionary force consisted of some 12,000 men, of which about half were European. The expedition sailed from Malacca in June 1811 leaving about 1,200 men behind who were sick and unfit for active service.[38] The force arrived off the coast of Java on 4 August of that year and the first landing was made near to the city of Batavia at a kampong (village) named Chillingching. The landing was twelve miles east of the town and Colonel Gillespie took up a position beyond the village and secured the road leading to Cornelis. The landing area was secured and although a patrol of the enemy cavalry attacked one of the advance posts it was driven off with a number of casualties. The 5th was a day of reconnaissance and it was recorded in Davidson's history of the 78th Highlanders that on that day the '. . . marines were greatly exhausted, and one of the rifle company of the 78th died on the road . . .'[39] The force marched to Batavia on the following day and halted at Tanjong Periot that night. From the camp many of the storage-houses and granaries in the city could be seen in flames. They had been plundered and set on fire by the local Malays and Javanese after the French troops had evacuated the town. The enemy had blown up the bridge leading to the city and on 8 August the Engineers constructed a temporary bridge allowing the British force to proceed. The burghers of Batavia immediately surrendered and informed the British that the French garrison had retreated to Weltevreeden.

One Malay was arrested on the morning of the 9th as he was about to set fire to some of the buildings in which powder and ammunition had been stored. He was summarily executed.

The disposition of the enemy force was at this time 3,000 men under command of General Jumelles at the cantonment of Weltevreeden, which was about three miles from Batavia and three miles beyond that position a large French force of some 10,000 men under General Jansens was securely entrenched at the camp of Cornelis.

The Staff of the British Forces was a combined one being drawn from both British and Indian units and was given by Wilson in his *History of the Madras Army* as follows:

'Major-General F. A. Wetherall, Second in Command.
Colonels Gillespie, Gibbs, and Lieutenant Colonel Adams,
H.M.'s 78th regiment [*sic*] commanding brigades.
Colonel P. A. Agnew, 21st M.N.I., Adjutant General.
Colonel Eden, 84th Foot, Quartermaster General.
Major P. VansAgnew, 24th M.N.I., Deputy Adjutant-General.
Major Burslem, 14th Foot, Deputy Quartermaster-General.
Captain Carroll, 69th Foot, Assistant Adjutant-General.
Lieutenant (James) Hanson, 14th M.N.I., Assistant Quartermaster-General.
Lieutenant Colonel Colin Mackenzie, Madras Engineers, Chief Engineer.
Captain (William) Farquhar, Lieutenant (William) Garrard,
Ensigns (Henry) Fullerton, (Duncan) Sim, (John) Coventry,
(Alexander) Anderson and (John George) Proby doing duty under the Chief Engineer.
Doing duty with Corps and Departments.
Captains (Horace) Durand, 5th M.N.I., and (Robert) Mears, 17th M.N.I., Lieutenants (Alexander) Stewart, 16th M.N.I., (John) Low, 24th M.N.I., (James) Bayley, 11th M.N.I., (Henry G.) Jourdan, 10th M.N.I., (Thomas) Clode, 25th M.N.I., (James) Dalgairns, 7th M.N.I., and (Robert John) Marr, 3rd M.N.I.'[40]

In addition two Surgeons and eleven Assistant Surgeons were attached to the staff.

On the 10th the British force under command of Colonel Gillespie advanced towards Weltevreeden and on reaching the cantonments found them abandoned, but, a little beyond, the rifle company of the 78th Regiment, under command of Lieutenant Pennycuik encountered one of the enemy piquets . The enemy picquet withdrew but the advance was halted about two miles from Cornelis. The situation is described by Davidson as follows:

'. . . His [enemy] right was defended by the canal called the Slokan his left was exposed, but the approach in front or flank was covered by a marsh and pepper plantations, and the road rendered impassable by a strong abattis of felled trees. His infantry, about 300 strong, with four field pieces, was drawn up behind this barrier, and opened a destructive fire of grape upon the head of the column as it advanced . . .'[41]

Captain Cameron who was leading his rifle company of the 78th Regiment was severely wounded and a number of his men killed and wounded. Captain Forbes and the light company then charged the obstacle but on meeting fierce resistance during which he lost fifteen out of thirty-seven men, he retired and then crossed a ditch to the left of the enemy. One of the British officers who had been detailed to command a cover party for the British guns was killed. The grenadier company of the 78th Regiment, together with a detachment of the 89th Regiment, attacked the enemy's guns and after a stiff resistance over-ran the position and killed all the enemy gunners bayonetting many of them in a hand-to-hand skirmish. The British force was now close to the walls of the fortified position of Cornelis and it was estimated that the enemy had lost about 500 men and that Brigadier-General Alberti had been seriously wounded. The casualties of the British force were mainly confined to the 78th and 89th Regiments and were given as Lieutenant Munro and seven soldiers killed and Captain Cameron and fifteen rank and file wounded, including two sergeants.[42]

On the 22nd the enemy made a sortie from Cornelis but although they initially captured two of the British advance gun positions they were driven off by the 78th Regiment under command of Major Lindsay and Captain Macleod. The 78th Regiment lost five men killed and five injured during this short but fierce encounter.

General Jansens considered that his position was sufficiently well fortified that he need only hold out on the defensive until the commencement of the rainy season when he calculated that the conditions would then compel the British to retire and end the siege. The camp of Cornelis was a rectangle of 1,600 yards long by 900 yards in width. It was strongly entrenched with the river Jacatra flowing on the western side and the canal Slokan running almost parallel to the east. Neither was fordable and also both had steep banks. The northern and southern ends of the camp were defended by deep ditches which could, when needed, be filled with water. The whole camp was also protected by seven redoubts and by numerous batteries and entrenchments. The redoubts had been constructed so that a cross fire could be maintained and so that they could mutually defend each other. It was estimated that the entire circumference of the defence work was about five miles and that some 280 guns were mounted as a defence.

The Commander of the British Force, Sir Samuel Auchmuty, ordered the British batteries to open fire with twenty 18-pounder guns and eight mortars and howitzers on the 24th, and a gun battle between the two forces commenced with the British gaining slight advantage as the standard of service of the guns was superior to that of the enemy.

There was a problem of delay and not wishing to be caught in the rainy season the General gave orders for the force to advance at midnight with Colonel Gillespie leading his troops as the advance guard. He was supported by Colonel Gibbs with the 59th Regiment and the 4th Battalion of the Bengal Volunteer battalion. The country was such that a night advance proved to be difficult and when daylight came it was found that the rear division under Colonel Gibbs had, in fact, lost its way. The advancing troops were challenged and having fixed bayonets the column advanced in the face of heavy grape and musketry fire but succeeded in taking possession of the No 3 redoubt. The grenadiers of the 78th Regiment managed to rush the bamboo bridge over the Slokan and were able to prevent the bridge from being destroyed. The left flank of the attack advanced and stormed the large No 4 redoubt and this was followed by the taking of No 1 redoubt, but as this fell it was blown up by what was said to have been the actions of some of the enemy officers who fired the magazine. The result was that both sides suffered casualties but in the meantime the attack against the No 2 redoubt was pressing ahead under the command of Lieutenant Colonel Macleod. The fighting continued with Lieutenants Hart and Pennycuik of the

78th Regiment being wounded and, although opposed by a determined enemy, the 78th pressed home their attack and succeeded in taking the enemy positions. The regiment then advanced along the high road under heavy grape and musketry fire but continued to take position after position. Lieutenant Colonel Campbell had both his thighs shattered by grape shot and despite medical attention died two days later.[43]

Whilst the 78th Regiment were taking the enemy positions to its own front, the 69th Regiment advanced along the Jacatra and drove the enemy from the defence positions there, however, this left two of the large redoubts still in enemy hands. The British forces regrouped and advanced towards these two enemy positions but as the British pressed home their attack the enemy withdrew. The whole attack had been completed in a space of three hours and although General Jansens was able to escape with a number of his men to Buitenzorg, some thirty miles from Cornelis, the enemy had lost about 2,000 men, of whom 1,000 were buried on the battlefield.[44] It was also estimated that some 6,000 prisoners were taken of which a large majority were European, and these included three Brigadiers, thirty field officers, seventy captains and 140 subalterns.[45]

The British lost nearly 900 men of whom eighty-five were officers. Davidson lists some of the casualties in his book and these are given as:

'22nd Dragoons	– Lieutenant Hutchins
14th Regiment	– Captain Kennedy, Lieutenant Coghlan
59th Regiment	– Captain Oliphant, Lieutenants Waring, Lloyd, Litton, Macpherson and Ensign Wolfe.
69th Regiment	– Lieutenant Colonel Macleod, Captain Ross, Lieutenant Hipkins.
78th Highlanders	– Lieutenant Colonel Campbell
5th Battalion of Volunteers	– Lieutenant Macdonald.
6th Battalion of Volunteers	– Lieutenant Murrall . . .'[46]

On 29 August the British force advanced to Buitenzorg but on reaching Camp Malayo it was learned that General Jansens had already evacuated the town and retreated to the eastern portion of the island. The advance was halted and the force returned to Weltevreeden. A small force was despatched by sea to Cheribon in order to intercept the General, but it arrived too late as General Jansens had passed through two days prior to this and was now proceeding to Samarang. On receiving this news General Sir Samuel Auchmuty decided to take up the pursuit in person and so embarked at Batavia on 5 September. He took a force of units drawn from both the British and Indian armies these being a troop of the 22nd Dragoons, the 14th and 78th Regiments of Foot, the grenadiers of the 3rd Volunteer Regiment from Bengal and six field-pieces together with artillerymen and pioneers, the whole force amounting to just under 1,000. The vessels used were the *Mysore* and the *Lowjee Family*. On 12 September the 14th Regiment and seven companies of the 78th Regiment landed at Samarang and were able to occupy the town without opposition. Once again General Jansens had retreated and was now located in a strong defence position in the hills of Serondole about six miles from the town of Samarang. A small reinforcement arrived on the 14th and it was then decided that the attack must proceed. The attack was launched at 2.00am on the 16th when the whole force under command of Colonel Gibbs of the 59th Regiment advanced and once again by the sheer determination of the British force the enemy positions were over-run and General Jansens was once again forced to retreat. On the night of 17 September, General Jansens sent a flag of truce proposing terms for the surrender of the island. The terms were rejected by Sir

Samuel Auchmuty who would accept nothing less than an unconditional surrender. He also warned that unless this was accepted immediately the British force would continue its advance. As a reply had not been received on the morning of the 18th, the troops continued their advance towards Soligata. After an advance of about seven miles a second flag of truce was sent and on this occasion the General agreed that the Governor should '... surrender himself and his army [as] prisoners of war, resign the sovereignty of Java, and all Dutch and French possessions in the East Indies into the hands of Great Britain, who should be left free with regard to the future administration of the island, the guarantee of the public debt, and the liquidation of paper money ...'[47]

As a result of the surrender Mr Thomas Stamford Raffles (later to become Sir Stamford Raffles) was appointed Governor of the island in recognition of his planning of the whole enterprise. Colonel Gillespie was appointed Commander of the Forces.

The Commander-in-Chief then arranged for the transfer of part of the cavalry and artillery, as well as the 69th Regiment of Foot, and the headquarters and a wing of the 89th Regiment of Foot back to India and on completion handed over command to Colonel Gillespie. The 78th Regiment of Foot, who had reached Samarang, embarked on the *Mysore* for Sourabaya, where they arrived on 28 September. The French flag was still flying over the Fort at the entrance to the Harbour but when the troops prepared to land and attack, the French at Fort Ludowick hauled down the French flag and surrendered. Davidson records in his history of the 78th Regiment that:

> '... The quarters occupied by the regiment at Sourabaya were very indifferent, and owing to the circumstance and the privations it had undergone during the campaign, together with the easy access which the men had to native liquors, they became very unhealthy, losing during some months between twenty to twenty-five men. A high rate of mortality existed until the latter end of the next year, when the regiment was moved into comfortable barracks at Denoyo, about three miles from the town of Sourabaya, where the number of deaths rapidly decreased, not being more than from three to five a month, and sometimes not so many ...'[48]

Whilst the 78th Regiment of Foot were stationed in Sourabaya the women and children of the regiment were permitted to join them from India.

On 23 May 1812 a general order was issued by the Commander-in-Chief in which he passed on to the troops the 'approbation' of His Royal Highness, the Prince Regent for the distinguished way in which the Java campaign had been waged and for the many acts of distinguished gallantry. Medals were granted to all the native troops that had taken part and to '... such of the superior European officers, as had distinguished themselves ...'[49] In the publication *British Battles and Medals* by Major L. L. Gordon, the allocation of the Java Medal is given as follows:

> '... This medal was issued by the H.E.I.C. to the native troops of the 3rd, 4th, 5th and 6th Bengal Volunteers, Madras Pioneers and the Governor's Bodyguard. One hundred and thirty gold medals were issued; 5,761 natives received silver medals as well as about 750 Europeans. Only those in the service of the Company were allowed to wear their medals ...'[50]

The British Army regiments were entitled to the Military General Service medal (1793–1814) with the clasp 'Java'. The qualification period for the clasp was from 14 to 26 August 1811, both dates inclusive. Gordon records the following as being entitled to the medal and clasp:

> '... 22 Light Dragoons; R.A. (10), 14, 59, 69, 78, 89 Foot; 3, 4, 6 Bengal Volunteers ...'[51]

There were very few naval and marine recipients of this medal and bar and these were allocated as follows:

HMS *Illustrious*	-	Captain Robert Festing RN
		Able Seaman Henry Bannister
		Marine Matthew F. Steele
HMS *Cornelia*	-	Lieutenant Edward Lloyd RN
		Master's Mate J. E. Cabburn
HMS *Scipion*	-	Leading-seaman Thomas Birtles
		Second Lieutenant George Gill, Royal Marines
		Private James Platt (Royal Marines)
HMS *Nisus*	-	Second Lieutenant William Calamy, Royal Marines
HMS *Bucephalus*	-	Second Lieutenant Samuel Garmston (or Garmstone) Royal Marines
HMS *Phoebe*	-	Marine Richard William Pascoe[52]

There are differences in the details of the issue of the medal, whereas Gordon lists eleven names, D. Hastings Irwin, in his book, *War Medals and Decorations* only lists nine Naval and Marine recipients, he omits Able Seaman H. Bannister and Leading-seaman Thomas Birtles. Irwin also shows Richard William Pascoe as being a Second Lieutenant.

In addition two British officers were awarded Military Gold Medals for their part in the Java campaign. They were Captain George Sayer RN and Captain Richard Bunce, Royal Marines.

Although in effect the Java campaign was now concluded, the force was again in action early in 1812. The Sultan of Palembang, on the island of Sumatra, attacked and destroyed the Dutch factory at Palembang and murdered all the inhabitants of the factory. Colonel Gillespie and a small force consisting of detachments of the Madras Horse Artillery and Bengal Artillery, together with the 59th and 89th Regiments of Foot and detachments of the Bengal Volunteers, arrived off the Palembang River during April 1812 and proceeded upriver towards the town which was a distance of about sixty miles. On the morning of 25 April news was received that the Sultan had fled and that the Sultan's troops were sacking the town and setting houses on fire. The British force moved to the palace and took command of that and gradually brought order to the area. The Sultan was formally deposed and the sovereignty conferred upon his brother. With peace restored the British force withdrew to Java arriving there at the end of May, having stopped to take possession of the island of Banca *en route*. Captain Robert Mears, 17th Regiment, Madras Native Infantry raised a native battalion in Java and was placed in charge of the island of Banca. Mears was later killed in the attack on a stockade which was being held by rebel forces under the ex-Sultan of Palembang.[53]

Although the British now controlled the island of Java, the Sultan of Mataram, one of the southern districts, roused local Javanese chiefs in an attempt to expel the British from the island. A British force was formed, consisting of detachments of the 22nd Dragoons, Madras and Bengal Artillery, 14th, 59th and 78th Regiment of Foot and also from the Bengal Volunteer units. The force reached Djojo-carta[54] on 17 June 1812. The principal fort, which was also the palace, was an oblong quadrangle and was '. . . enclosed by a good rampart with a bastion at each angle, the whole surrounded by a broad wet ditch. It was garrisoned by about 15,000 men, and defended by nearly 100 guns . . .'[55] The estimated number of defenders varies, some authors give the number as 17,000 whilst others (including Sir Thomas Raffles) put the figure as much less.

The palace was attacked on 20 June and after a fiercely fought battle the palace was taken with a loss of ninety-nine killed and wounded and with a great number of the enemy killed. Two officers of the Madras Horse Artillery were commended for conspicuous gallantry; they were Captain Henry Thomas Rudyerd and Lieutenant Charles William Black.[56]

With this action the British force entered into a period of duty as an occupation force and encountered no further resistance to the new administration by the Lieutenant Governor, Mr Thomas Stamford Raffles. The 14th Regiment of Foot remained in Java as the garrison regiment until the peace of 1814. In that year Lord Castlereagh, the British Foreign Secretary negotiated the Anglo-Dutch Convention and, with a King once again on the throne of Holland and the country free from the French, it was agreed that the former Dutch colonies would be returned to them, with the exception of the Cape of Good Hope, Ceylon and Guiana. However it was not until 1816 that the Dutch returned to Java and regained control of the administration.[57]

After the conclusion of the fighting in Java and the Spice Islands, a number of commendations were published in various orders including commendations for gallant service to the detachment of H.M.'s 22nd Dragoons; the Madras Pioneers, under command of Lieutenant G. M. Stewart (of the 24th Regiment, Madras Native Infantry); and Lieutenant Colonel Colin Mackenzie of the Madras Engineers. One special mention was made in Wilson's book on *The History of the Madras Army* in respect of Gunner John O'Brien of the Madras Horse Artillery, who acted as a despatch rider between the British Forces during the final action to capture Djojo-carta. O'Brien was thanked in general orders and was awarded a gold medal. An order published at Welte Vreeden on 27 May 1813 carried the following paragraph '. . . The Commander of the forces regrets that the medal which it was his intention personally to present to John O'Brien in commemoration of his gallant conduct near Djojokarta [*sic*] has not yet been received from Bengal; it will, however, be forwarded to Madras through the regular channel . . .'[58]

For his part in the planning and the subsequent administration of Java, Thomas Stamford Raffles, whose first wife Olivia had died in Java on 26 November 1814 at the age of 43, was knighted by the Prince Regent on 29 May 1817.

As a result of the campaigns in Java and the Spice Islands the following Battle Honours were awarded:

Battle Honour	*Unit*	*Date awarded*
Amboyna	Madras European Regiment (later 1st Bn The Royal Dublin Fusiliers)	12 March 1841
Ternate	Madras European Regiment	12 March 1841
Banda	Madras European Regiment	12 March 1841
Java	14th (Buckinghamshire) Regt of Foot	28 May 1818
	59th (2nd Nottinghamshire) Regt of Foot	28 May 1818
	69th (South Lincolnshire) Regt of Foot	28 May 1818
	78th (Highland) Regt of Foot (Ross-shire Buffs)	28 May 1818
	89th Regt of Foot	28 May 1818
	HEIC Governor General's Body Guard	GGO 43/1829
	Madras Pioneers (later 2nd Q.V.O. Sappers & Miners)	20 Dec 1820[59]

Following the campaign a number of memorials were erected in both England and India, some in memory of individuals and others in respect of regiments that participated in the campaigns. One of the most impressive was erected at Barrackpore Park, about 200 yards west of the Government House. The cenotaph was erected by '. . . Gilbert, Earl of Minto, Governor-General of British India, as a tribute of personal feeling and respect to the memory of the brave . . .' The cenotaph was in memory of those who lost their lives in both the Java and the Mauritius campaigns of 1810 and 1811 and the following officers were listed on the plaque dedicated to the Java campaign:

Lieutenant Munro of H.M. 78th Regiment	10 August 1811
Lieutenant Colonel Charger H.M. 69th Regiment	22 August 1811
Lieutenant Shephard, Madras Pioneers	"
Ensign McLeod, Madras Pioneers	"
Lieutenant Patton, Royal Artillery	24 August 1811
Lieutenant Farnaby, Bengal Artillery	"
Lieutenant Colonel McLeod, H.M. 69th Regiment	26 August 1811
Major Campbell, H.M. 78th Regiment	"
Captain Kennedy, H.M. 14th Regiment	"
Captain Olpherts, H.M. 59th Regiment	"
Captain Ross, H.M. 69th Regiment	"
Lieutenant Hutcheen, H.M. 22nd Dragoons	"
Lieutenant Waring, H.M. 59th Regiment	"
Lieutenant Lloyd, H.M. 59th Regiment	"
Lieutenant Litton, H.M. 59th Regiment	"
Lieutenant Hopkins, H.M. 69th Regiment	"
Lieutenant Fergusson, 20th Madras N.I.	"
Ensign Wolfe, H.M. 59th Regiment	"
Lieutenant Murrall, 24th Bengal N.I.	23 September 1811
Captain Shaw, 13th Bengal N.I.	28 September 1811
Lieutenant McLean, H.M. 14th Regiment	20 June 1812'[60]

The fort at Amboyna retained the name of Fort Victoria even after the withdrawal of the Company's troops and its return to Dutch control. The *Singapore Free Press* newspaper published the following report in its edition published on Thursday morning 7 April 1836 (vol 1, no 27):

> '. . . During three weeks in the month of October [1835] a thick sulphury mist lay over the Islands [Molucca Islands] and on the 1st November at about 3 o'clock in the morning an un-commonly strong earthquake was felt at Amboina [Amboyna] and several of the surrounding islands . . . this was followed on the 4th of the same month by several new shocks . . . it is but too probable that there must have been an eruption of the Volcano of Banda . . . The earthquake [at Amboyna] has been dreadful in its consequences: fifty-eight men, women and children lost their lives by the falling of the Barracks in Fort Victoria. Sixty-six persons in all have been killed . . .'

Although the Native regiments employed in the Java campaign were made up of volunteers these regiments would have been based on the standard establishment approved for the 'Bengal Native Army'. The authorised establishment consisted of:

Colonel Commandant	1	Lieutenant Colonels	2
Majors	2	Captains	7
Captain/Lieutenant	1	Lieutenants	22
Ensigns	10	Sergeants	2
Subadars	20	Jemadars	20
Havildars	100	Naiks	100
Drummers and Fifers	40	Sepoys	1,600

Staff

Adjutant and Quartermaster (for the Regiment)	1
Adjutant (one for each battalion)	2
Paymaster	1
Sergeant Major	1
Quartermaster Sergeant	1
Drum Major	1
Fife Major	1
Drill Havildars	2
Drill Naiks	2
Surgeon	1
Surgeon's Mate	2
Native Doctors	2

The establishment was sub-divided into two battalions, each consisting of two grenadier and eight battalion companies.[61]

There is one aspect of the British occupation of Java that has received little attention from military historians and that is the Sepoy Conspiracy of 1815 when a number of the men of the 4th Light Infantry Battalion, Bengal Army entered into a conspiracy with the local Javanese leaders and attempted to mutiny. At the time of the outbreak the whole affair was kept secret in order to avoid the spread of dissention within the Company's other units and also to avoid further insurrection by the local rulers.

The disaffection with the 4th Light Infantry Battalion commenced as early as 1813. It has been recorded that the Javanese at first dreaded the Sepoys as they were '... insolent and oppressive ...'[62] and it was because of their dread of these troops that Mr Thomas Stamford Raffles decided that the Volunteer Battalions should be used on garrison duties. He allocated half a battalion for garrison duties at each main centre. The 4th Light Infantry Battalion was kept on these duties during the years following the initial occupation of the Island, however the actual strength at outposts varied.[63] In July 1813, another volunteer unit, which had been named the Java Light Cavalry Regiment, which had only recently arrived from Bengal, was relieved of garrison duties and this move led to disaffection among the Bengal light infantrymen.[64] In addition, the fact that the Bengal troops were not permitted to have their families with them and were expected to remit part of their meagre pay back to Bengal, combined with the fact that men were having difficulty in actually remitting the money back to their relatives,[65] all led to a general feeling of unrest within the Light Infantry Battalion. The Battalion was also short of European officers as, of the seven allocated to the 4th Light Infantry Battalion prior to the Java mutiny only three officers remained,[66] the others having been withdrawn to fill civil administration posts on the island.

These factors all tended to lead to the general unrest and discontent in the Central Javanese garrisons and the situation further deteriorated when, in August 1814, news reached Java that Napoleon had been defeated at Leipzig during October of the previous year. The Dutch residents on Java expected the British to withdraw and soon rumours circulated within the 4th Light Infantry Battalion that the British intended to sell the Sepoy Battalions to the Dutch and of course abandon the men in Java.

Some of the Sepoy officers in Surakarta made contacts in the court of the Sunan and were later found to have been accepting gifts from the Javanese nobles. By September 1815, the local Javanese rulers had entered into correspondence with the Sepoys at the garrisons of Weltevreden and Surabaya as well as Klaten and Bayalali in an attempt to bring about an uprising in Central Java. In addition a number of deserters had been permitted to secretly settle by some of the local rulers and one Sepoy officer, Subadar Fakir Singh had agreed to use his influence to subvert the '. . . Malay Battalions in Batavia . . .'[67] The Sunan was unhappy about the impending massacre of the British officers but wished to ultimately gain control of the whole of Central Java with the help of the Sepoys, who he then hoped would either return to their homes in Bengal or settle on the west coast and thus cease to be a threat to his future kingdom.

The rising was planned to start at the headquarters of the 4th Light Infantry Battalion at Yogyakarta. Groups of Sepoys began to meet secretly on the ramparts at night and on 21 October 1815 the European Sergeant Major was warned by his bearer that there was trouble in the battalion and he was later again warned and told to secure the door of his room at night.

One conspirator, Mata Deen,[68] met about twenty-nine Sepoy officers and men at one o'clock in the afternoon of 24 October and it was generally agreed that the British officers should be murdered between 6.30 and 7.00 pm that evening. The conspirators met again at about 7.00 pm and decided to delay the uprising for two days in the hope that more Sepoys would join the mutiny. There then followed further meetings and the outbreak was again delayed. On this occasion the delay was extended until 29 October when, after the murder of the British officers, it was intended that the mutineers would seize the Sultan's treasury and after transferring the contents to the Residency that building would be converted and become a Hindu temple.

With the number of meetings being held, it was inevitable that the conspiracy would be discovered and Lieutenant Steel,[69] the acting commanding officer, became aware of what was happening and paraded the Sepoys and warned the men that anyone convicted of being involved in the plot would be executed by being '. . . blown away . . .' from the mouth of a cannon. The threat had the desired effect and two of the ring-leaders, Mata Deen and Jemadar Dai Ram, changed their plans. Captain Robert Cock[70] returned to Yogyakarta and held an inquiry lasting from 9 to 16 November. As a result of this inquiry, during which many men of the battalion gave evidence against the conspirators, the ring-leaders and twenty others were sent under guard to Serondol, near Senarang, to await a Court Martial which was held in January 1816.

With the prompt action by the British officers, the situation had been saved and it was seen to be expedient to bring the Sepoy battalions back to strength with regard to British officers. Moves were made to improve the security in Central Java and a new Corps of twenty-five mounted troops was formed to guard the palace and to ensure that further conspiracies were impossible. These troops were recruited, in the Central Javanese court towns, from Dutch and German residents who had Indonesian mothers.

The Sepoys who had been convicted of conspiracy were swiftly dealt with; seventeen were shot by a firing squad at Samarang, and about fifty were sent back to Bengal 'in irons'.

ANNEXURE A
LIST OF OFFICERS OF THE HONORABLE EAST INDIA COMPANY'S ARMIES WHO VOLUNTEERED AND SERVED IN THE JAVA CAMPAIGN

(Officers are listed in order of precedence within their parent regiment)

1st Volunteer Battalion, Bengal Army		*Parent regiment*
Captain James Garner		15th Regiment, Bengal Native Infantry
Lieutenant William D. Turner		"
Lieutenant Joseph Garner		"
Lieutenant Hugh Davidson		"
Lieutenant David Crichton		"
Lieutenant Gilbert Cowper		"
Ensign H. F. Wroughton		"
Major William Burton	(in Command)	25th Regiment, Bengal Native Infantry
Captain Henry Carter		"
Captain Mossem Boyd		"
Lieutenant William H. Kemm		"
Lieutenant Brown Roberts	(Adjutant)	"
Lieutenant John C. B. Parke		"
Lieutenant Gardner Boyd		"
Lieutenant John Thwaites		"
Lieutenant James Vyse		"
Lieutenant George Stubbins		"
Ensign Henry C. Medwin		"
Ensign James Parsons		"
Ensign Adonijah Smith		"
Ensign J. Neale Wilson		"
Assistant Surgeon Watson		Bengal Medical Establishment

2nd Volunteer Battalion, Bengal Army		
Captain James R. Lumley	(in Command)	8th Regiment, Bengal Native Infantry
Captain Henry Manley		"
Lieutenant the Hon. J. T. Aylmer		"
Lieutenant Andrew O'Shea		"
Lieutenant Matthew C. Paul		"
Lieutenant Francis Heron		"
Lieutenant Charles E. Turner		"
Lieutenant James Fagan	(Adjutant)	"
Lieutenant Alexander Wilson		"
Lieutenant John Lucas Earle		"
Lieutenant Thomas Palmer		19th Regiment, Bengal Native Infantry
Lieutenant Edward Pettingall		"
Lieutenant E. W. Broughton		"
Ensign Charles Thoresby		"
Lieutenant Joseph Green	(Quartermaster)	25th Regiment, Bengal Native Infantry
Lieutenant Andrew Allen	(Acting Adjutant)	"
Ensign Robert McMullin		Not posted to a regiment
Assistant Surgeon George Lambe		Bengal Medical Establishment
Assistant Surgeon Edward Muston		"

Madras Volunteer Battalion		
Captain William G. Waugh		Madras European Regiment
Lieutenant David Agnew		3rd Regiment, Madras Native Infantry
Lieutenant Robert John Marr		"
Captain W. M. Robertson		4th Regiment, Madras Native Infantry
Ensign James H. Logan		5th Regiment, Madras Native Infantry
Lieutenant Pringle Fraser		7th Regiment, Madras Native Infantry
Lieutenant James Myers		"

Madras Volunteer Battalion		*Parent Regiment*
Lieutenant Benjamin R. Hitchins		7th Regiment, Madras Native Infantry
Ensign William Low		8th Regiment, Madras Native Infantry
Lieutenant James Noble		15th Regiment, Madras Native Infantry
Ensign Robert B. Dawes		,,
Lieutenant Josias Torriano		16th Regiment, Madras Native Infantry
Lieutenant Alexander Cowan		17th Regiment, Madras Native Infantry
Captain John Mackenzie		20th Regiment, Madras Native Infantry
Lieutenant William Thomas Sneyd		,,
Lieutenant John Newall		,,
Lieutenant James Crokat		,,
Lieutenant George Grant		24th Regiment, Madras Native Infantry
Lieutenant Benjamin Blake		25th Regiment, Madras Native Infantry
Surgeon George Alexander		Madras Medical Establishment
Assistant Surgeon Archibald Campbell		,,
4th Light Infantry Battalion, Bengal Army		
Lieutenant Thomas Henry Paul		5th Regiment, Bengal Native Infantry
Lieutenant Duncan G. Scott		,,
Lieutenant Alexander M'Donald		7th Regiment, Bengal Native Infantry
Lieutenant Alexander Stewart		14th Regiment, Bengal Native Infantry
Lieutenant Henry Hall	(Adjutant)	16th Regiment, Bengal Native Infantry
Lieutenant Robert Gunn		17th Regiment, Bengal Native Infantry
Lieutenant William Kedlie		19th Regiment, Bengal Native Infantry
Lieutenant John Tulloch		22nd Regiment, Bengal Native Infantry
Lieutenant Robert Cock		26th Regiment, Bengal Native Infantry
Lieutenant Thomas Thackeray		,,
Ensign Robert Vetch		,,

Bengal European Regiment

The following officers served with the regiment at Amboyna.

Major Samuel Kelly
Captain Sir Thomas Ramsay, Bart.
Captain H. Blankenhagen
Captain Edmund Morris
Lieutenant Foster Walker
Lieutenant Thomas Kirchoffer
Lieutenant Alexander Brown
Lieutenant Charles Smith
Lieutenant P. S. Van Swinden
Lieutenant Edward Fitzgerald
Lieutenant Robert McKerrill
Lieutenant Robert Ledlie
Lieutenant John Fulton
Lieutenant Henry P. Carleton
Ensigns Francis Crossley, James Harrison and George Wray
Assistant Surgeons Patrick Halket and John Eckford.

By 1814 the number of officers serving in the Java campaign had increased and the following officers were then serving in the area with the Volunteer units, or Garrison, as shown.

Java Light Cavalry Regiment (stationed on Java)		*Parent Regiment*
Lieutenant Samuel I. Bacon		1st Regiment Bengal Native Cavalry
Cornet Hubert de Burgh		2nd Regiment Bengal Native Cavalry
Cornet Frederick Peret		3rd Regiment Bengal Native Cavalry
Cornet George Burges		5th Regiment Bengal Native Cavalry
Cornet George Roxburgh		6th Regiment Bengal Native Cavalry
Major Lucius R. O'Brien	(Commanding)	8th Regiment Bengal Native Cavalry
Cornet G. A. Kempland		"
Amboyna Corps		
Captain H. Blankenhagen	(Commanding)	Bengal European Regiment
Lieutenant John Fulton	(Adjutant)	"

Bengal European Regiment	*Location*
Captain Sir T. Ramsay, Bt	Amboyna
Captain John Stuart	Java (Assistant Commissary General)
Captain Edmund Morris	Amboyna
Lieutenant William Home	Java
Lieutenant James Auriol	Java
Lieutenant Foster Walker	Amboyna
Lieutenant Thomas Kirchoffer	Amboyna (later Fort Adjutant Ternate)
Lieutenant Charles C. Smith	Java
Lieutenant Thomas Watson	Java
Lieutenant Charles Smith	Banda
Lieutenant George Bolton	Java
Lieutenant P. S. Van Swinden	Amboyna (Brigade Major)
Lieutenant Samuel Watson	Java
Lieutenant Edward Fitzgerald	Amboyna
Lieutenant Robert Ledlie	"
Lieutenant William Burroughs	"
Lieutenant Metcalf S. Hogg	Java (Acting Adjutant and Quartermaster to the Division)
Lieutenant Joseph Orchard	Amboyna
Lieutenant John Irwin	Java
Lieutenant Henry P. Carleton	Amboyna
Lieutenant Francis Crossley	Amboyna (Fort Adjutant)
Lieutenant John Cockburn	Java
Lieutenant James Harrison	Amboyna
Ensign George Wray	"
Ensign Benjamin Ashe	Java
Ensign Alexander Irvine	Java
Ensign Thomas Joseph Goding	Java
Ensign James Marshall	Amboyna
Ensign William Davison	Java
Ensign James Easson	Java
Ensign Joseph Bunyon	Java
Ensign William Richard Lee Faithfull	Java
Surgeon John Henderson	Java
Assistant Surgeon Patrick Halket	Amboyna
Assistant Surgeon John Eckford	"

Parent Regiment	*Name*	*Serving with:*
Bengal Native Infantry		
1st Regiment	Lieutenant Thomas C. Jackson	6th Volunteer Battalion, Java
2nd Regiment	Lieutenant James C. Grant	3rd Volunteer Battalion, Java
"	Lieutenant W. G. Mackenzie	Assistant to the Resident at Amboyna
"	Lieutenant George W. A. Lloyd	Java Infantry Volunteers
"	Lieutenant George Richard Pemberton	3rd Volunteer Battalion, Java
"	Lieutenant Alexander Francis P. Macleod	"
"	Ensign Stephen Swayne	6th Volunteer Battalion. Java
3rd Regiment	Captain Denis H. Dalton	Commanding Volunteer Light Infantry Battalion, Java
"	Captain Bensley Walker	5th Volunteer Battalion, Java
4th Regiment	Captain Alexander Campbell	4th Volunteer Battalion, Java
"	Captain James Clarke	"
"	Lieutenant Patrick M. Hay	Adjutant 4th Volunteer Battalion, Java
"	Lieutenant Thomas Hepworth	Java Infantry Volunteers
"	Lieutenant Peter Grant	4th Volunteer Battalion, Java
"	Lieutenant Charles Christie	"
"	Ensign William Jover	Java Infantry Volunteers
5th Regiment	Lieutenant William F. Wilson	Volunteer Light Infantry Battalion, Java
"	Lieutenant Thomas Trist	3rd Volunteer Battalion, Java
6th Regiment	Captain Charles Poole	6th Volunteer Battalion, Java
"	Captain Simon Fraser	Light Infantry Volunteer Battalion, Java
"	Lieutenant Robert Pringle	Adjutant 6th Volunteer Battalion, Java
7th Regiment	Lieutenant Henry B. Rawlins	4th Volunteer Battalion, Java
"	Lieutenant Henry Shadwell	Java Infantry Battalion
"	Lieutenant John H. Aspinwall	Quartermaster 4th Volunteer Battalion, Java
8th Regiment	Lieutenant Alex Black	Java Infantry Volunteers
"	Ensign Richard Burney	"
9th Regiment	Captain William Brooks	3rd Volunteer Battalion, Java
"	Lieutenant William P. Hardwicke	Java Infantry Volunteers
10th Regiment	Lieutenant John H. Littler	6th Volunteer Battalion, Java
"	Lieutenant Patrick Dudgeon	Quartermaster, 6th Volunteer Battalion, Java
11th Regiment	Lieutenant Scipio E. Richards	Quartermaster, Volunteer Light Infantry Battalion, Java
"	Lieutenant John Dunn	4th Volunteer Battalion, Java
"	Lieutenant John Macartney	"
"	Lieutenant Thomas H. Rouse	"
12th Regiment	Captain Henry Hodgson	"
"	Lieutenant Alexander M'Leod	Java Infantry Volunteers

Parent Regiment	*Name*	*Serving with:*
Bengal Native Infantry		
12th Regiment	Lieutenant Llewellin Conroy	Java Infantry Volunteers
”	Lieutenant Charles Edward Davis	4th Volunteer Battalion, Java
13th Regiment	Lieutenant Colonel James Dewar	Commanding, 3rd Volunteer Battalion, Java
”	Captain G. T. D'Aguilar	6th Volunteer Battalion, Java
”	Lieutenant Robert Seymour	”
”	Lieutenant George Byrant	”
14th Regiment	Captain Thomas Evans	5th Volunteer Battalion, Java
”	Lieutenant Woodward Bidwell	Quartermaster, 3rd Volunteer Battalion, Java
”	Lieutenant Alexander Stewart	5th Volunteer Battalion, Java
”	Lieutenant William Cotes	3rd Volunteer Battalion, Java
15th Regiment	Captain John Leys	Volunteer Light Infantry Battalion, Java
”	Lieutenant George P. Green	3rd Volunteer Battalion, Java
”	Lieutenant Gilbert Cowper	Java Infantry Volunteers
”	Ensign Owen Phillips	3rd Volunteer Battalion, Java
16th Regiment	Lieutenant Gerard J. Hendy	Volunteer Light Infantry Battalion, Java
”	Lieutenant Anthony Lambert Swanstone	5th Volunteer Battalion, Java
17th Regiment	Lieutenant Archibald Montgomery	Quartermaster, 5th Volunteer Battalion, Java
”	Lieutenant Robert Gunn	Adjutant, Volunteer Light Infantry Battalion, Java
”	Lieutenant Alexander Dick	5th Volunteer Battalion, Java
”	Lieutenant Frederick Anstice	4th Volunteer Battalion, Java
18th Regiment	Captain William Collyer	5th Volunteer Battalion, Java
”	Captain George Veal Baines	”
”	Lieutenant James Pearson	Volunteer Light Infantry Battalion, Java
”	Lieutenant Colin M'Donald	5th Volunteer Battalion, Java
19th Regiment	Lieutenant Godfrey P. Baker	Volunteer Light Infantry Battalion, Java
20th Regiment	Major Udney Yule	Java
”	Lieutenant Thomas Travers	Town Major, Batavia
”	Ensign John S. Marshall	Java
21st Regiment	Captain Jerimiah M. Johnson	Paymaster to the troops at Java
”	Lieutenant Thomas Fiddes	4th Volunteer Battalion, Java
”	Lieutenant Henry Dwyer	”
”	Lieutenant Hugh Sibbald	”
”	Lieutenant James Oram Clarkson	Java Infantry Volunteers
22nd Regiment	Major Peter Grant	Commanding 4th Volunteer Battalion, Java
”	Captain Hugh Griffiths	Commanding 5th Volunteer Battalion, Java
”	Lieutenant Charles T. Higgins	Amboyna

Parent Regiment	*Name*	*Serving with:*
Bengal Native Infantry		
22nd Regiment	Lieutenant John Tulloch	Volunteer Light Infantry Battalion, Java
”	Lieutenant Henry C. M. Cox	Adjutant, 5th Volunteer Battalion, Java
”	Ensign William H. Wake	Java Infantry Volunteers
23rd Regiment	Lieutenant Alexander Horsburgh	5th Volunteer Battalion, Java
”	Lieutenant Thomas Williams	”
”	Ensign Hugh L. Barnett	”
24th Regiment	Captain Robert C. Garnham	3rd Volunteer Battalion, Java
”	Lieutenant Brook Watson	6th Volunteer Battalion, Java
”	Lieutenant John Cruickshank	Adjutant, 3rd Volunteers Battalion, Java
25th Regiment	Lieutenant John R. Knight	Java Infantry Volunteers
”	Lieutenant David Mason	”
26th Regiment	Lieutenant Robert Cock	Volunteer Light Infantry Battalion, Java
”	Lieutenant Peter P. Morgan	5th Volunteer Battalion, Java
”	Lieutenant Thomas Frobisher	”
”	Lieutenant Archibald E. Charters	5th Volunteer Battalion, Java
”	Lieutenant William Percival	Java Infantry Volunteers
27th Regiment	Lieutenant Arthur A. Homer	6th Volunteer Battalion, Java
”	Lieutenant Newton Wallace	”
Bengal Regiment of Artillery	Major Edward W. Butler	Java
”	Captain John David Smith	Java
”	Captain Henry Stark	2nd Troop Horse Artillery, Java
”	Captain James H. Brooke	3rd Troop Horse Artillery, Java
”	Captain/Lieutenant John P. Boileau	2nd Troop Horse Artillery, Java
”	Captain/Lieutenant Samuel Shaw	Com. of Stores, Java
”	Captain/Lieutenant Charles Harris	Java
”	Captain/Lieutenant John Rodber	3rd Troop Horse Artillery, Java
”	Lieutenant Samuel Parlby	”
”	Lieutenant Johnathan Scott	Java
”	Lieutenant George E. Gowan	1st Troop Horse Artillery, Java
”	Lieutenant John Curtis	2nd Troop Horse Artillery, Java
”	Lieutenant John Broadhurst	Amboyna
”	Lieutenant James C. Hyde	3rd Troop Horse Artillery, Java
”	Lieutenant Charles P. Kennedy	1st Troop Horse Artillery, Java
”	Lieutenant Gabriel N. C. Campbell	”
”	Lieutenant John James Farrington	Java
”	Lieutenant Allan Cameron	Java

ANNEXURE B
MEDAL ROLL OF THE MILITARY GENERAL SERVICE MEDAL (1793–1814) WITH THE BAR 'JAVA'

The medal was authorised by General Order of 1 June 1847, and the actual medals and bars, of which there were twenty-nine, were issued on application in 1848. This therefore implies that not all those who were entitled would have applied, as in the meantime at least a percentage of the participants would have died. The details of the medal are as follows:

Obverse: The diademed head of Queen Victoria with 'VICTORIA REGINA' and the date '1848'.

Reverse: Queen Victoria standing on a dias about to place a laurel wreath on the Duke of Wellington, who is kneeling on his left knee. Beside the dias is a diminutive representation of the British Lion dormant. The following inscriptions are also included: Top half – 'TO THE BRITISH ARMY'. Lower – '1793 – 1814'.

Size: 1.42 inch in diameter.

Ribbon: 1.25 inch in width. Crimson with 1/8 *inch wide dark blue borders.*

Suspension: By a plain straight swivelling suspender.

Designer: W. Wyon, RA.

Naming: In indented large Roman capitals.

Bar: The bar for 'Java' was awarded for the period 14 to 26 August 1811.

The following is a list of those who qualified for the medal and who are listed on the Medal Rolls for the MGS Medal held at the Public Record Office, Kew. Those listed were shown as being entitled to the medal *and* the bar 'Java'.

22nd Light Dragoons

Ballinghall	George	Assistant Surgeon
Chase	W.C.	Cornet
Eden	John	Lieutenant
Middleton	Charles	"
Oakes	R.M.	"
Slegg	Richard	"
Smith	J.B.	"

Other ranks

Barlow	William	Private
Black	William	"
Barrett	John	"
Barrett	William	"
Bradshaw	William	"
Cherrington	John	"
Cheetham	William	Corporal
Crowder	Evan	Private
Dawson	Richard	"
Fowles	James	Serjeant Major
Gowan	James	Private
Gunton	Robert	"
Hailstone	George	"
Hammond	William	"
Higgins	John	"
Johnson	William	"
McLean	William	"
Munro	Thomas	"
Middleditch	Thomas	"
Milne	William	"
Nice	Benjamin	Troop Serjeant Major
Perry	Joseph	Private
Ramsay	William	Serjeant
Robinson	Robert	Private
Shaw	Thomas	"
Smith	Allen	"

22nd Light Dragoons – Other ranks

Smith	Thomas	Serjeant
Stephens	Thomas	Private
Walford	John	"
Watson	William	"
Whittle	Samuel	"
Williams	John	"
Windsor	Richard	"

Royal Artillery

Byers	John Sparks	2nd Captain
Colebrooke	William	Captain

Other ranks

Arkinstall	Joseph	Private	4 Bn RA
Coarer	Thomas	"	6 Bn RA
Hughs	William	"	4 Bn RA
Hutchinson	Thomas	Gunner	6 Bn RA
McKenzie	Alexander	"	6 Bn RA
Pike	John	Private	6 Bn RA
Roylance	Thomas	"	4 Bn RA
Smith	George	"	4 Bn RA
Starman	James	Gunner	4 Bn RA
Wells	Thomas	Private	6 Bn RA

14th Foot

Hall	Thomas	Lieutenant
L'Estrange	Edward	Ensign
Rochfort	Gerald	Lieutenant
Watson	Jacob	Captain

Other ranks

Abbotts	Joseph	Private
Amey	John	"
Andrews	Thomas	"
Armstrong	William	"
Bacon	Thomas	Sergeant
Bartlett	William	Private

14th Foot – Other ranks

Baxter	John	Private
Beeby	William	”
Bowler	John	”
Bradley	Joseph	Serjeant
Broom	John	Private
Burke	John	”
Carr	James	”
Conway	John	”
Crick	Daniel	”
Curtis	John	”
Cullington	Robert	Sergeant
Dane	James	Private
Davy	William	”
Davison	William	Corporal
Dawson	John	”
Douglas	William	Private
Dunkley	Joseph	”
Elbro	John	”
Fitzpatrick	John	”
Foster	Edward	”
Foster	John	”
Foulds	Joseph	”
France	Robert	”
Freer	Robert	”
Gibbard	William	”
Griffiths	William	”
Gurney	William	”
Halford	John	”
Harrobin	William	”
Henry	James	Sergeant
Holliday	James	”
Hollis	William	Private
Holt	Richard	”
Howe	William	”
Howlett	John	”
Huckling	Benjamin	”
Humphries	John	”
Hunt	James	”
Hunt	John	”
Jeffs	Thomas	”
Jennings	James	”
Jones	John	”
Joseph	Henry	”
Joyce	Joseph	”
King	Samuel	”
Masters	David	”
McCollough	James	Sergeant
Milton	John	”
Montgomery	John	Private
Newitt	William	”
Orme	Richard	”
Roonie	Thomas	”
Russell	William	”
Shinney	William	”
Smith	Jabez	”
Smith	John	”
Smith	Robert	”

14th Foot – Other ranks

Starr	George	Private
Sturges	Joseph	Sergeant
Taylor	Thomas	Private
Towe	John	”
Townley	Samuel	”
Whilpton	David	”
Woodhouse	Joseph	”
Woodland	Samuel	”
Worthey	John	”
York	John	”

59th Foot

Brohier	John	Lieutenant

(also spelt Broheir)

Chadwick	James	Lieutenant
Gordon	Duncan	Ensign
Graham	David	Captain
Low	John	Lieutenant

24th Regt MNI attached to 59th.

MacMullen	S. (MD)	Surgeon

(shown as M'Mullen in Army List)

Stewart	Archibald H.	Ensign

Other ranks

Ashmore	John	Private
Ashworth	John	”
Barnes	Thomas	”
Barrett	John	Corporal
Booker	Robert	Private
Boland	John	”
Bowden	Joseph	Sergeant
Brown	William	Corporal
Buckmaster	Robert	Private
Butler	John	”
Byrne	Dennis	Sergeant
Bywater	Mark	Private
Burke	William	”
Callican	John	”
Campbell	Charles	”
Carrington	William	”
Cartwright	Joseph	”
Clarke	James Bass	”
Cleary	John	”
Cording	Thomas	”
Dilks	Henry	”
Dowman	William	”
Doyle	Thomas	Quartermaster Sergeant
Dudley	John	Sergeant
Findlay	John	Private
Finn	Owen	”
Flint	Anthony	”
Green	Charles	”
Green	William	”
Hackett	John	”
Hagerty	Thomas	”
Haines	Patrick	”
Harrison	William	”
Hayes	John	”

59th Foot – Other ranks

High	Charles	Private
Hildyard	Robert	"
Hogan	Thomas	"
Holmes	George	"
Holohan	Thomas	"
Horton	James	"
Humphries	Thomas	"
Jeffries	Richard	"
Jones	Edward	"
Jones	Samuel	Sergeant
Kirk	Joseph	"
Lampion	James	Private
Lane	Job	"
Leaver	John	"
Lee	John	"
Light	Thomas	"
Marryatt	Richard	"
Mather	Noah	"
Mayo	Robert	"
McCarthy	Dennis	"
McDonald	Kenneth	"
McHoffy	William	"
McKenzie	Alexander	"
Meade	John	"
Melson	William	"
Morgan	William	"
Murdock	John	"
Neilson	Charles	Corporal
Nutt	John	Private
Nutt	William	Drum Major
Palmer	Samuel	Private
Parker	Samuel	"
Preston	Samuel	"
Radley	George	"
Rea	John	Corporal
Read	James	Private
Reavell	George	"
Richards	Henry	Corporal
Ruscow	Charles	Private
Scott	Edward	"
Sheppard	Richard	"
Slack	John	"
Smith	Donald	"
Smith	Thomas	"
Smith	William	"
Stanaway	Thomas	"
Stephenson	Sently	"
Taylor	Matthew	"
Tell	William	"
Todd	William	"
Townsend	Edwin	Sergeant
Turner	Charles	Private
Vessey	John	"
Walker	William	"
Wallace	Peter	"
Watts	Benjamin	"

59th Foot – Other ranks

Wheatcroft	Matthew	Private
White	James	"
Wilcox	John	"
Willis	John	"
Wilson	William	"
Woodruffe	John	"
Yates	Joseph	"
Young	James	Sergeant

69th Foot

Baker	Robert	Lieutenant
Durnford	George	Lieutenant
attached 69th Foot		
Foulston	John	Lieutenant
(name also spelt Foulstone)		
Gunning	Matthew	Lieutenant
Jenour	Matthew	"
Leslie	John H. H.	Captain
Lowrie	Charles	Lieutenant
Macpherson	Duncan	"
Mitchell	Thomas P.	"
Sprowle	Edward	"

Other ranks

Angle	Richard	Private
Armstrong	James	"
Austin	George	"
Baker	Samuel	"
Bateman	Thomas	"
Betteson	John	"
Bilton	William	"
Birch	Charles	Corporal
Bradbury	John	Private
Brannen	Daniel	Sergeant
Briggs	William	Private
Bristowe	George	"
Brown	John	Corporal
Bryan	Patrick	Private
Byrne	James	"
Chadwick	Samuel	Sergeant
Chittle	James	Private
Clay	Benjamin	"
Connolly	Michael	"
Cousins	Andrew	"
Coyle	Charles	"
Craigie	Robert	"
Dempsey	Patrick	"
Duckett	Job	"
Dunsheath	Hugh	"
Egelstone	Thomas	"
Fields	John	"
Foyson	Benjamin	"
Framington	John	"
Gibson	James	"
Gilbert	Richard	"
Gilbert	Thomas	"
Graham	Samuel	"

69th Foot – Other ranks

Gray	George	Corporal
Grimes	Robert	Private
Hackett	Martin	Sergeant
Hallom	Jonathan	”
Hancock	John	Private
Hanson	William	”
Hardwick	Robert	”
Harvey	Robert	”
Hays	James	”
Headon	David	Corporal
Hennan	Hugh	Private
Henney	George	”
Hill	Joseph	”
Hill	Thomas	”
Hollis	Thomas	”
Houghton	Richard	”
Howlett	William	”
Jackson	David	”
Johnson	Joseph	Sergeant
Johnson	James	Private
Jones	Edward	Sergeant
Knee	Samuel	Private
Kelly	Silvester	”
Kennaway	William	”
Keowne	Michael	”
Keir	William	”
Knight	George	”
Leary	John	”
Lees	George	”
Lewis	John	”
Liddle	John	”
Logan	Henry	”
Marr	George	”
McGahey	Patrick	”
McGuire	Michael	Sergeant
McKeowin	Charles	Private
McQuade	Cullan	”
Melvin	Samuel	”
Mills	James	”
Mothershead	David	Corporal
Orchard	James	Private
Peadon	Alexander	”
Phillips	Richard	”
Pope	Henry	”
Pratt	Benjamin	”
Pritchard	Richard	Corporal
Quinn	Bernard	Private
Quinn	Hugh	”
Robinson	John	”
Shepperd	Thomas	”
Simpers	George	”
Stoker	Joseph	”
Swindles	Daniel	”
Swingler	George	”
Talbot	Richard	”
Taylor	John	”

69th Foot – Other ranks

Taylor	Thomas	Private
Thomas	John	Drummer
Tracey	Michael	Private
Waddell	Matthew	”
Wayland	Matthew	”
Wilson	William	”

78th Foot

Cameron	Alex M.	Lieutenant
Douglas	Henry N.	”
Fraser	James	” and Adjutant
Hart	Richard	Lieutenant
Pennycuick	John C.	Ensign
Robertson	Charles	Lieutenant

Other ranks

Anderson	Alexander	Sergeant Major
Anderson	William	Private
Ashbourne	George	”
Bain	Alex	Sergeant
Barclay	Alex	Private
Beeson	Thomas	”
Bennie	Robert	”
Bowman	John	”
Calder	Donald	”
Cameron	John (1)	”
Cameron	John (2)	”
Cameron	John (3)	”
Campbell	Alex	”
Campbell	Angus	Sergeant
Campbell	Donald	Corporal
Campbell	John	Private
Campbell	Roderick	”
Cummins	John	”
Duff	Donald	”
Dunn	Peter	”
Elder	John	”
Erskine	Robert	”
Ferguson	Hugh	”
Fraser	Donald	Sergeant
Fraser	Forbes	Private
Fraser	Simon	”
Graham	Donald	Sergeant
Graham	Peter	”
Grant	Alexander	Private
Grant	Peter	”
Grote	Alex	”
Gunn	John	”
Harrier	William	Corporal
Hassock	William	Private
Hay	Robert	”
Henderson	Alex	”
James	Roderick	”
Keith	John	”
Keith	William	”
Laidlow	James	”
Manuel	Thomas	”

78th Foot – Other ranks

Matheson	William	Private
McArthur	Thomas	,,
McAuley	John	,,
McBain	Donald	,,
McBain	John	,,
McDonald	Alex	,,
McDonald	George	,,
McDonald	John	,,
McGough	John	,,
McIntosh	Donald	,,
McIver	Donald	,,
McLean	Donald	,,
McLeod	Alex	,,
McLeod	Donald	,,
McLeod	John	,,
McLeod	Murdoch	,,
McKay	Alexander	,,
McKay	Donald	,,
McKay	John	,,
McKenzie	Hugh	,,
McKenzie	John (1)	,,
McKenzie	John (2)	,,
McKenzie	William	,,
McQuaig	John	,,
McQueen	Ewen	,,
McRae	Alex	Sergeant
McRae	Farquhar	Private
McRae	Finlay	,,
McRae	John	,,
Menton	David	,,
Miller	John	,,
Miller	William	,,
Moore	Thomas	,,
Munro	Charles	Corporal
Munro	Hugh	Private
Murray	William	Corporal
Pyatt	James	Corporal
Reid	Robert	Private
Riddle	Peter	Corporal
Robertson	Alex	Sergeant
Robertson	Angus	Private
Ross	Donald	,,
Ross	James	,,
Ross	Robert	,,
Scott	John	,,
Shaw	Alex	,,
Smith	Malcolm	,,
Smyth	Michael	,,
Stalker	David	,,
Stewart	Charles	,,

78th Foot – Other ranks

Stewart	John	Private
Sutherland	John	,,
Sutherland	John	Sergeant
Sutherland	William	Private
Urquhart	Benjamin	,,
Weir	James	,,
Weir	John	,,
Wood	James	,,
Wybrough	Joseph	,,

89th Foot

Dalgety	Alexander	Lieutenant
McKechnie	Alexander	Assistant Surgeon
Young	Plomer	Lieutenant

Other ranks

Albrooke	Josiah	Private
Bradley	Cornelius	,,
Bradley	John	,,
Carroll	Christopher	,,
Carroll	Patrick	,,
Connell	James	,,
Dixon	James	Corporal
Dougherty	Michael	Private
Duke	Thomas	Sergeant
Eagen	Laughlan	Private
Eason	Joseph	,,
Gallagher	Michael	,,
Gee	John	,,
Gott	John	,,
Hayes	Michael	,,
Horan	Patrick	,,
Hughes	Owen	,,
Kennedy	James	,,
Kerr	James	,,
Leger	John	Sergeant
Lynam	Martin	Private
Matthews	Daniel	,,
McCann	Peter	,,
McCoy	Dennis	,,
McLoughlin	John	,,
Moraney	John	,,
Murphy	John	,,
Pearce	James	,,
Phelan	Thomas	,,
Power	Michael	,,
Quinn	James	,,
Spence	George	,,
Stack	Thomas	,,
Thompson	Robert	,,

Staff Officers

Blackiston	John	Lieutenant	ADC to Sir Samuel Auchmuty
Taylor	Thomas	Captain	24th Dragoons (employed as ADC)
Tylden	John	Captain	23rd Foot
Wetherall	George A.	Captain	1st Foot (employed as AAG & ADC)

HEIC SERVICE PERSONNEL AWARDED THE BRITISH MILITARY GENERAL SERVICE MEDAL WITH 'JAVA' CLASP

Anderson	Alexander	Ensign	Madras Engineers
Annesly	James	Surgeon	Medical Staff, Madras Army
(East India Register shows name as 'Annesley')			
Baillie	George	Assist Surgeon	Medical Staff, Madras Army
Baker	Godfrey P.	Lieutenant	19th Bengal Light Infantry
Bowen	Herbert	Captain	6th Vol Bn, Bengal Army
Christie	Charles	Lieutenant	4th Vol Bn, Bengal Army
Dick	Alexander	Lieutenant	5th Vol Bn, Bengal Army
Dickson	William	Lieutenant	Bengal Army, ADC
Farrington	John James	Lieutenant	Bengal Artillery
Fiddes	Thomas	Lieutenant	4th Vol Bn, Bengal Army
Fielding	William George A.	Lieutenant	Governor General's Bodyguard, Bengal
Frobisher	Thomas	Lieutenant	5th Vol Bn, Bengal Army
Harrison	James	Lieutenant	Madras Horse Artillery
Horsburgh	Alexander	Lieutenant	2nd Vol Bn, Bengal Army
Hunter	George	Lieutenant	4th Vol Bn, Bengal Army
Marshall	John S.	Ensign	20th Regt Bengal Native Infantry
MacDonald	Christopher	Sgt Major	Bengal Light Infantry Vols
McKenzie	William	Assist Surgeon	Madras Horse Artillery
Norris	William	Lieutenant	3rd Vol Bn, Bengal Army
Pearson	James	Lieutenant	Light Infantry, Bengal Army
Seymour	Robert	Lieutenant	6th Vol Bn, Bengal Army
Sim	Duncan	Ensign	Madras Engineers
Thompson	John	Assist Surgeon	Medical Staff, Bengal Army
Tulloch	John	Lieutenant	22 Regt, Bengal Light Infantry
Dunlop	James	Corporal	Madras Horse Artillery
Denton	James	Corporal	Madras Horse Artillery

ROYAL NAVY AND ROYAL MARINE PERSONNEL AWARDED THE MILITARY GSM WITH 'JAVA' CLASP

Birtles	Thomas	Seaman
Colburn	John	Masters Mate
Clarke	James	2 Lieutenant Royal Marines
Calamy	William	2 Lieutenant Royal Marines
Cupey	John	Private, Royal Marines
Donnely	Lawrence	Lieutenant, Royal Marines
Field	Francis	Seaman
Festing	Robert	Captain, Royal Navy
Fox	Dennis	Private, Royal Marines
Giles	G.	Lieutenant, Royal Marines
Morgan	William	Private, Royal Marines
Norris	John	Seaman
Pascoe	R. W.	Lieutenant, Royal Marines
Platt	James	Private, Royal Marines
Reynolds	Jerimiah	Seaman
Simons	H. C.	Seaman
Steele	M. F.	2 Lieutenant, Royal Marines
Gamestone	I.	Lieutenant, Royal Marines

The above list of Royal Naval and Royal Marine personnel who were entitled to the Military General Service Medal, with the 'Java' clasp have been included in Medal Roll reference (PRO) WO/100/1. Three other seamen qualified and are included in Medal Roll (PRO) WO/100/11 and are:

Banister	Harry	Seaman
Harris	Charles	Seaman
Spear	William	Seaman

The following list, given alphabetically, shows additional awards made of the medal and clasp as a result of '... applications received from those resident in the Colonies ...' (Medal Roll (PRO) WO/100/10).

Boyd	William	Corporal	69th Foot
Bailey	William	Private	69th Foot
Barrett	John	"	69th Foot
Brown	James	"	69th Foot
Brannen	Dennis	"	69th Foot
Crowe	Samuel	"	22 Light Dragoons
Cooper	Thomas	"	69th Foot
Cooke	John	"	69th Foot
Cartwright	Joseph	"	69th Foot
Dickens	Charles	"	22 Light Dragoons
Damerum	William	"	69th Foot
Donaghue	John	"	89th Foot
Francis	John	"	69th Foot
Grice	William	Sergeant	22 Light Dragoons
Hind	James	Private	14th Foot
Highfield	Thomas	"	69th Foot
Hill	Edward	"	89th Foot
Joseph	John	"	69th Foot
Leach	John	Corporal	22 Light Dragoons
Murphy	John	Private	14th Foot
Neale	Patrick	"	14th Foot
Newbold	John	"	69th Foot
Pegg	John	"	22 Light Dragoons
Roach	James	"	89th Foot
Rosia	Elisha	"	69th Foot
Rudge	Joseph	"	69th Foot
Richards	George	"	69th Foot
Spooner	John	"	69th Foot
Thompson	Richard	"	22 Light Dragoons
Wilkins	Thomas	"	69th Foot
Ward	William	"	69th Foot

An annotation on Medal Roll (PRO) WO/100/11 records the following deletion from the 'Java' medal roll: "... Dunn, Peter, Sergeant. 78th Regt. Bar for Java to be removed ...".

The foregoing lists have been compiled from the Military General Service Medal rolls held in the Public Record Office, Kew under references WO/100/1 to WO/100/11. The totals of the entries differ slightly from both previously published works dealing with the award of the MGS Medal with the clasp 'Java' which have appeared in the following publications: *British Battles and Medals* by Major L. L. Gordon and *The Military General Service Medal, 1793–1814* by Colonel K. O. N. Foster. A comparison of numbers is given as follows:

Regiments	*Medal Rolls included above*	*Gordon (pages 40–47)*	*Foster*
22 Light Dragoons	45	54	54
Royal Artillery	12	10	11
14th Foot	80	74	75
59th Foot	105	104	104
69th Foot	124	105	108
78th Foot	106	107	107
89th Foot	40	41	40
Staff Officers	4	4	4
Medical Staff	1	1	—
HEIC	26	41	41
RN and R Marines	21	12	1
Comparative totals	564	553	545

The foregoing does not take into account any awards of the Naval General Service Medal which have been adequately covered by Captain K. J. Douglas-Morris in his publication on that subject.

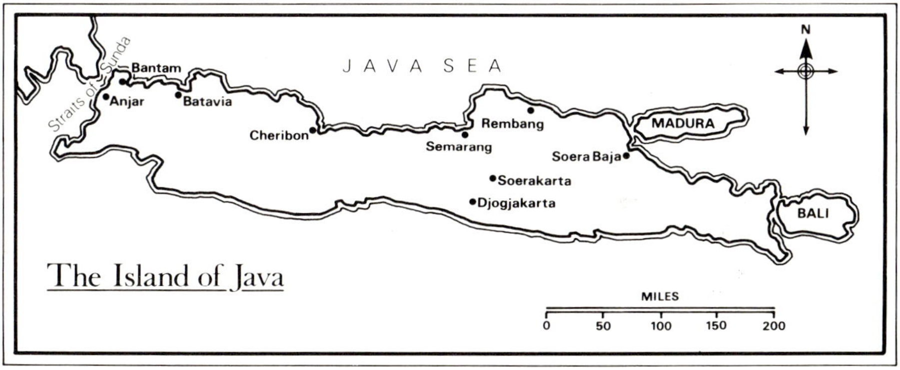

NOTES TO CHAPTER FOUR

1. Lieutenant Colonel W. J. Wilson, *History of the Madras Army,* (Government Press, Madras, 1882), vol 2, p 257.
2. Ibid, p 258.
3. Ibid.
4. John Crawfurd, *A Descriptive Dictionary of the Indian Islands and Adjacent Countries,* (Oxford University Press, 1971), p 11.
5. George Masselman, *The Money Trees – The Spice Trade,* (World's Work Ltd), p 121.
6. Crawfurd, op cit, p 33.
7. Ibid.
8. Wilson, op cit, vol 2, p 259.
9. Ibid, pp 259 and 260.
10. East-India Register and Directory, 1803, pp 148 and 160. Lieutenant Colonel Daniel McNeile.
11. Wilson, op cit, vol 3, p 32.
12. Ibid.
13. Ibid, p 34.
14. Ibid, p 35.
15. Ibid, p 95.
16. Ibid.
17. Colonel H. Yule and A. C. Burnell, *Hobson-Jobson,* (John Murray, London, 1903). Puckallies – a puckauly, water carrier. In the Madras Drill Regulations for 1785 there was shown, for each battalion, an allowance of 'ten puckalies'.
18. Wilson, op cit, vol 3, pp 97 and 98.
19. Crawfurd, op cit, p 27.
20. Ibid.
21. K. C. Tregonning, *The British in Malaya,* (The University of Arizona Press, Tucson, 1965), p 134.
22. Ibid, p 135.
23. East-India Register and Directory, 1809, p 180.
24. Wilson, op cit, vol 3, p 314.
25. Ibid, p 315.
26. Yule and Burnell, op cit, pp 193, 217, 742 and 743.
27. Wilson, op cit, vol 3, p 316. The wounded were one Corporal, four European privates, one marine and four seamen.
28. East-India Register and Directory, 1810. Captain Charles W. Yates' parent unit was the 13th Regiment, Madras Native Infantry.
29. Wilson, op cit, vol 3, pp 317 and 318.
30. Ibid, p 318.
31. Ibid, p 319.
32. Ibid, p 320.
33. East-India Register and Directory, 15 Aug 1816, p 446.
34. Ibid.
35. Crawfurd, op cit, pp 165–167.
36. Rowland Brinckman, *Historical Records of the 89th Princess Victoria's Regiment,* (Gale and Polden, 1888), p 10.
37. Major H. Davidson, *History of the 78th Highlanders (Ross-shire Buffs),* (W. & A. K. Johnston, 1901), p 75.
38. Ibid, p 76.
39. Ibid.
40. Wilson, op cit, vol 3, pp 323 and 324, and East-India Register and Directory for 1812.
41. Davidson, op cit, p 77.
42. Ibid, p 78.
43. Ibid, p 81.
44. Ibid.
45. Mills and Wilson, *History of British India.*
46. Davidson, op cit, p 82.
47. Ibid, p 84.
48. Ibid, pp 84 and 85.

49. Wilson, op cit, vol3, p328.
50. Major L. L. Gordon, *British Battles and Medals,* (Gale and Polden, 1962), p23.
51. Ibid, pp 37 and 38.
52. Ibid p38; and Thomas Carter and W. H. Long, *War Medals of the British Army 1650–1891,* (Norie and Wilson, 1893), pp76, 152 and 153. Despite a search in the India Office Records and Library and the Public Record Office, etc, no trace has yet been found of the HEIC Java Medal Rolls. It is thought that these are in all probability held in one of the many archives in India.
53. Wilson, op cit, vol3, p329. Captain Robert Mears was killed on 15 September 1812.
54. Djojo-carta is now known as Jokyakarta, also as Jogjakarta. Crawfurd shows alternative spelling as being Yugyakarta.
55. Wilson, op cit, vol3, p329.
56. Ibid, and the East-India Register and Directory, 1812, pp 228 and 229.
57. Tregonning, op cit, p146; and Maurice Collis, *Raffles,* (Faber and Faber, 1966), pp 84 and 114.
58. Wilson, op cit, vol 3, p331.
59. N. B. Leslie, *The Battle Honours of the British and Indian Armies, 1695–1914,* (Leo Cooper, 1970).
60. C. R. Wilson (ed), *List of Inscriptions on Tombs and Monuments in Bengal,* (Government Press, Calcutta, 1896), pp 163 and 164. Wilson has mistakenly listed Lieutenant-Colonel Christopher Clarges of the 69th Foot as 'Lieutenant Colonel Charger'.
61. Lieut F. G. Cardew, *A Sketch of the Services of the Bengal Native Army. To the year 1895,* (Government Press, Calcutta, 1903), p68.
62. India Office Library and Records. Bengal Secret and Political Consultations, vol279. Major J. M. Johnson to Mr Thomas Stamford Raffles, 16 December 1815.
63. IOLR, G 21/68. General Return of the Army, 1 October 1812. During a period in 1812–1813, Surakarta was garrisoned by a detachment from the 14th Regiment of Foot (strength given was 63 men). The garrison strengths of Klaten and Yogyakarta, which were provided by troops of the 4th Light Infantry Battalion, was 71 at Klaten and 267 at Yogyakarta.
64. IOLR, G 21/33.
65. IOLR, Bengal Secret and Political Consultations, vol 279, 4 May 1816, No 25.
66. IOLR, Bengal Secret and Political Consultations, vol 278.
67. The Malay Battalions mentioned were probably the Amboynese and Javanese Corps that had been raised and, by 1816, had a strength of 1,200. IOLR, G 21/65, Memorandum respecting Java, subheading 'Military Establishment'.
68. Mata Deen was chosen to be the leader of the party designated to murder the European officers.
69. Lieutenant James Steel, 21st Regiment Bengal Native Infantry serving with the Light Infantry Volunteer Battalion in Java. East-India Register and Directory for 1816, p78.
70. Lieutenant Robert Cock, 26th Regiment Bengal Native Infantry, serving with the Volunteer Light Infantry Battalion, Java.

CHAPTER FIVE

The Malacca Garrison 1795–1830

Malacca has a long history of occupation which started when the Portuguese expedition, led by Diogo Lopes de Sequeira, set out from India in 1509 and captured Malacca. Soon after having established a post at Malacca the Portuguese were attacked by the forces of Sultan Mahmud, the 'King of Malacca'.[1] The attack was not wholly successful as although some of the Portuguese were killed, Sequeira escaped and returned to India whilst eight others eventually escaped and made their way to Sumatra where they were later found by Alfonso de Albuquerque. Following this first setback, the Portuguese mounted another expedition, under the command of Alfonso de Albuquerque, and attacked the town and port again in 1511 and brought the town under control and set up a permanent settlement there and built a massive fortress which was known as 'A Famosa' (The Renowned).[2] The Portuguese considered the capture of Malacca to be one of their most important conquests in the Far East and the attack on the town proved their superiority in battle. Having once captured the town and established their fortress, which was first a construction of a palisade and a moat, they commenced a construction of stone which was recorded as having a tower five-storeys high. Ruy de Brito, the first Captain of Malacca had the keep completed in 1514 and used the keep as a residence. He had apartments on the second floor and visiting dignitaries were also housed in the keep, as were the administrative offices. The Portuguese suffered frequent attacks from the Malays that they had driven from the town and were also besieged by the Dutch during 1606 and 1608.[3] In 1613 it was recorded that the fortress had been expanded by successive Portuguese 'Captains of Malacca' so that now the area had been extended and bounded by '. . . extensive outer walls with six large Bastions entirely surrounding the old inner Fortress or Keep and the hill [now known as St Paul's Hill]. The seaward side extended from the riverside to a sharp point at the site of the pier and then returned to the Bastion of Santiago . . .'

An account of the fortress of Malacca was written by Barretto de Resende and is thought to refer to his visit to Malacca in 1638. His account reads '. . . The Bastions contain 41 pieces of Artillery of 12 to 44 pounds iron shot. All are of bronze with the exception of nine iron pieces . . . There are in the town two hundred and fifty married whites[4] who would possess 200 black captives of different races, all competent to carry arms . . .' He goes on to describe the fortress 'A Famosa' '. . . The fort within this town where the Captain resides is five stories high; the Captain lives on the second storey, which is square like the tower, each wall being 20 paces wide. The other appartments are set aside for the Captain's guests, and for storing ammunition. On the first floor four thousand candys of rice were stored . . .'[5]

The Dutch finally captured the Settlement from the Portuguese in 1641 after a blockade and siege lasting nine months. The Dutch force amounted to 1,500 men with as many Malay auxiliaries. The storming party numbered 650 and on their surrender the Portuguese garrison was found to number a mere 200 Europeans and some 400 natives.[6]

The Dutch remained in occupation of the Settlement for the next 154 years. The

Fortress had been badly damaged during the siege and subsequent attack but the Dutch renovated and extended the fortification. The old Gateway, which is known as Porta de Santiago, and still stands at the foot of St Paul's Hill, was built by the Dutch and is recorded in the report written by the Dutch Governor, Balthasar Bort, in 1678. The Dutch also built a flanking fortification on the southern side of the town. This small fort was built on St John's Hill in approximately 1750 on a site that had originally been the chapel of St Joa Baptista. The chapel had been destroyed by the Achinese when they attacked the town when it was under Portuguese control. The Achinese established a battery on the hill after they had destroyed the chapel and bombarded the town from that point. It was therefore important to the Dutch that this site should be occupied by them to avoid any repetition of the bombardment of the town from that hill. The Dutch held the Settlement without undue trouble until the surrender to the British in August 1795. It was essential, due to the conflict in Europe, that the British should safeguard their interests in the East Indies and it was therefore vital that Malacca should be under British control. An expedition was despatched, under the command of Captain Newcombe of the *Orpheus* and, with the forty-four gun ship *Resistance* under command of Captain Edward Pakenham, set out from Madras with a military force of four European battalions.[7] When Captain Henry Newcombe anchored in the bay the massive fortress, which by then had been enlarged and had nine bastions, looked impregnable but was in fact only held by '. . . a mere 269 men, of whom over 200 were unenthusiastic Javanese, Amboinese, or other Malaysians . . .'[8]

After some negotiations, which proved to be negative, it was decided that on 17 August a night landing would be attempted with the landing force under command of Major Archibald Brown. The landing did not go smoothly as in the darkness the troops landed at the wrong point and encountered jungle that was so thick that the landing had to be abandoned. The troops remained in their boats whilst an officer waded through the surf until he came to open ground and then called the boats inshore and the troops disembarked. They then advanced through coconut trees towards the fort. At daybreak their presence was discovered and shots were fired from the fort. During the entry into the bay the British ships were fired upon by a Dutch ship which had run aground and being unable to escape opened fire. The fire was returned by the *Resistance* and the Dutch ship struck her colours. After the brief attempt at defence by the ship and the fort honour appeared to have been satisfied and the Dutch Governor surrendered.

The capitulation brought the British the problem of manning the massive fort. It was agreed that the British troops would man the fort, and that the Dutch would raise a sum of money within the Settlement for the upkeep of the garrison. The British Commandant received the keys of the fort and all military stores were placed under his control. One item in the agreement which set the tone of the occupation was that '. . . the British and Dutch flags were to be displayed on proper occasions, on two flag-staves in the fort . . .'[9] It was generally felt that although Holland was under the control of the French, the Dutch were not the enemy and it was accepted that Malacca had been occupied to protect British interests. The Dutch colonies had been called upon to collaborate with the British, consequently when Malacca was occupied the Dutch administration continued to control the Settlement with the Commanding Officer of the British garrison taking the seat as 'Second in Council'.[10] The initial agreement lasted only a short time as Commodore Peter Rainier set out from Madras on 15 October 1795 with an expedition to take possession of Dutch Amboyna. The expedition first called at Prince of Wales Island (Pulo Pinang) and then Malacca. Rainier was not at all satisfied with the way the British settlement was being

died on 13 August.[20] In 1817 the 20th Regiment lost yet another officer at Malacca when on 26 March Lieutenant Henry P. Short died.[21]

A careful search has been made in the old fort Cemetery at Malacca for the graves of the six officers who died between the years 1812 and 1818 but no trace has been found.

On 21 September 1818, Malacca was handed back to the Dutch by Major William Farquhar, who was still serving as the British Resident, and accepted by the Dutch Commissioners Admiral Walterbeck and Timmermann Thyssen. Although repossession had been agreed as early as 1816, due to its deterioration as a result of the British demolition of 'A Famosa' the Dutch were not in a hurry to reoccupy. However, despite this, the British Government asked for restoration expenses for the period from 2 November 1816 to 21 September 1818 during which time the Dutch should have been in occupation. The expenses were said to have amounted to '... 28,022 Spanish Dollars and 87 cents ...'[22]

Major William Farquhar published a Garrison Order on 19 September 1818 detailing the actual handover ceremony and procedure:

> '... Monday next, 21st instant, being the day appointed for restoring this settlement to His Netherlands Majesty, the following ceremony will be observed on the occasion – The British colours to be hoisted at sun-rise. All the troops in garrison off duty to parade at their respective barracks, at seven o'clock in the morning, from whence they will remove to the vicinity of the Flag-staff on St Paul's Hill, when they will take up position in open ranks on the right of the Dutch troops. The civil and military officers not on duty with corps to assemble at the Government House in full dress at seven o'clock. The British Resident and Dutch Commissioners, with their respective staffs, public officers, &c, will proceed in procession to the spot allotted [to] the troops in the vicinity of the Flag-staff. On their arrival there, they will be received by the troops with presented arms. Drums beating two rolls.
>
> The British Proclamation will then be read by the Resident and the same repeated in the Chinese and Malay languages, after which the master attendant will begin to lower the union on the royal salute from the battery, and continue to do so gradually until it reaches the ground. The troops, during the lowering of the British flag, will again present arms. Drums beating God save the King.
>
> The Dutch men-of-war in the roads will also fire a royal salute whilst the British flag is descending. The British troops will then take up a new position on the left of the Dutch line, when the Proclamation in Dutch will be read, and explained by the Commissioners or their secretary.
>
> The Dutch colours will then be hoisted full mast, under a royal salute from the British battery, as likewise from the Dutch squadrons.
>
> The troops paying the same compliment of presenting arms on the hoisting of the Dutch colours as they did on the lowering of the British. Drums beating a Dutch march. After which the troops will march off to their respective barracks. The ceremony being thus concluded, the whole of the garrison guards will be immediately relieved by the Dutch troops ...'

After the ceremony the British garrison departed and moved by ship to the settlement on Pulo Pinang.

Malacca remained a Dutch possession until 1 March 1825 when it was ceded to Britain under the Anglo-Dutch Treaty of 1824. It was, in effect, exchanged for the Honourable East India Company's settlement at Bencoolen on the west coast of Sumatra. At this time it became a dependency of Bengal, but was transferred to the Penang administration in 1826.

Grave of Lieutenant Joseph William Colquhoun, who died on 4 July 1826, at Fort Road Cemetery, Malacca. Photograph by Mrs J. Harfield, 1977.

Newbold records in his history that at the time of the handover the total population of '... Malacca and its territory including Naning, at 37,706 souls of whom the greater portion are Malays ...'[23]

From 1826 the garrison troops at Malacca were provided by detachments from the Native Infantry regiment stationed at Prince of Wales Island. Singapore Garrison Order No 1 of 1827, which was a repeat of the General Order published at Fort Cornwallis (Pulo Pinang) on 12 January 1827, showed Lieutenant Edward W. Snow of the 24th Regiment, Madras Native Infantry being appointed as Cantonment Adjutant at Malacca.[24] It was during 1826 that yet another officer of the Bengal Native Infantry died at Malacca, he was Lieutenant Joseph William Colquhoun who died on 4 July 1826.[25]

In February 1827, an order was published appointing Lieutenant Alex E. Baillee of the Madras Artillery as the '... Junior Officer in command [of] the Detachment of Artillery ...' and ordered him to take charge of the Ordnance department at Malacca.[26] Lieutenant Baillee was relieved at Malacca by Lieutenant George Balfour, also of the Madras Artillery, in May 1827 who was ordered to '... proceed by the earliest opportunity to Malacca to which station he is now appointed ...'[27] A further examination of the orders revealed that Lieutenant Balfour was granted an allowance for carrying out the duties of 'Officer in charge of the Ordnance Depot at Malacca'. The actual order authorised the charge to be the '... same as those of Lieut Baillee ...'[28]

In August 1828 four stores lascars were sent from Singapore to Malacca to work in the

Ordnance Depot under command of Lieutenant Balfour. During the same month Lieutenant William Edward Brooshooft of the 35th Regiment Madras Native Infantry arrived in the Settlement from Prince of Wales Island on furlough on a sick certificate.[29] Brooshooft was to return to Pulo Pinang and was to be killed in a tragic accident at the then famous Penang Waterfall on 1 March 1830.[30].

In addition to officers from Pulo Pinang and Singapore visiting Malacca on leave it was also considered to be suitable as a family station. There are no records showing the precise number of married Europeans on station during the early years and although an examination has been made of the gravestones that remain in the Fort Road Cemetery at Malacca no trace has been found of any graves belonging to members of families. However the East-India Register and Directory for 1828, second edition, does show that Sophia, the youngest daughter of 'Mr Conductor T. Green of the Madras Establishment' died at Malacca on 24 November 1827.

The year 1829 passed quietly for those stationed at Malacca and there were no deaths recorded for that year among the personnel serving on the military strength of the East India Company. On 9 February 1830 Lieutenant Francis Pelly of the Bombay Engineers died at the Settlement. On 1 May of that year Malacca was combined with Prince of Wales Island and Singapore to become a single Residency under the control of Fort William.

NOTES TO CHAPTER FIVE

1. Francisco de sa de Menses, *The Conquest of Malacca,* (University of Malaya Press, 1970), pp ix–x.
2. The Reverend Piyasilo Samanera, *Historical Guide of Malacca,* (Published Malacca, 1970).
3. John Crawfurd, *A Descriptive Dictionary of the Indian Islands and Adjacent Countries,* (Oxford University Press, Kuala Lumpur, 1971), p248.
4. 'Married whites' – 'the married men whose duty it was to populate the settlement'. *Journal of the Straits Branch of the Royal Asiatic Society,* no 60, 1911.
5. Candy – a weight used in Southern India. It varies but is generally accepted as 500 pounds.
6. Crawfurd, op cit, p248.
7. K. C. Tregonning, *The British in Malaya,* (University of Arizona Press, USA, 1965), p131.
8. Ibid.
9. T. J. Newbold, *British Settlements in the Straits of Malacca,* (Oxford University Press, Singapore, 1971), p123.
10. Tregonning, op cit, p132.
11. East-India Register and Directory for 1802.
12. Newbold, op cit, p126.
13. Straits Settlements Factory Records, volume 12. London to Penang, 18 April 1805.
14. Tregonning, op cit, p144.
15. Hodson Index, National Army Museum, London. The Index shows Charles Porteous as being promoted to Captain on 30 November 1804. He was stationed at Malacca from 1811. He moved to Bencoolen, on the West Coast of Sumatra in 1814, and died there on 8 April 1816.
16. East-India Register and Directory for 31 August 1812.
17. Major V. C. P. Hodson, *Officers of the Bengal Army, Part II,* (Constable & Co, 1928). Thomas Jones was born in 1784. He was a Cadet in 1798 and arrived in India on 24 February 1800. He became an Ensign on 11 January 1800; Lieutenant on 29 May 1800; Captain on 24 January 1809. He served as Adjutant of 1/2nd Regiment Bengal Native Infantry from 1807 until March 1810. He was at the capture of Java in 1811. He died, unmarried, at Malacca, on 28 October 1812.
18. Hodson, op cit, p563. John Jolliffe was born in 1787. Became a cadet in 1808. He arrived in India on 24 July 1809 and was appointed Ensign from 27 September 1809; Lieutenant on 16 December 1814. He served with the Java Infantry Volunteers from 1814 and was still attached to them at the time of his death. He may well have been 'in transit' at Malacca when he died on 2 June 1815.
19. East-India Register and Directory – 1816 (second edition).
20. East-India Register and Directory – 1817 (first edition).

21. East-India Register and Directory - 1819 (first edition).
22. Newbold, op cit, p 129.
23. Ibid, p 135.
24. National Reference Library Singapore. Singapore Garrison Orders 1826–1827. Garrison Order No 1 dated 29 January 1827.
25. Lieutenant Joseph William Colquhoun was born at Luss, Co Dumbarton on 3 February 1805. He was the second son of Lieutenant Colonel Ludovic Colquhoun and Barbara Camilla, the daughter of the Reverend Dr Joseph McIntyre. He was baptised on 15 February 1805. He joined the Honourable East India Company as a Cadet in 1820 appointed Ensign with effect from 9 March 1821 and was posted to the 2/19th Regiment, Bengal Native Infantry (2nd Nassauri Battalion) in 1823. Promoted to Lieutenant on 11 July 1823 he was then transferred to the 16th Regiment, Bengal Native Infantry. In May 1824 the regiment was re-numbered as the 32nd Regiment, Bengal Native Infantry. Lieutenant Colquhoun was granted leave on a sick certificate on 18 February 1825 and was still on sick leave, in Malacca, when he died on 4 July 1826. A Will, dated 'Malacca 26th February 1826', was proved on 5 December 1826.
26. Singapore Garrison Orders 1826–1827. Garrison Order dated 20 February 1827.
27. Singapore Garrison Orders 1826–1827. Garrison Order dated Sunday 20 May 1827.
28. National Reference Library, Singapore. Penang Orders and Notices. Entry 'Malacca dated 21st July 1828'.
29. Penang Orders and Notices. Entry 'Military dated 24 August 1828'.
30. Lieutenant William Edward Brooshooft was appointed Lieutenant on 1 May 1824 and from that date served with the 35th Regiment, Madras Native Infantry. He moved to Prince of Wales Island garrison with that regiment. From 27 August 1824 he was appointed 'Quartermaster, interpreter and paymaster' for the regiment a post he held until early 1830. He died on 1 March 1830 as a result of an accident.

The Evans–Velge tomb at St Paul's Church cemetery at Malacca. Photograph by Mrs J. Harfield, 1977.

CHAPTER SIX

The Founding of Singapore and its Defences

The occupation of Singapore island by the Honourable East India Company dates from 1819, and before dealing with the setting up of the Settlement and its defences it is relevant to the study of the island to examine previous occupation of the area. Traces of ancient man, in the form of stone implements, have been discovered across the Causeway between Singapore Island and the State of Johore, and in Malaya in Upper Perak. The finds were estimated as probably being 250,000 years old. Further finds of stone tools of a later age show that cave dwellers existed in Malaya during the period 4,000 to 10,000 years ago, and a fragment of Greek pottery was discovered in a Neolithic burial site, dating about fourth or fifth century BC, in its correct stratification layer.[1]

From Central Asia, during the period 2500–1500 BC, there was a great migration when the forebears of the present-day inhabitants of South East Asia and the Indonesian Islands travelled southwards and left many identifiable items. Stone Age round axes, of uncertain date, have been found on Singapore Island at Tanjong Kerang, and also on the beach of Pulau Ubin, an island some one and half miles off the northern coast of Singapore Island.[2]

Recorded history of Singapore Island can be traced to the Tamil invasion of the Malay empire Sri Vijaya in AD 1025. Tumasik (meaning Sea Town)[3], the Javanese name by which Singapore was then known, was listed amongst the conquests mentioned on a commemorative monument which had been excavated at Tanjore, in Madras. During the period 1250–1300 Tumasik, or Temasek as the name was sometimes spelt, was an independent port of call and for more than a century it flourished under its own monarch. In the early fourteenth century it is recorded in ancient Chinese documents that a Siamese naval expedition failed to capture Tumasik, or as it was then sometimes called by travellers from the Indian sub-continent, 'Singapura'. The name has been taken from two Sanskrit words 'Singa Pura', meaning 'Lion City'. In the work *A History of Singapore* by H. F. Pearson, the derivation of the name is given as follows:

> '. . . About the year 1150, Prince Buana, with his Malay followers set out in prahus from Palembang. They went down the river . . . [and] . . . into the open sea. Then he continued north towards Malaya. He visited many strange islands until at last he came to the island of Riouw. The island was ruled by a Queen, who made Prince Buana welcome. But soon he became tired of the pleasures of the Queen's court. He made up his mind to go hunting again, and he said goodbye to the Queen. Once again Prince Buana set off with his fleet of prahus. He landed on an island and led his men in search of deer. The chase brought them close to a big rock. He climbed it and looking across the water saw another island which had a long sandy beach. "Come" he cried to his men, "we will see if we can find more sport on that island." When they had crossed the sea, Prince Buana led the way inland, and was just beginning to feel disappointed because there

> were no animals to be seen when suddenly, in his path stood the biggest and most fierce-looking animal he had ever seen. Before anyone could move it slipped away and disappeared. "What beast was that?" asked Prince Buana. An old man, who remembered stories he had heard in his youth told him it was a lion. Prince Buana was so pleased to find such a fine animal that he decided to build a city on the island because of his meeting with the animal, he called his city "Singapura" which means the City of the Lion . . ."[4]

The city, which thrived as a centre for trade under the Sri Vija Empire, was occupied by Majapahit forces from Java in 1376 and then came under the protection of Thailand until it was occupied by Parameswara, or Iskandar Shah, a Prince of Palembang during the Javanese war of succession.[5] In 1391 he was expelled by the Thais and fled to Malacca where he established the dynasty that spread Islam and succeeded in dominating the trade of the area until the arrival of the Portuguese at Malacca.[6] Sultan Iskandar Shah died in AD 1413.[7]

Singapore is again mentioned as a port, although a minor one at the end of the fifteenth century when the Governor of the Port was required to make the journey to Malacca to pledge his loyalty to the Sultan of Malacca. The unfortunate Governor was late arriving at the Sultan's court at Malacca and was executed as a traitor.[8] It remained a small trading port and is again mentioned in the early seventeenth century when the Portuguese became masters of Malacca and the Malay dynasty fled and had established itself at Johore. Singa Pura is then recorded as having a Harbour Master,[9] but due to its tragic history it was never a popular place to the Malays and so was never developed beyond being a small trading port at the mouth of the Singapore River. In 1703 the Sultan of Johore offered Singa Pura to Alexander Hamilton, who observed that it was no place for a private person but would probably be a good place for a trading company to settle a colony.[10]

The foregoing is only a very brief note on the history of Singapore Island prior to the arrival of Sir Stamford Raffles in 1819 and is not intended to be a comprehensive history as that is more than adequately dealt with in other published histories. Singapore is shown on some early maps; Jan Huyghen van Linschoten published maps in his work which was dated 1595, and on the map of the East Indies shows Singapore as 'Sincapura'. The English edition of his work was published in 1598.[11] A chart of the southern part of the Malacca Strait by Captain J. Lindsey, dated 1798, shows 'Sincapour' and 'Sincapour or Governors Straits'[12] and in 1806 the East India Company published a chart, redrawn from a chart by James Horsburgh, which showed 'Sincapour Island'.[13]

John Crawfurd describes Singapore island in his book *A Descriptive Dictionary of the Indian Islands and Adjacent Countries*, published in 1856, as follows '. . . Singapore in its greatest length from east to west, is 25 miles long, and in its greatest breadth 14, having an area of 206 square geographical miles . . . To the north it is bounded by the territory of Jehore . . ."[14]

With the return of the Settlement at Malacca to the Dutch in 1818 the Honourable East India Company then had a need for a port, or harbour, at the eastern end of the Strait of Malacca. In 1819 Sir Thomas Stamford Raffles left the settlement at Bencoolen on the west coast of Sumatra and formed an expedition at Prince of Wales Island (Pulo Pinang) with a set task of finding a suitable site for a port and settlement. His small fleet consisted of the *Nearchus* (Captain William Maxfield), *Minto* (Lieutenant J. S. Criddle), *Mercury* (Captain R. Beaumont), *Indiana* (Captain James Pearl), the brig *Ganges* (F. J. Bernard) and the schooner *Enterprise* (Captain R. Harris). In addition two survey ships joined the fleet, these were the *Discovery* and the *Investigator*. These two vessels arrived at the Carimons on the afternoon of 19 January and the survey of the islands commenced at once. The islands are about twenty miles from Singapore (on a clear day they can easily be seen from Mount Faber on Singapore Island). The main island, Great Carimon, is twelve miles long and five miles wide, with two peaked mountains in the centre, the highest being two thousand feet. Little Carimon is two miles long with a hill 500 feet high. The islands were convenient for the Indiamen on their way to trade with China but it was recommended that the two anchorages available were unsuitable and 'hardly defensible'. On 26 January the two vessels sighted the *Nearchus*, *Mercury* and *Ganges*. Major William Farquhar, Madras Engineers, who had formerly been the Resident at the Settlement at Malacca, persuaded the group to return to the Carimons where he landed with other officers of the party. Despite the objections raised by the survey party Farquhar was delighted with the site and was convinced that a town with every facility for trading could be built. That evening Sir Stamford Raffles arrived with the *Indiana* and the *Enterprise* and was recommended that he should see an alternative site. The expedition left the Carimons and arrived off the coast of Singapore Island at about 4.00 pm on 28 January. No landing was made that day but a party from the native settlement was received on board.

Wa Hakim (a fifteen-year-old boy) later recorded that on the following day Raffles came ashore with one other white man (this was Major William Farquhar) and a Sepoy who was armed with a musket. At this time there were about a hundred small huts and houses at the mouth of the river, the largest of which was the Raja's house. It has been estimated that this house stood approximately where the present day Dalhousie monument stands. In addition to the Malays living in the houses there were some thirty families living on boats on the Singapore River, although in some cases part of the family did live ashore.

The party were entertained at the Raja's house and given rambutans and other kinds of local fruit. At about 4.00 pm Raffles and his party re-embarked. That same afternoon Major Farquhar sailed for Rhio with instructions to negotiate with the Raja Muda who exerted control over Singapore. The following day Raffles concluded his provisional treaty with the local Raja (Temenggong). Under the treaty the East India Company were authorised to establish a post, or settlement, and in return the Temenggong would receive an annual payment of 3,000 dollars. So long as the 'English' remained to protect him the Temenggong agreed to have no further relations with other nations that might call at the port.

In the account of the founding of the Colony by the local historian Abdullah bin Abdul Kadir, a conversation between Major William Farquhar, who later was to become the Settlement's First Resident, and the Temmenggong, whose full title was Datoo Temenggong Sree Maharajah Abdul Rahman, is reported as follows – Farquhar had enquired as to the name of the nearby hill and was told '. . . For ages it has been called Forbidden Hill . . .' On asking the reason for this name, it was explained that local legend had it that dating back in ancient times the Kings had their palace built on the summit and it was forbidden for anyone to ascend the hill except by the ruler's command or summons. The Temenggong also told Farquhar that '. . . behind it is a stream known as the Forbidden Stream, for it was the place where the consorts and the wife of the King used to bathe, and no one was allowed to approach . . .'[15] Some support is given to the legend as in 1822 the remains of vast buildings were discovered on the summit of the hill.[16]

The day following the first landing by Sir Stamford Raffles, troops and their baggage were landed and tents were erected in the area now covered by the Court House, the church

Plate I. View of Mount Erskine and Pulau Tikus, Penang, 1817–1818.
Courtesy of the Penang State Museum.

Plate II. View of the Government Bungalow, Penang, 1817–1818.
Courtesy of the Penang State Museum.

Plate III. View of Haliburton's Hill, Penang, 1817–1818. Courtesy of the Penang State Museum.

Plate IV. The Padang, Singapore. Painting is from the site of the Scandal Point Battery and shows an officer and a sepoy. Courtesy of The Director, National Library, Singapore.

Plate V. Singapore River and Government Hill (later Fort Canning Hill), 1843–1847.
Courtesy of The Director, National Library, Singapore.

Plate VI. Singapore – looking south from Government Hill (later Fort Canning Hill), August 1846. Right foreground shows an officer of the 21st Regiment Madras Native Infantry, the unit was serving at Singapore when the picture was painted. Courtesy of The Director, National Library, Singapore.

Plate VII. Singapore Waterfront, from the sea, in 1861. From a lithograph by W. H. McFarlane of Edinburgh, based on an oil painting by W. Gray. Note – Fort Canning is shown on the top of the hill on the right of the picture. Courtesy of the National Museum Singapore.

Plate VIII. The Indian General Service Medal, (1854–1895), with clasp 'Perak'. Obverse and Reverse. Photograph by G. Parselle.

and the Padang. Raffles then gave orders for a fort to be built on the hill at the back of the kampong, which was of course Forbidden Hill, also known at that time as Bukit Larangan. The fort was to be capable of mounting 8- or 10-pounder artillery pieces; to have a magazine of brick or stone and barracks for thirty European artillery personnel as well as temporary accommodation for the remainder of the garrison, should such an emergency arise.[17] There is no record of this fort having been built or even started on Forbidden Hill, but there is evidence of a small fort having been partially constructed near the mouth of the Singapore River.

Raffles wrote to his friend William Marsden on 31 January 1819 describing the remains of earlier fortifications that had been found on the island. He wrote '... the lines of an old city and of its defences are still to be traced ...' The traces of the old defences are also mentioned by John Crawfurd, who became the second Resident of the Settlement. He visited the settlement in January 1822, *en route* to Siam and recorded in his diary his description of the ancient ruins of 'Tumasik' '... the walled town was bounded to the east by the sea, to the north by a wall and to the west by a salt creek. This enclosed space is a plain ending in a hill of considerable extent. The whole is a kind of triangle of which the sea side is about a mile in length ...' The wall was about sixteen feet in breadth and '... at present about eight or nine feet in height, runs very near a mile from the sea coast to the base of the hill until it meets a salt marsh. As long as it continues in the plain it is skirted by a little rivulet running at the foot of it and forming a kind of moat; and where it attains the elevated side of the hill there are apparent the remains of a dry ditch. On the western side which extends from the termination of the wall to the sea, the distance, like that of the northern side is very near a mile ...' The report also commented that there were no traces of embrasures or loop-holes. Crawfurd also describes the finding of what were once cultivated fruit trees growing wild on Forbidden Hill. These trees included durian, rambutan, duku and shaddock, etc, which had all grown to a great size. Abdullah, the local historian, also recorded this find saying '... the men came across many fruit trees, as large as durian trees, six feet or more around the trunk ...'[18]

Farquhar was appointed Resident and ordered a flagpole to be erected on the top of Forbidden Hill. The Temenggong's men refused to accompany him as the local superstition had it that the hill was haunted. Farquhar then ordered some of the Sepoys to accompany him and also ordered that a gun should be taken to the top of the hill. Fortunately the side of the hill was only sparsely covered with vegetation and so the gun was brought to the top whereupon Farquhar ordered the gun to be loaded. He then fired twelve rounds in succession himself, no doubt to impress the local inhabitants and to try to break the superstitions connected with the hill. He then had a flagpole erected and the bushes cut down and a path made to the top of the hill.

On 6 February Major Farquhar was appointed as Resident and Commandant of the garrison and it was on that day that the Treaty was formally made and the British flag was officially hoisted over the new settlement.[19] On that date Raffles wrote an official letter to Major Farquhar 'Resident and Commandant, Singapore' setting out his duties, etc. He specified his military requirements as follows:

> '... Along the coast in the vicinity of the Settlement one or two strong batteries for the protection of the shipping, and at Sandy Point a redoubt and to the east of it a strong battery for the same purpose.
>
> The entrenchment of the Cantonment by lines and a palisade, as soon as the labour can be spared from works of more immediate importance.

These defences, together with a Martello tower on Deep Water Point, which it is my intention to recommend to the Supreme Government, will in my judgment render the Settlement capable of maintaining a good defence. The principle on which works were charged for at Malacca, is to be considered as applicable to this station, and it is unnecessary for me to urge on you the necessity of confining the cost of these works within the narrowest limits possible. As the construction of them, however, will necessarily demand a greater portion of care and superindendence than the performance of your duties will permit you to devote to them I have appointed Lieutenant Ralfe of the Bengal Artillery to be the assistant Engineer. This officer will likewise have charge of the ordnance and military stores, and for the duties attendant on both these appointments conjoined I have fixed his salary at Spanish dollars 200 per mensem, to commence from the 1st instant, and subject to the confirmation of the Supreme Government.

As you will require the aid of a Staff officer to conduct the duties of the garrison, I have thought proper to authorise the appointment of a cantonment adjutant on the same allowances lately authorised at Malacca. As this officer may be considered your personal staff, I shall not make any permanent arrangement regarding it, but have appointed Lieutenant Dow to the temporary performance of its duties.

The indent for ordnance and stores which you have handed to me shall be transmitted to Bengal without delay, and I request you will lose no time in the erection of storehouses for their reception. An application for the number and description of troops which you have recommended to form the garrison of the residency will accompany the indent, together with an application for provisions equal to their supply for 12 or 15 months . . . A store house for the Commissariat department is at present of indispensible necessity and you will accordingly be pleased to erect a house of this description, of such materials as can be procured, and as soon as you may find practicable. A magazine built of such materials, for the military stores, would be subject to some risk; and I therefore confide to your professional judgment [Farquhar was, of course, an officer of the Madras Engineers] the adoption of such measures for their security as you may judge most expedient under the circumstances . . .'[20]

The letter gave Farquhar instructions in respect of the disposal of the various ships that were left at Singapore by Raffles. Raffles recommended that the ship *Mercury* was suitable for re-employment for the purpose of conveying '. . . troops and stores from P. of W. Island . . .' The letter also included details of the various accounts that the Resident was required to keep and these were listed as follows:

'. . . The accounts of the residency are those which detail the receipt and disbursement of the public money. These are principally:

(i) An account particulars of military disbursements in which every military abstract and disbursement is clearly and correctly entered.

(ii) A general account particulars, which will comprize the particulars of every disbursement of whatever nature and containing also, under the head of 'Military Establishment', a correct copy of No1 . . .'[21]

The work on the defences of the Settlement proceeded slowly. A small fortification was built at the mouth of the Singapore River and a Battery was built on a small knoll just above the shoreline on the coast between the mouth of the Singapore River and Sungei Kallang, just south of the present day St Andrew's Cathedral. Lieutenant Henry Ralfe, of the Bengal Artillery was in charge of the construction. This was the main defence position of the

Settlement until the construction of Fort Fullerton. After this fort had been built at the mouth of the Singapore River and taken into service, the battery at Scandal Point, as the site had become known to the local European residents, was reduced in size and used only for the firing of salutes. The site provided a cool meeting place in the evening for the European residents during which time current local events were discussed and it was through this usage that the site became known as 'Scandal Point'.

Forbidden Hill became known as Bukit Bandara (Flag Hill), and sometimes as Bukit Tuan Bonham, but from 1822, after Raffles had built his house upon it facing the harbour and sea, it became known as Government Hill. Raffles described his house as being '. . . 150 feet front and 59 feet deep [and] was finished in two weeks from its commencement . . .'[22] At the time of the building of the house Raffles decided, after a suggestion by Doctor Wallich, to build a Botanical Garden at the north-east side of the hill. A plot of forty-eight acres was used and this flourished until 1829 when it was abandoned as a measure of economy. Raffles left Singapore in June 1823 never to return and in fact his total period of residency in the settlement was just over a year. The settlement was developing fast by this time as when Raffles and Farquhar first landed the estimated total of inhabitants at that time was '. . . about one hundred and fifty Malays, half fishermen and half pirates . . .'[23] but by 1823 the population was recorded as being 10,683, of which only seventy-four were Europeans.[24]

From 1823 until 1826 the Settlement at Singapore was a dependency of Bengal but after this date it was incorporated with Prince of Wales Island (Pulo Pinang) and Malacca and then became part of the Straits Settlements. During this early period of occupation the method of travelling from the south, where the town was located, to the north of the island was by boat. A bridle path was started in 1821 which followed roughly the route of the present-day Serangoon Road. By 1822 it had reached Woodsville Corner but did not reach the Serangoon River until 1827 and at this time the road was started to Bukit Timah.

An early Christian cemetery was established high on the hill near to Raffles' residence. This early cemetery was discontinued in 1823 and all traces of it have been obliterated by various rebuilding programmes, the first of which was the major project, the construction of the fort which was to be known as Fort Canning. There is no record of any military personnel being buried in the early cemetery but the burial register has not been found so that evidence is inconclusive. Charles Burton Buckley, in his history of Singapore, records that a number of gravestones were moved from the first cemetery and placed in the new cemetery which had been opened on the lower slopes of Fort Canning Hill (the cemetery is now generally known as the 'Fort Canning Cemetery'). There does not appear, however, to be a record of the number of stones transferred nor any record of inscriptions, except for that of 'John C. Collingwood of the Ship Susan', who died in 1821. In the *Journal of the Straits Branch of the Royal Asiatic Society* for the year 1912 it was reported that Collingwood's tombstone had been found in the lower ground.[25]

John Crawfurd, the second Resident of the Settlement took over Raffles' house on Government Hill and extended and improved it at his own expense. It remained the residence of the Resident, and later the Governor, until 1858. The name generally used for the hill throughout the period 1822 to 1858 was Government Hill.

In May 1823 the Sepoy soldiers' accommodation was moved and the new barracks were set up in Sepoy Lines, which was adjacent to Pearls Hill. Just prior to the move of the troops to their new accommodation, Major Farquhar challenged Raffles' authority, as the Lieutenant Governor, to act independently of the Resident on the numerous items of detail that had been entrusted to him by the Supreme Government. Farquhar also challenged Raffles' right to criticize him directly to the Commander-in-Chief in India on military matters. Raffles, furious at this challenge, quite illegally dismissed Farquhar as Resident and Commandant. In the third week of May, Major Farquhar was further surprised when he was informed of the appointment, by the Supreme Government, of Mr Surgeon John Crawfurd, as the new Resident of the Settlement. The situation was obviously beyond repair and Tregonning in his book *The British in Malaya* comments on Raffles as follows '... As a lieutenant-governor who ordered and acted over his head, as a civilian who criticized the military actions of a soldier, as a superior who practiced nepotism and who imposed humiliations, Raffles in many ways must have been an obnoxious and overbearing person. Clever and ambitious men who fight hard for success often are ...'[26] Farquhar in fact remained in the Settlement for a further six months after Raffles had left, finally leaving in December. He called at Malacca and Penang *en route* to India. In Malacca his twenty-year rule as Resident was still remembered and he was met with affection during his short visit. In Calcutta he appealed against his dismissal and was vindicated. He sailed for London and retirement, being granted furlough from 8 September 1824. He was promoted to Colonel in 1829 and died, at the age of sixty-eight, in 1839.

By 1826 the garrison had settled into its new accommodation and was following a routine of parades, drills, musketry training and guards.

Singapore Garrison Orders for 1826 give examples of the orders that were published by the 'Acting Resident' and are shown as follows:

> '*Singapore Sunday 19th November 1826*
>
> The Guards to be relieved by Detachments tomorrow morning at Sunrise. The Detachment of Golundauze to furnish the Native Officer for the Main Guard.
>
> Officer for the ensuing week Lieutenant Long 25th Regt ...'[27]

This was Lieutenant Samuel Long of the 25th (or Marine) Bengal Native Infantry.[28] On Wednesday 29 November 1826 the Commandant published the following order:

> '... The monthly muster of troops and establishment at the station for November, will be taken on Friday morning the 1st ultimo at Sunrise ...'[29]

The entries for December of 1826 were varied commencing on 1 December with an appointment:

> '... Assistant Surgeon Lindesay authorised to proceed to Calcutta on urgent private business. Assist. Surgeon Montgomerie to be "in charge of troops" for his 3 month period of absence ...'

The order published on Sunday 3 December notified the officer of the week, which was again Lieutenant Long of the 25th Regiment.

Orders for 5 December gave instructions to the Artillery stationed in the Settlement who were ordered to fire '... A salute of 13 guns ... on the embarkation of the Resident of His Netherland Majesty's Settlement [at] Rhio ...'

In the following week the order convening a Court Martial was published:

> '*Singapore Wednesday 13th December 1826.*
>
> *Station Orders by the Commandant*
>
> A European Station Court Martial to assemble tomorrow morning at 10 o'clock at the adjutants bungalow, 25th Regiment NI Volunteers for the trial of John Syfax, Fife Major, 25th Regiment NI Volunteers and such other prisoners as may be brought before the Court ... President Captain C. E. Davis, 58th Regt NI, Lieutenant Long and Lieutenant Tierney, 25th Regiment NI Volunteers, Members. Lieutenant and Inter-

preter Thompson, 25th Regiment NI Volunteers will be pleased to attend Court . . .'[30]

The result of the Court Martial was published on 15 December:

'Singapore Friday 15th December 1826.

Station Orders by the Commandant

The Proceedings of an European Station Court Martial of which Captain C. E. Davis 58th Regiment NI being approved of, the Court is desolved, and the following particulars published for general information.

Crime. John Syfax Fife Major 25th Regt NI Volunteers.

Confined to the Main Guard on the following charges:

1st For being drunk and incapable of performing his duty on the evening of Tuesday the 12th instant.

2nd For refusing to return to his Lines when sent for by the Sergeant Major of the Regiment: for running away and endeavouring to evade the Sepoys sent to search for him, and otherwise behaving in a manner disgraceful to his character as one of the Regimental Staff of the Corps [at] Singapore.

13th December 1826 By order of the Commanding Officer.
Signed S. F. Hannah, Lieut.
Adj 25 Regt NI Volunteers.

Finding and Sentence – The Court having duly weighed the whole of the evidence brought before it find the Prisoner John Syfax Fife Major 25th Regiment NI Volunteers Guilty of the whole and every part of the charges brought against him and it therefore sentences him, the aforesaid John Syfax to be reduced to the Rank and Pay of a Drummer.

(Signed) C. E. Davis. Captain
58th NI President.
Approved and confirmed.
(Signed) W. Vincent Major
Commanding the Troops . . .'[31]

The foregoing shows that military justice was, in the East Indies in the early nineteenth century, swift. In this instance the crime was committed on one day and by the evening of the following day a Court Martial had been convened, the trial held and sentence passed.

The artillery whilst not being involved in firing against an enemy had plenty of practice at the Singapore Garrison. Two orders published during the last weeks of 1826 give orders for the firing of the guns, these orders were published as follows:

'Singapore Sunday 17 December 1826

A salute of 13 guns to be held in readiness to be fired on the landing of His Excellency the Commander-in-Chief of Her Most Catholic Majesty serving at the Phillippine [*sic*] Islands . . .' and

'Singapore Sunday 31st December 1826

A Royal Salute to be fired tomorrow morning at sunrise in honour of the New Year. [and in the same order] Guards to be relieved at Sunrise tomorrow.

Officer for the ensuing week – Lieutenant Long 25 Regt . . .'[32]

At this time in the garrison it appears that guards were changed weekly, as were the Garrison Duty Officer. An examination of the orders for the latter part of 1826 reveals that two officers of the 25th Regiment, Bengal Native Infantry did the duty in turn, they were Lieutenant Samuel Long and Lieutenant John Tierney. The reason for this is not known as these two officers were not the junior officers of the regiment at the time nor was either the senior subaltern.

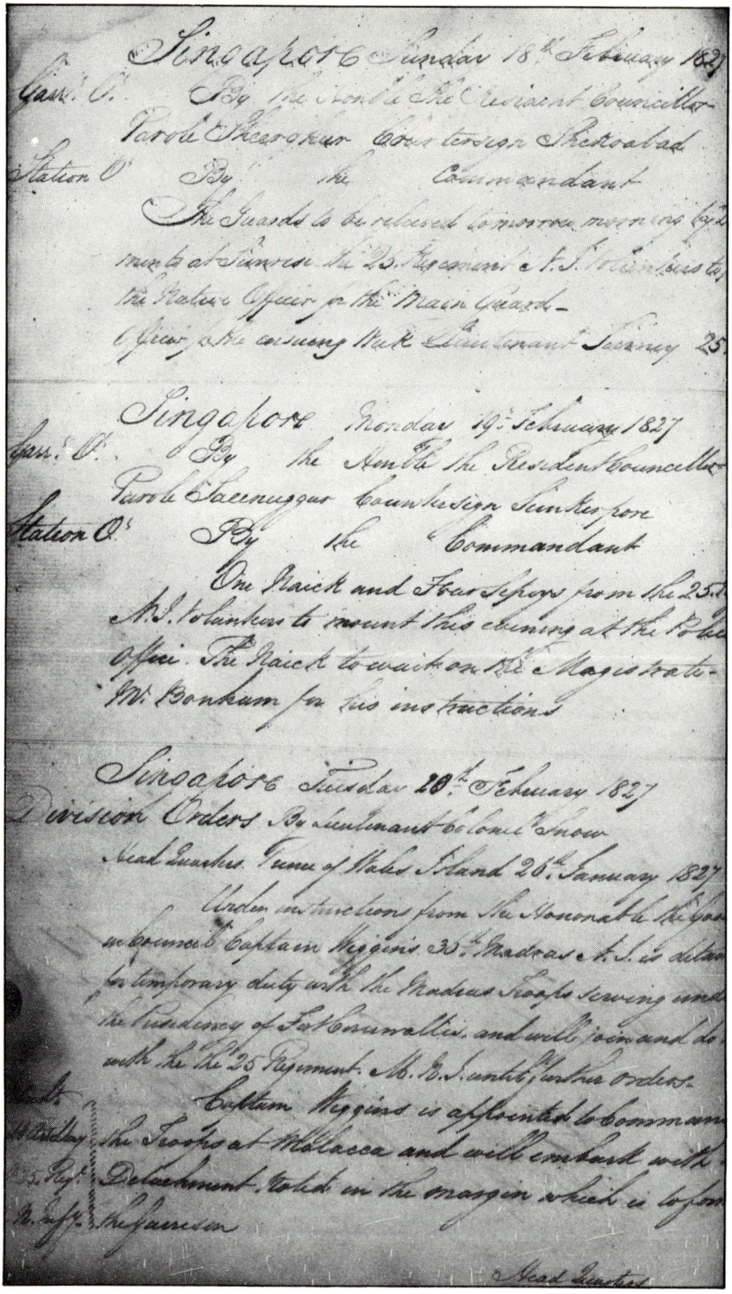
Singapore Sunday 18th February 1827
Garr. O. By the Honble the Resident Councillor
Parole Sheerpoor Countersign Shekoabad
Station O. By the Commandant
The Guards to be relieved tomorrow morning by 2 ments at Sunrise the 25 Regiment N.I. Volunteers to the Native Officer for the Main Guard —
Officer of the ensuing Week Lieutenant Jeremy 25

Singapore Monday 19th February 1827
Garr. O. By the Honble the Resident Councillor
Parole Bacenuggur Countersign Seenkerpore
Station O. By the Commandant
One Naick and Four Sepoys from the 25 N.I. Volunteers to mount this evening at the Police Office. The Naick to wait on the Magistrate Mr. Bonham for his instructions

Singapore Tuesday 20th February 1827
Division Orders By Lieutenant Colonel Snow
Head Quarters Prince of Wales Island 26th January 1827
Under instructions from the Honorable the Gov in Council Captain Higgins 30th Madras N.I. is detai for temporary duty with the Madras Troops serving under the Presidency of Fort Cornwallis, and will join and do with the the 25 Regiment M.N.I. until further Orders.
Captain Higgins is appointed to command the Troops at Malacca and will embark with Detachment noted in the margin which is to form the Garrison
Havildar 25 Regt. N. Infy.
Head Quarters

Singapore Station Orders, February 1827.
Courtesy of the Director, National Library, Singapore.

The year 1827 is well documented as not only are the Garrison Orders available but also copies of the Singapore newspaper the *Singapore Chronicle and Commercial Advertiser.* Although this newspaper started publication in 1826 no copies have been traced for the year 1826 and the date of the first paper available was Thursday 4 January 1827 in which it was reported that '. . . Six hundred Madras troops had arrived at Penang to relieve the 65th Bengal Native Infantry. The Bengal troops, now here, are also to be removed immediately and replaced by Sepoys from Madras . . .'[33]

An order was published on 9 January 1827 calling for returns to be submitted and read '. . . The Honorable the Governor in Council is pleased to direct that all Reports, Returns etc to be henceforth made to Lieutenant Colonel E. W. Snow of the 17th Regiment of Madras Native Infantry as Commanding Officer of the Troops serving under Prince of Wales Island, Singapore and Malacca . . .' This order was signed and dated at Penang on 25 December 1826.[34] This general order was quickly followed by a 'Division' order issued by Lieutenant Colonel Snow which showed that he was indeed taking firm control of his new command. The order read '. . . The Officer Commanding at Malacca and Singapore, will transmit to Head Quarters by the earliest opportunity returns for December for the Troops, Ordnance, Military Stores accounting for every Military Man at their respective stations. A return of all Officers and men on Guard explaining the particular duties of Officers employed on the Staff is also to be forwarded . . .'[35] In the meantime Lieutenants Long and Tierney were continuing to act as Garrison Orderly Officers on alternate weeks.[36]

Further changes were to be made, no doubt due to the arrival of Lieutenant Colonel Snow, and General Order No 1 of 1827 was published on 29 January of that year which saw radical changes in the staff appointments in the Settlements. It is of interest to consider the order in full:

'. . . General Orders, Fort Cornwallis the 12th January 1827.

The Honorable the Governor having been pleased to appoint Lieutenant Colonel Snow to Command the Troops at Fort Cornwallis, the same is notified in General Orders.

The appointment of Station Staff is abolished from this date and a Station Brigade Major is appointed to assist Lieutenant Colonel Snow Commanding the details of the Fort and Cantonment.

The Honorable the Governor in Council is pleased to make the following appointments: Captain Lake of the Madras Engineers to be Inspector of Stores, Works and Estimates of Prince of Wales Island, Malacca and Singapore.

Captain Ralfe of the Bengal Artillery to be Town Major and Military Secretary to the Honorable the Governor.

Lieutenant Hodge of the 1st Madras Artillery to be Executive Officer of Public Works at Prince of Wales Island.

Lieutenant Jackson of the Bengal Artillery, to be Executive Officer of Public Works at Singapore.

Lieutenant Risdon of the 8th Regiment Madras Cavalry, to be Cantonment Brigade Major at the Presidency Cantonment.

Lieutenant Snow of the 24th Regiment Madras NI to be Cantonment Adjutant at Malacca . . .'[37]

The order was however amended in as much as Captain Lake's appointment was changed. The new appointment was given as '. . . to be denominated [*sic*] Inspector General of Works at Prince of Wales Island, Malacca and Singapore instead of Inspector of Stores,

Works and Estimates . . .' in addition he was to be '. . . retained on the Honorable the Governors Personal Staff as his Aide de Camp . . .'[38]

On the last day of January a Court Martial was convened at Singapore to try a native soldier. The order read:

'. . . A Native Station Court Martial convened to try Gazie, Lascar of the Ordnance Commissariate Department. The 25th Regt NI Volunteers to provide One Subadar [*sic*] and three Jemadars and the Detachment of Golundaz one Jemadar to form the Court. Superintending Officer for the above – Lieutenant Tierney 25 Regt Interpreter for the above – Lieutenant Thomson 25 Regt NI . . .'[39]

The prisoner on this occasion was found to be 'Not Guilty' and was 'Released from confinement'.

On 20 February 1827 further changes were made by the new military commander. An order was published at Singapore which read:

'Division Orders by Lieutenant Colonel Snow.
Headquarters Prince of Wales Island 26th January 1827.
Under instruction from the Honorable the Governor in Council Captain Wiggins 36th Madras NI is detached for temporary duty with the Madras Troops serving under the Presidency of Fort Cornwallis and will join and do duty with the 25 Regiment M.N.I. until further orders.

Captain Wiggins is appointed to Command the Troops at Malacca and will embark with the detachment noted in the margin which is to form the Garrison . . .'[40]

The exact strength of the Garrison is not known as the entry in the margin of the document had been damaged and was unreadable.

The Singapore Garrison continued to perform normal duties, guards and parades with little of note happening to relieve the daily routine. The Orders published during the next few months of 1827 give an indication of the types of parades, etc, that were ordered. The monthly muster parades continued, the muster parade for the beginning of March being published as follows:

'*Singapore, Tuesday 27th February 1827*
Station Orders. The monthly Muster of Troops and Establishments at the Station for February will take place on Thursday morning, the 1st Proximo at Sunrise . . .'[41]

On 9 March the Routine order for the disbursement of pay and allowances was published, the order following the set pattern for such orders, which read:

'*Singapore Friday 9th March 1827*
Pay and Allowances of the Troops and Establishments at the Station in arrears for February and advance for March 1827 will be issued tomorrow morning at the Treasury on Government Hill . . .'[42]

The order is interesting on two counts as it confirms the use of the name 'Government Hill' and also shows that the troops were paid at two monthly intervals.

The garrison guard duties continued to be changed weekly, at Sunrise, which is confirmed again by a routine order which gives the following details:

'*Singapore Sunday 25th March 1827*
Station Orders by the Commandant.
The Guards to be relieved by Detachments tomorrow morning at Sunrise, the 25th Regiment NI Volunteers [to] furnish the Native Officer for the Main Guard. Officer for the ensuing week Lieutenant Tierney 25 Regt . . .'[43]

Maintenance of the garrison buildings was also catered for in Station Orders and on

Wednesday 28 March 1827 an order was published dealing with that subject:

'... A Committee of Survey to assemble tomorrow morning at 6 o'clock at the Main Guard Room to report on repairs required to that building.
President Captain Davis 58th Regt NI
Lieutenants Hannah and Long 25 Regt NI Vols Members ...'[44]

Two days later a further garrison order was published which indicated that the Committee had completed their report. It read '... The Committee of Survey of which Captain Davis 58 Regt NI is President is dissolved ...'[45]

During the month of April the garrison infantry unit was changed and the orders show the gradual progression to the point of handover. On 15 April the following order was published:

'*Singapore Sunday 15th April 1827*
Major Mallandain of the 35th Regiment Madras Native Infantry having reported his arrival on HCS Hastings with a Detachment of Troops destined for the relief of those at present serving at Singapore. The Troops are accordingly to be landed tomorrow morning at daylight on Boats to be furnished by the Master Attendants Department and will proceed to occupy the Tents to be erected for them, in the front of the Singapore Institution house.

Major Vincent 25 Regiment NI Volunteers will be pleased to make over the Command of the Troops tomorrow morning to Major Mallandain as Senior Officer.

Present States of Troops and Establishment at the Station to be forthwith sent to the Station Staff Officer.

General Order by the Hon'ble the Resident Councillor
The Salute of 13 Guns from HCS Hastings to be returned by one of the Garrison forthwith ...'[46]

On the following day a further order was published which set out further details concerning the actual handover of command and the various other offices:

'*Singapore Monday 16th April 1827*
Station Orders by Major Vincent.
In confirmatory to Garrison Orders of yesterday all Reports of the Station to be made to Major Mallandaine [*sic*].

Lieutenant R.N. Campbell of the 4th Regiment, Madras Native Infantry, the officer approved by General Order of the Honorable the Governor in Council under date the 26th February last, to the situation of Cantonment Adjutant Singapore having arrived Captain Davis will be pleased to deliver over to Lieutenant Campbell the office of Cantonment Adjutant together with the Commissariat Stores at as early a period as practicable. On the final transfer of the Commissariat Stores to Lieutenant Campbell Captain Davis is permitted to return to Bengal per HCS Hastings.

The Resident Councillor takes this opportunity of recording the impression felt by himself and he understands equally by all his Predecessors in the Government of Singapore of the efficient Services and Zeal of Captain C. E. Davis, of the 58th Regiment Native Infantry, who in addition to his Military Duties has greatly distinguished himself by the Ability Activity and Correctness with which he has conducted the details of the offices of Commissariat, Civil and Military Pay Departments for nearly the past seven years ...'[47]

Captain Charles Edward Davis, formerly of the 29th Regiment Bengal Native Infantry had in fact been appointed as Cantonment Adjutant, Singapore in 1820 and had become a much respected member of the European community.

Major Mallandaine published a Garrison Order convening a Board of Officers for the purpose of checking the Commissariat stores with Captain James Tennant, 35th Regt MNI as President and Lieutenants Thomson and Long of the 25th Regt NI Volunteers to serve as members. The check of stores was obviously completely satisfactory as a further order was published allowing Captain Davis to depart. The order read:

> '*Order by Major Mallandaine Singapore 18 April 1827*
>
> Captain C. E. Davis 58th Regiment Bengal Native Infantry, late Cantonment Adjutant at Singapore is attached to the Head Quarters Division of the 25th Regiment NI Volunteers and will accompany the Detachment of that Regiment about to proceed to Calcutta on the HC Ship Hastings . . .'[48]

and on the following day the order giving details of departure of the 25th Regiment NI Volunteers was published:

> '*Singapore 19th April 1827*
>
> *Garrison Orders by the Hon'ble the Resident Councillor*
>
> The continuation of the Garrison order under 17th instant the undermentioned officers proceeding to Bengal with the Troops on the HC Ship Hastings will send forthwith to the proper officer for examination and payment Bills for Lodging Money [to] the 25th instant inclusive the probable date of Departure of the Detachment.
>
> Major Vincent 25th Regiment NI
> Captain Davis 58th Regiment NI
> Lieutenant Thomson 25th Regiment NI
> Lieutenant Hannah 25th Regiment NI
> Lieutenant Long 25th Regiment NI . . .'[49]

Not only was the departure well published in Garrison orders it was also covered by the local press and on Thursday 26 April 1827 the following report was published in the *Singapore Chronicle*:

> 'SINGAPORE. Thursday 26th April 1827. The HASTINGS, Captain Dawson, arrived here on the 15th instant, with Madras Troops from Penang, and sailed again this morning for Calcutta with two Companies of the 25th Bengal Native Infantry. One Company of the Bengal Sepoys still remain under the Command of Lieut Tierney, but will leave this station probably in the CAMBRIDGE in which the Hon the Governor is almost daily expected. The Officers who have accompanied the troops in the HASTINGS and whose departure is viewed with general regret by a society to which they have in no common degree endeared themselves are Capt C. E. Davis, Major Vincent and Lieuts Thomson, Hannah and Long . . .'

The local newspaper at this time also published lists of passengers arriving and departing from the Settlement and therefore gives an accurate cross check against the Garrison Orders for the arrival and departure of the military. In the newspaper for 26 April 1827 the following arrivals and departures were published:

> '*Arrivals*
>
> 15th April 1827 HC Ship Hastings, departed from Pinang 9th April.
> Major Mallandaine and family,
> Captain Turner and family,
> Capt Chauvel and family,
> Dr Warrand and family,
> Capt Tennant,
> Lts Campbell, Balfour and Trotter.

Departures

25th April 1827 HC Ship Hastings, for Calcutta.
Major Vincent
Capt and Mrs Davis and family,
Lts Thomson, Hannah and Long and
Dr Innes . . .'[50]

With the changeover of command at the Garrison there were, as is normally expected, a number of changes in organisation and these are reflected in a series of orders published by the new 'Senior Officer' and a sample of these orders are shown to illustrate the type of order and wording used during the early nineteenth century in the East Indies:

'Garrison Orders by Major Mallandaine.

Singapore 25th April 1827, Wednesday.

Medical Officers in charge of Corps and Detachments at this Station will be pleased to send in States of their sick every Monday morning by 10 am to the Officer Commanding Troops.

The Guards will be relieved tomorrow at 5 o'clock.

When any Military Building requires repair of a temporary nature the Commanding Officer will direct in Station Orders the Executive Officer to perform the duty in the manner laid down by Government . . .'

'Singapore 27th April 1827 Friday.

The Muster of the Troops and Establishment will be taken by the Cantonment Adjutant on Tuesday morning next the 1st proximo at 6 o'clock on the General Parade . . .'[51]

The orders reflect a general tightening up of discipline within the garrison which included checking on sick parades, the early parade time for the change of guard duties, moved from 6 o'clock in the morning to 5 o'clock; and the requirement of the Garrison Adjutant to take the Muster parade. This was only the beginning of the orders as Major Mallandaine also published a directive on dress for his Garrison officers which read:

'. . . The costume laid down for officers by the Commander-in-Chief of the Madras Army is your guide, and the Brigadier expects you will not permit officers to appear in Public in any fancy dress whatever. Every consideration and attention has been paid to the comfort of officers costume, and you are held responsible that everybody under your command pays due attention. Colonel Snow is more particular on this point as it is reported as a local custom for Gentlemen to appear at Public parties in white jackets which cannot be tolerated in the Army of Fort St George . . .'[52]

It would appear that in this instance Mallandaine republished an order that had been given to him from his superior at Prince of Wales Island. Within a week another order was published which would undoubtedly have caused a certain amount of discontent among the rank and file of the garrison. The order read:

'Garrison Order by Major Mallandaine

Singapore 3rd May 1827 Thursday

The Troops will hold themselves in readiness to Parade in Review Order at the shortest notice to be drawn up in open column of Sub divisions right in front, on the road leading to the Master Attendants Office and fronting the Parade, the artillery on the right with its leading sub-divisions drawn up opposite the Bungalow at the present occupied by the Madras Artillery, the 25th Regt Bengal NI next in Order, then the 35th Regt Madras N Infantry.

A portion of the Artillery to be in readiness to return the salute of the Ship CAMBRIDGE as likewise be prepared to fire a salute of 17 guns on the landing of the Honorable the Governor of Prince of Wales Island and its Dependencies. A Bugler from the 35th Regiment MN Infantry be immediately sent to Major Mallandaine's quarter for the purpose of sounding the Assembly for the Troops.

An Honorary Guard of 2 Havildars, 2 Naick [*sic*], 24 Privates under a Native Officer to mount immediately at the House of Mr Napier at Campong Glam. The Officers to report to the Cantonment Adjutant for instructions . . .'[53]

On the following day a further order was published by Major Mallandaine which covered various aspects of garrison activities.

'*Singapore 4th May 1827 Friday*

The detachment Madras Artillery and Detachment 25th Regiment Bengal NI will parade tomorrow morning at Gun-fire in marching order for the purpose of being inspected by the Officer Commanding the Division . . .'[54]

The final phase of the handover from Bengal to Madras army units took place during May 1827 and various orders were published to achieve the completion of the handover. The orders on 4 May 1827 dealt with the handover of the Bengal Artillery detachment and read:

'. . . Lieutenant Jackson being relieved from all Garrison duty the Detachment of Bengal Artillery to be made over to the charge of Lieutenant Tierney, 25th Bengal NI Volunteers and to be held in readiness for embarkation . . .'

Lieutenant Balfour Madras Artillery[55] assumed command of the Madras Artillery from that date and also had charge of the Park Establishment and the Magazine Department.

Details of the embarkation of the rear party of the 25th Regiment Bengal Native Infantry were published on Monday 14 May 1827:

'. . . The Detachment of the 25th Regt Bengal Infantry under the Command of Lieut Tierney, will embark on board the ship David Clarke for Prince of Wales Island on Wednesday next the 16th instant. The Commissariat Officer will ship Provisions for the Party calculated for a voyage of 5 days . . .'[56]

and on 17 May the following instruction was issued which dealt with the handover of the Artillery detachment.

'*Garrison Order by Major Mallandaine*

Agreeable to instructions received from the Hon'ble the Governor in Council. Lieutenant Balfour of the Madras Artillery will receive charge of the Detachment [of] Bengal Golundauze remaining at Singapore. Lieutenant Tierney will hand over to Lieutenant Balfour all Books, Papers &c relating to the Detachment . . .'[57]

During the same month a Divisional Order was published dealing with the rations for the European Artillery stationed in the Singapore Garrison which reveals an interesting view of the daily diet of soldiers stationed in the East Indies at that time. The order reads:

'*Headquarters Singapore 7th May 1827*

Division Order by Lieutenant Colonel Commandant Snow.

The European Artillery in Singapore are to receive the same rations with respect to quantity as directed at Prince of Wales Island. Until arrangements can be effected for the procuring a regular supply of Fresh Provisions the Commissariat Officer will issue Rations as follows:

Sunday – Fresh Pork or Beef and Bread,
Monday – Salt Beef and Rice,
Tuesday – Salt Pork and Bread,

Wednesday - Fresh Beef or Pork and Bread,
Thursday - Salt Beef and Rice,
Friday - Fresh Pork or Beef and Bread,
Saturday - Salt Beef and Rice . . .'[58]

There are very few instances of promotions and appointments of other ranks being recorded in the documents of the East Indies but within the Singapore Garrison Orders for 1827 there were two such orders issued during the changeover period and both were issued by Major Mallandaine. The first dated 4 May 1827 gives two appointments:

'. . . Garrison Serjt Major Richard Bailey is appointed Barrack Serjeant at Singapore to complete the establishment. Corporal William Well, of the Madras Artillery to act as Cantonment Sergeant Major at Singapore . . .',

but this was amended by a further Garrison order which was dated 8 May which ordered the following change: '. . . Corporal Well of the Madras Artillery to act as Garrison Sergeant Major and Barrack Sergeant at Singapore, subject to the confirmation of Government . . .'. For some reason, which is not recorded Richard Bailey was returned to Calcutta on the ship *Cambridge*.

The local newspaper recorded the departure of Lieutenant Tierney and his detachment of the 25th Regiment Bengal Native Infantry as follows:

'*Departures*

David Clark. Departed 20th May 1827 for Penang and Madras. Lt Tierney . . .'[59]

In the same paper Lieutenant Baillie was notified as having arrived on 20 May on board the vessel *Colonel Young*. On the same day as his arrival Major Mallandaine published an order setting out his duties and appointment:

'*Singapore 20th May 1827 Sunday*

Lt Baillie Madras Artillery having reported his arrival will agreeably to Division Orders of 12th inst receive charge of the Madras Artillery and Bengal Golandauz from Lt Balfour who is now directed to proceed by the earliest opportunity to Malacca to which station he is now appointed . . .'[60]

At about this time there was an epidemic on the island and from the evidence of orders it can be assumed that there was a considerable increase in the number of soldiers who were hospitalised as on 24 May 1827 the following 'Station Order' was issued:

'. . . Mr Acting Conductor Hall will provide two of the Public Tents for the accommodation of the sick of the 35th Regiment Madras NI as near to the Hospital as it can conveniently be done . . .'[61]

The only record of a European death during this period is that of the infant daughter of Lieutenant C. Turner who died on 29 May 1827. The obituary appeared in the *Singapore Chronicle* – 'Died on Tuesday the 29th May of fever and dysentry, Isabella, infant daughter of Lieut C. Turner of the 35th Regt, Madras Infantry, aged 14 months and 18 days . . .'[62]

During the month of June further changes were made in the garrison and these are, as before, recorded in the Station Orders issued by the senior officer who was still Major Mallandaine. The order for 13 June 1827 read:

'. . . Sub-Conductor Hall will be pleased to deliver over the charge of the Magazine and Military Stores to Lieutenant Baillie, Madras Artillery.

A Committee will assemble at the Arsenal tomorrow morning at 10 o'clock for the purpose of seeing the same delivered over. Reports in Triplicate to be sent to the Commanding Officer.

Captain Blundell – President Lieut Trotter 35th Regt Member.
Lieut Turner 35th Regt Member

Lieut Baillie is directed agreeably to Garrison Order of the 12th instant to select the Artificers therein required and forward a Roll of them to the Cantonment Adjutant's Office.

Mr Sub-Conductor Hall with the Non-commissioned Staff Bengal Golundauz and the remaining part of the Artificers are directed to hold themselves in readiness to proceed to Calcutta at the shortest notice.

Mr Sub-Conductor Ryan not now required at this station he will be prepared to return to Prince of Wales Island.

Captain Blundell Madras Artillery will be pleased agreeably to Garrison Order of 12th June to select 4 eligible men to fill the vacancies of the Non-Commissioned Staff Appointments and send in recommendations to the Officer Commanding the Station . . .'[63]

The recommendations which had obviously been submitted by Captain Blundell to the Officer Commanding were approved and published in an order dated 17 June 1827. The four appointments were listed and the soldiers nominated as follows:

Commissariat Sergeant	– Sergeant Daniel Cunningham.
Magazine Sergeant	– Sergeant William Trapp.
Engineer Sergeant	– Corporal Walter Narros.
Laboratory Corporal	– Gunner Dennis McSweeney.

Whilst all the various changes were being made within the Singapore Garrison by the new Officer Commanding the actual defences of Singapore were being examined in detail by Captain Edward Lake, Madras Engineers, who held the appointment of '. . . Inspector of Works &c, Prince of Wales Island . . .' Lake proceeded to Singapore to report on the works and fortifications and it has been suggested that he partially based his proposals on the recommendations given to him by Lieutenant Phillip Jackson, Bengal Artillery. Jackson had been stationed at Singapore from 21 January 1822 and had been appointed as the officer in charge of the artillery unit stationed in the Settlement. In October of that year Sir Stamford Raffles appointed Jackson as the Assistant Settlement Engineer and this appointment was later confirmed by John Crawfurd. He continued in that appointment until 1 July 1827. In the ensuing years the title had been changed to that of Executive Officer on 18 December 1824 and he had also accepted the additional duty of Surveyor of Lands as from January 1823. The idea which Jackson is said to have passed to Lake originally came from Colonel William Farquhar, who apart from being the First Resident was also a Madras Engineer officer.

Jackson had also taken up the appointment of Assistant to the Officer in Charge of the Residency in 1826 and when Captain Lake appeared and gave instructions for Lieutenant Jackson to spend some time each week on field survey work in connection with the proposed defence plan, Jackson produced a 'Sick certificate' to the effect that it was not advisable for him to be out in the sun.[64] Jackson was permitted to proceed to Europe on a Medical Certificate.[65] The order allowing him to relinquish his post was published as follows:

'*Singapore 19th June 1827*

Lieutenant Goldingham, Madras Artillery, Assistant to the Surveyor at Prince of Wales Island is appointed to relieve Lieutenant Jackson as Executive Officer at Singapore and will report forthwith to this station . . .'[66]

The order was also published in the *Singapore Chronicle*[67] and Lieutenant Jackson was reported in the newspaper dated Thursday 5 July 1827 as departing on 'The Hero of Malowne on 1st July 1827 bound for Mauritius . . .'[68]

Despite the conflict between Captain Lake and Lieutenant Jackson the report on the defences was completed and as the report contains much detail it is produced in full:

REPORT ON THE BEST MEANS OF FORTIFYING SINGAPORE
(by Edward Lake Inspector General May 1827)

1. The Town of Singapore is favourably situated for defence, surrounded as it is on two sides by the sea, on one by a broad and deep creek, having a large part of the land front defended by a morass, which it would be easy to lay underwater when occasion required it, the only difficulties which present themselves are the hills on the N.E. quarter occupied by the Residency, and a range of hills to the Southward. Even these difficulties might be surmounted, by levelling some of the hills, and by a judicious and defilade-ment of the Works, but unfortunately the ground laying between the creek and the fresh water stream, which would be indispensable for the purpose of a covert way, glacis, and Esplanade, has been granted to individuals, and is covered with valuable and substantial houses, which could not be removed but at a vast expense. The ground on the sea faces has also been granted away and it would therefore be necessary to remove from the sea sufficient for the works on those sides, and the expense of this together with that of purchasing the ground above alluded to, will I presume be an insurmountable objection to any project for fortifying the town throughout.

2. But although these considerations may induce the Government to forego any attempt to fortify the town so as to render it equally secure against an enemy by sea or land, they will probably think it indefensible not to protect it from an attack by shipping, and to prevent a landing (which if effected would involve the speedy destruction of the Town), on any point between St John's Island and Tanjong Katong. It will also I presume be the intention of Government to have a fortified post sufficiently large and strong to give protection to the public stores and treasure, and to a certain extent to that of individuals, and to form a defensible point of general rendezvous in the event of an enemy effecting a landing and it will also doubtless be thought desirable that there should be a place of refuge for the shipping should an enemy's ship of war come into the Roads. By a disposition which I shall proceed to point out, I think all these objects may to a considerable extent be attained and a general and connected line of defence formed. To recapitulate then shortly the objects of the plan I am about to propose; they are, first, to protect the Town from attack by sea or bombardment by shipping.

3. To protect the coast from a landing between St John's Island and Tanjong Katong, 3rdly to construct a work for the protection of the stores and for the defence of the Garrison in the event of the enemy effecting a landing, 4th to make a secure harbour for the shipping in time of war. To protect the Town from attack by sea or bombardment by shipping I propose to effect this by a casemated Redoubt on Battery Point, and by a half Mean Battery on the S.W. angle of the Town, the position of which will be more clearly shewn by reference to the plan of the town and environs, which accompanies this Report, and on which it is marked No. 1.

22. Whichever of these points is selected (I should myself give the preference to Captain Pearl's) I should recommend all the works to be constructed of earth, leaving them to be erected with brick, in case it is thought advisable to do so at a future day, and a body of 300 convicts employed thereon would very soon complete them.

23. On receiving the decision of the Hon'ble the Governor in Council as to which hill is to be appropriated for this purpose, I shall be prepared to furnish the requisite plans of the works necessary for its defences.

24. I take this opportunity of recommending that any new streets which may be formed in the town, should if possible, be laid out in such a way as to be enfiladed by the work on whichever hill it may be placed.

25. Harbour. The Roadstead in which ships lay while receiving and discharging cargo, is open, and at no great a distance from the shore to be defended from thence. It therefore becomes a question what is to be done for the protection of shipping that may resort here in time of war. Fortunately an inlet of the sea affords between Sandy Point and the Island, a secure harbour for boats, junks, and vessels of less than . . . tons, which will be perfectly defended by the projected redoubt on Sandy Point and by the batteries No. 3 and 4. The Singapore creek would also afford shelter equally secure for a large number of small craft, about 2½ miles South West of the Roads is a harbour containing water sufficient for vessels of the largest burthen, formed by the island of Singapore on one side, and that of Blakang Mati or St George's Island on the other. In the middle of the entrance from the roadstead is a small island called Pulo Ayer Brani or Gage Island, which divides the channel into two, that nearest Singapore being 1,000 yards across, and the small one 350 yards. The entrance at the West end of the harbour is 500 yards wide. I do not feel myself qualified to pronounce an opinion on the capability of the harbour in a Naval point of view, but if it is found equal to expectations that have been raised regarding it in this respect, there is no doubt that it may be made impregnable against the strongest naval force an enemy might bring against it.

26. In my project for the defence of the harbour it must be considered that if an enemy were to land and make themselves Masters of Blakang Mati, it would give them command of the harbour, and therefore that island should be fortified to prevent that contingency, as well as to protect the entrance of the harbour from attack by sea. I have marked in the plan, the position of the batteries which would be sufficient for this latter purpose, and it will be immediately seen by reference thereto that no enemy's ship would venture to pass such narrow channels so defended.

27. It would be premature in me to offer detailed plans of the Works necessary to protect this harbour until it is ascertained whether it be calculated to hold the ships that would probably resort here in time of war, and whether it is the intention of Government to make it secure against an attack by sea or land. I have the honour however to annexe to this report, the copy of a letter from the Executive Officer regarding it, and I shall only observe in addition to his remarks, that not only a communication with the town by carts could be kept up, but that there would be no difficulty in carrying a canal from the Singapore creek to the new harbour if it were thought necessary.

28. In closing this report I trust that the Hon'ble the Governor in Council will take into consideration that the plan of defence it contains is offered as being the best that occurs

to me at a moderate expense if the importance of the object be considered under the circumstances of the greater part of the ground about Singapore having been granted away and appropriated, and not as the best which the localities and situation of the place would have been capable of affording if a project of defence had been determined on, upon the first occupation of the Island, and if the position of the Town, and the issues of grants had been regulated with reference to that.

Inspector Genls.
Office Singapore
May 1827

Edward Lake
Inspector Genl.

(True Copy)
Saml Best
Captn Engineers

(True Copy)
W. J. Butterworth
Governor

(True Copy)
W. M. R. Sturt Major
Offg Secy to the Govt of India
Military Dept.

Captain Lake's report was never implemented and the only positive action taken on the report was the building of Fort Fullerton at the mouth of the Singapore River. For one reason or another the report was not sent on to Bengal until 1829. It was during the following year that Major Lake, as he was by then, departed from Singapore with his wife and family on the vessel *Guildford* on 29 December 1830. The *Singapore Chronicle* for Thursday 30 December 1830 showed '. . . Major & Mrs Lake and Masters A. & W. Lake departing on the "Guildford" bound for London . . .'[69] The vessel was reported to have cleared Sunda Head but was never heard of again. It presumably sank somewhere in the Indian Ocean. With the loss of Major Lake the report appears to have been filed and the question of building adequate defences for the Settlement was shelved until 1843.

From an order published in July 1827 it is clear that the accommodation provided for the Sepoys was made of attap. An order by Major Mallandaine dealt with one of the problems of this type of building:

'*Singapore 23rd July 1827 Monday*
As from the combustible materials etc of which the Sepoy Huts are constructed and their proximity to the place of Arms, in case of fire it would be extremely difficult to secure the Arms of the Detachment. The Commanding Officer therefore deems it necessary for their preservation that the End Houses of each line should be pulled down . . .'

The *Singapore Chronicle* which was published fortnightly not only gave details of the movements of officers to and from the garrison at Singapore but also proved to be useful for details of the social side of the garrison. In the edition printed on 16 August 1827 a notice appeared notifying the birth of a son to Major Mallandaine. The phrasing of the birth notice is of interest as this format is no longer used, being

'*BIRTH*
On the 10th instant, the Lady
of Major Mallandaine, Commanding
Singapore, of a son . . .'

During September 1827 there were further changes recorded within the garrison. On 10 September the following order appeared:

'. . . Lieut Beresford of the Artillery having reported his arrival with a Detachment of Golundauz and Gun Lascars at this Station they are received on the strength of the

Garrison from yesterdays date. Lieut Beresford will refer himself to the Officer Commanding the Artillery.

The Detachment of European Artillery at this Station will hold themselves in readiness to proceed to Prince of Wales Island under Lieut Beresford at the shortest notice. Embarkation Returns in triplicate to be sent to the Cantonment Adjutant as soon as possible . . .'

The detachment did in fact remain at Singapore for a further two weeks and the order for embarkation was published on 25 September:

'*Station Movement Orders by Major Mallandaine*

The Detachment of European Artillery under Lieut Beresford will embark tomorrow morning at 6 o'clock on board the Ship "Grace" for the purpose of proceeding to Penang . . .'

From the pages of the *Singapore Chronicle* it is again possible to keep an account of the social events of the Settlement which would have been fairly restrictive and would have consisted of picnics, balls and band concerts. However in the paper published on Thursday 25 October there was a notice concerning the marriage of the Adjutant of the 35th Regiment, Madras Native Infantry. 'Married at Singapore on Wednesday 10th inst by the Rev R. Burn, Lieut S. R. Hicks, Adjt 35th Regt, MNI, to Sarah Kennedy only Daughter of Lieut Colonel Comdt C. Farran of the Madras Army . . .',[70] and at the end of the year, which has been remarkably well documented, a notice was published in respect of a birth. The published notice is interesting in that it illustrates a variation of the regiment's title:

'*Birth.*
On Sunday night at 12 o'clock the
2nd instant the Lady of Lieut Charles
Turner 35th Regt Madras Volunteers,
of a son . . .'[71]

In 1828 the Garrison troops at Singapore were changed. The Penang Military Orders dated 3 January of that year ordered two Companies of the 25th Regiment, Madras Native Infantry to embark on the *Speke* to relieve four companies of the 35th Regiment, Madras Native Infantry at Singapore.[72] The *Speke* brought the relief force from Penang and returned on 25 January 1828 with the companies of the 35th Regiment MNI and the following officers and families:

'Major and Mrs Malandaine [*sic*]
Capt Tennant,
Lt and Mrs Turner,
Lt and Mrs Chauvel,
Lt Trotter . . .'[73]

The records for 1828 and 1829 are sparse and there would appear to be little change in the routine of the Garrison. The new Officer Commanding does not appear to have issued numerous orders as did his predecessor, Major Mallandaine. Major Mallandaine and his family applied for a passage to Madras (this was from Prince of Wales Island) which was granted on 23 April 1829.

During 1829 a replacement detachment of Gun Lascars arrived at Singapore on 4 May and it is recorded that on 25 May '. . . 19 Sick Sepoys under Lieutenant Gosling . . .' were sent from Singapore back to Madras on the HCS *Hastings*.[74]

In the same month a request was made, from Singapore, to Prince of Wales Island for the provision of 'Facings' for the military clothing of the 25th Regiment then stationed at

Singapore, and also during the month a report was submitted by the 'Inspector General' on the failure of the retaining wall at the Saluting battery at Singapore. Orders were issued for a Contractor to rebuild the wall.

Officers were permitted to proceed on furlough to England and on 20 July Lieutenant Richard H. James, 35th Regiment, Madras Native Infantry, who was at that time serving in Singapore, applied for permission to proceed to England on furlough '. . . from Singapore . . .' and was granted the usual allowance of 1,500 Rupees for passage money.

Captain Lake features in the *Singapore Chronicle* during December 1829. He had been on duty in Malacca and returned on 16th November, the entry in the newspaper reading:

> '*Arrivals*
> 16th November 1829. HC Schooner Zephyr. Departing from Malacca on 12th inst. Major Lake . . .'[75]

From this it is assumed that he had recently been promoted but probably to a temporary rank as his final entries in the East India Registers show him as being a 'Captain'. In the newspaper published on 3 December 1829 he is referred to as 'Captain Lake', the item being a birth notice which read:

> '*Birth*
> On the 9th ultimo, the Lady of
> Captain Lake, of a daughter . . .'[76]

Although no record has been found of the death of his daughter it is believed that she may well have died during 1830 as she is not shown on the shipping list of the *Guildford* when the family departed from the Settlement on the ill-fated voyage in December 1830.

A further batch of invalid Sepoys were evacuated during February 1830 from Singapore to Penang and travelled on the Company vessel *Penang Merchant*. There is no record of there being any casualties among the European officers, soldiers or families during the early part of that year.

Singapore was combined with Malacca and Prince of Wales Island on 1 May 1830 and then became a single Residency under the control of Fort William.

NOTES TO CHAPTER SIX

1. Information obtained from the 1957 Exhibition of Stone Implements held at the National Museum (then known as the 'Raffles Museum'), Singapore.
2. Ibid.
3. *Guide to Singapore and Spotlight on Malaya*, ninth edition, (Papineau Studios Advertising, Singapore, 1955).
4. H. F. Pearson, *A History of Singapore*, (University of London Press).
5. Ibid.
6. *Guide to Singapore*, op cit, p145.
7. Ibid, p149.
8. Walter Makepeace (ed), *One Hundred Years of Singapore*, (J. Murray, 1921), vol1.
9. Ibid.
10. Ibid.
11. Jan Huyghen van Linschoten, *The Reys-Gheschrift vande Navigation der Portugaloyers in Orienten*, (James Wolfe, London, 1598).
12. C. A. Gibson Hill MA, *Singapore Old Strait and New Harbour 1300–1870*, (Government Printing Office, Singapore, 1956), plate2.
13. Gibson-Hill, op cit, p74.
14. John Crawfurd, *A Descriptive Dictionary of the Indian Islands and Adjacent Countries*, (reprinted Kuala Lumpur, Malaya, 1971), p395.

15. Makepeace, op cit.
16. Ibid.
17. Ibid.
18. Ibid.
19. Charles Burton Buckley, *An Anecdotal History of Old Times in Singapore, 1819–1867*, (published 1902, reprinted by University of Malaya Press, 1965), p33.
20. Ibid, pp 43 and 44.
21. Ibid, p44.
22. Sophia Raffles, *Memoirs of the Life and Public Services of Sir Stamford Raffles*, (London, 1830).
23. R. Montgomery Martin, *History of the British Colonies*, (Cochrane and McCrone, London, 1834), vol 1, p430.
24. Ibid.
25. H. A. Stallwood, 'The Old Cemetery on Fort Canning, Singapore', *Royal Asiatic Society, Straits Branch, Journal LXI*, (1912).
26. K. C. Tregonning, *The British in Malaya*, (University of Arizona Press, Tucson, 1965), p160.
27. Singapore National Library, Microfilm No66. Singapore Garrison Order 1826–1827. 19 November 1826.
28. East-India Register and Directory, 1826, (Bengal), p68.
29. Singapore Garrison Orders 1826–1827, 19 November 1826.
30. Ibid, 13 December 1826.
31. Ibid, 15 December 1826.
32. Ibid, 31 December 1826.
33. Singapore National Library. *Singapore Chronicle and Commercial Advertiser*, No73, 4 January 1827.
34. Singapore Garrison Orders 1826–1827, 9 January 1827.
35. Ibid, 9 January 1827.
36. Ibid, 14, 21 and 28 January 1827.
37. Ibid, 20 February 1827. Details of the officers mentioned in the order are:
Captain Edward Lake, Madras Engineers.
Captain Henry Ralfe, Bengal Artillery. Ralfe joined the Bengal Artillery in 1810. 'In March 1815, Capt Ralfe was obliged, on account of ill health, to proceed to Europe. He again, in March 1818, left England for Bengal, and in the following November was selected for the command of a detachment of artillery, and directed to place himself at the disposal of Sir T. S. Raffles. He, accordingly, left Bengal . . . was subsequently appointed to the command of the Artillery, Assistant Engineer, and to the charge of the military stores at the Settlement formed on the Island of Singapore, where he remained 'till November 1821, when he was permitted, on account of ill health, to return to Bengal, from whence in a few months, he was, from the same cause obliged to proceed to England . . .' (*The Services of General and Field Officers of the Indian Army*, (London, 1826), vol3, p289). Ralfe returned from furlough in 1826.
Although the order showed Lieutenant Hodge as being of the '1st Madras Artillery' the officer concerned was 'Lieutenant Peter Pender Hodge, 1st Regiment, Madras Native Infantry'.
Lieutenant Philip Jackson, Bengal Artillery. (The entry in the East-India Register shows Jackson as serving at 'Sincapore' – using the old spelling for the Island).
Lieutenant Peter Risdon, 8th Regiment Light Cavalry Bengal Army.
Lieutenant Edward Winterton Snow, 24th Regiment, Madras Native Infantry.
38. Singapore Garrison Orders 1826–1827, 29 January 1827.
39. Ibid, 31 January 1827.
40. Ibid, 20 February 1827. Officer listed was Captain Henry Wiggins of the 36th Regiment, Madras Native Infantry.
41. Ibid, 27 February 1827.
42. Ibid, 9 March 1827.
43. Ibid, 25 March 1827.
44. Ibid, 28 March 1827.
45. Ibid, 30 March 1827.
46. Ibid, 15 April 1827.
47. Ibid, 16 April 1827.
48. Ibid, 18 April 1827.

49. Ibid, 19 April 1827.
50. *Singapore Chronicle and Commercial Advertiser,* No81, 26 April 1827.
51. Singapore Garrison Orders 1826–1827, 25 April 1827.
52. Ibid, Division Order dated 27 April 1827.
53. Ibid, 3 May 1827.
54. Ibid, 4 May 1827.
55. East-India Register and Directory. Officers mentioned are:
1st Lieutenant George Balfour, Madras Artillery.
Lieutenant John Tierney, 25th (or Marine) Regiment, Bengal Native Infantry.
56. Singapore Garrison Orders 1826–1827, 14 May 1827.
57. Ibid, 17 May 1827.
58. Ibid, 7 May 1827.
59. *Singapore Chronicle and Commercial Advertiser,* No83, Thursday 24 May 1827.
60. Singapore Garrison Orders 1826–1827, 20 May 1827.
61. Ibid, 24 May 1827.
62. *Singapore Chronicle and Commercial Advertiser,* No84, Thursday 7 June 1827.
63. Singapore Garrison Orders 1826–1827, 13 June 1827.
64. Gibson-Hill, op cit, p83.
65. Singapore Garrison Orders 1826–1827, 18 June 1827.
66. Ibid, 19 June 1827.
67. *Singapore Chronicle and Commercial Advertiser,* No85, Thursday 21 June 1827.
68. Ibid, No86, Thursday 5 July 1827, and Gibson-Hill, op cit, p83.
69. *Singapore Chronicle and Commercial Advertiser,* No177, Thursday 30 December 1830. The route frequently followed by vessels proceeding from Singapore to England was to sail south-east past the island of Banca. They would then pass through the Sunda Strait between Sumatra and Java and then turn north-west to call at Bencoolen (not all vessels made this port of call) before crossing the Indian Ocean.
70. Ibid, No94, Thursday 25 October 1827.
71. Ibid, No98, Thursday 20 December 1827.
72. Singapore National Library (Microfilm No S2). Penang Orders and Notices: January–December 1828.
73. *Singapore Chronicle and Commercial Advertiser,* No101, Thursday 31 January 1828.
74. Penang Orders and Notices, 1829, Military, 25 May 1829.
75. *Singapore Chronicle and Commercial Advertiser,* No148, Thursday 19 November 1829.
76. Ibid, No149, Thursday 3 December 1829.

CHAPTER SEVEN

The Naning Campaign of 1831 and 1832

The small inland state of Naning was located between the Malacca territory and Rambau with a population of about 5,000[1] in 1829. The state first came under control of a foreign power in the early sixteenth century when the Portuguese controlled Malacca. When the Dutch captured Malacca in 1641 they also took control of the Naning territory by treaty. The Dutch control was minimal as Dutch officials generally tried to avoid visiting the Naning state having some distrust of the Naningites. There were rumours of constant unrest in the state and rumours that during 1644 the Naning chiefs plotted to mount a raiding party and attack and set fire to Malacca town. The Dutch-Naning treaty of the 1640s required the Naning chiefs to pay an annual tribute of one-tenth of their crops to the Dutch at Malacca and they were required to journey to Malacca town annually to pay homage to the Dutch Governor.[2]

During the early eighteenth century the title of Penghulu became hereditary in the family of Abdul Said. In 1746 the annual payment of one-tenth of the state's crops had been commuted to being an annual payment of 400 gantangs of paddy.[3] The quantity was not thereafter varied although by 1831 it was estimated that 400 gantangs really only equated to about one-thousandth part of the total Naning production. It was therefore a token payment.

The British captured Malacca from the Dutch in 1795 and in 1801 the British concluded a treaty with the Penghulu of Naning. The treaty required the Penghulu to hold allegiance to the Honourable East India Company; to continue to pay a tenth of its produce to the Company's representative at Malacca, but this was commuted to the 400 gantangs of paddy; the Penghulu, or one of his Chiefs, to pay an annual homage to the Company's representative at Malacca; the Company reserved the right to appoint the Penghulu when the office became vacant; all tin was to be brought to the Company's go-downs at Malacca. Naning obtained tin from Sri Menanti, Sungai Ujong and Rambau etc; the compulsory delivery to the Company at Malacca of all surplus Naning pepper at a fixed rate of $12 per bahara[4] and, finally, the Naningites were forbidden to trade with anyone else. All trading from the state had to be conducted through the Company.[5]

There were many Company officials who saw the treaty as completely binding and that it made the Naning territory an integral part of the Malacca territory. During the initial British occupation of Malacca, that is from 1795 to 1818, the controls were not rigidly adhered to and the Penghulu continued to rule the state as he had done during the Dutch rule of Malacca. There appears to be no evidence that, during this initial period, the Penghulu or any of his chiefs ever came to Malacca to pay annual homage.

In 1807 the British Resident, Major William Farquhar, Madras Engineers, ordered that the Penghulu should no longer pass the death sentence on offenders within his own state.

The fact that a treaty was concluded indicates that Naning should not be regarded as part of the Malacca Territory and Samuel Garling, the Resident Councillor of Malacca, wrote a

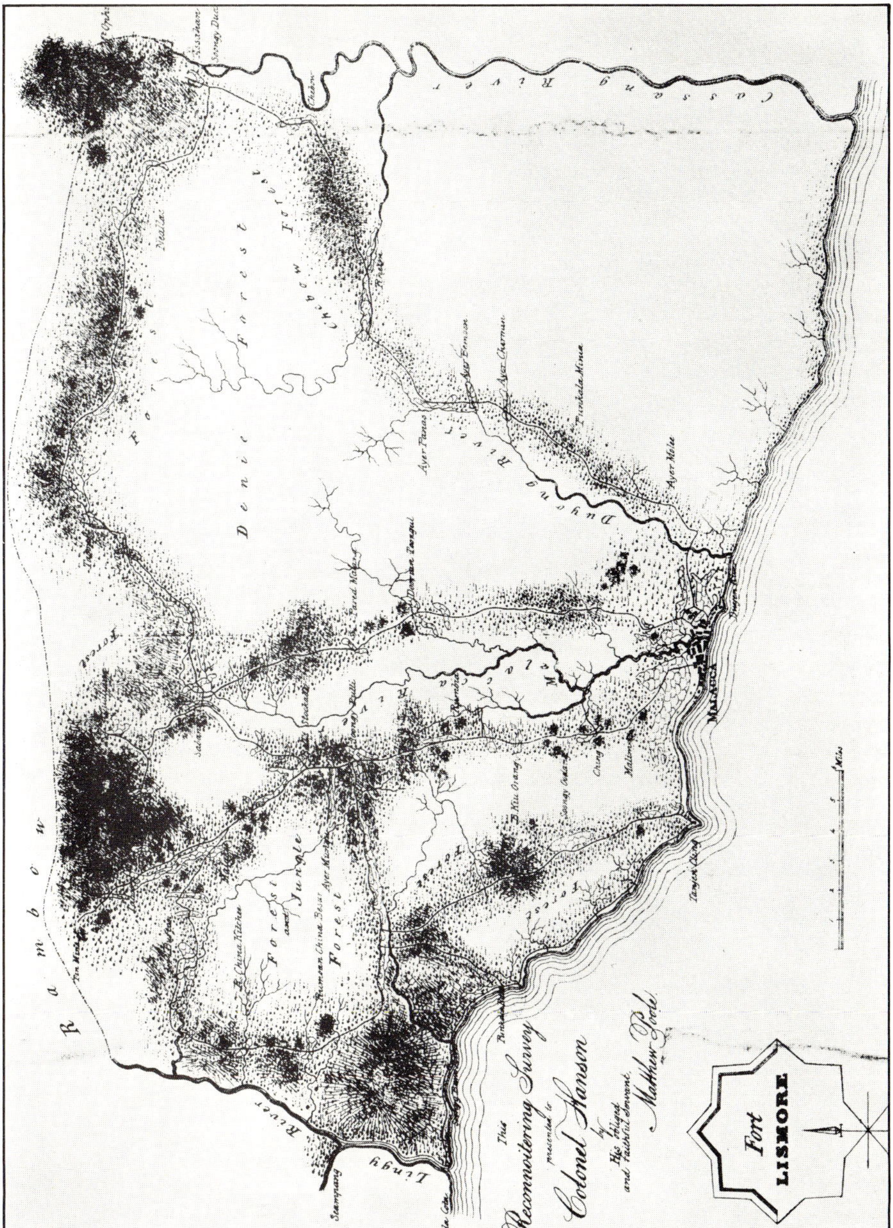

Map of Malacca and Naning. Reproduced from an original copy of the *Malaysian Peninsula* by P.J. Begbie. Courtesy of the National Library, Singapore.

minute in 1828 to the effect that he considered that Naning was an independent state and set out his reasons for this conclusion. Certainly the people of Naning considered themselves to be independent and that their ruler, or Penghulu, was their only lord.

The Superintendent of Lands at Malacca, William Thomas Lewis,[6] convinced the Governor at Penang, Robert Fullerton, that the '. . . Penghulu of Naning was no more than a glorified village headman and described the Company's relationship with him in terms of that between "a liege lord and vassal" . . .'[7] In 1828 Lewis visited Naning with instructions to levy the 'tenth' there in return for the allocation of small pensions to the Penghulu and the Sukus.[8] An offer was made to increase the annual tribute from 400 to 1,600 gantangs of paddy in return for their continued independence and jurisdiction over their own people. Lewis is reported to have avoided giving an answer and with the uncertainty of the political state of Naning the agents sent by Lewis to collect the 'tenth' met with opposition in Naning and had to return to Malacca empty-handed.

The matter was referred to the Penang Government but the problem was held in abeyance pending the visit of the Governor Robert Fullerton to Malacca.[9] The delay and hesitation convinced the Penghulu that there was a fair chance that the Company's resolve over the Naning problem was now wavering. Governor Fullerton arrived in Malacca and asked the Penghulu to meet him but Abdul Said refused to attend any meeting.[10] In July, Thomas Church the Deputy Resident at Malacca visited Naning to hold a census and reported that he had been received in '. . . a most friendly manner and with every token of respect . . .'[11] However the 'tenth' was, at that time, still unpaid and had Governor Fullerton sent in troops to collect the payment the whole matter might well have been resolved. He referred the matter to the Supreme Government in India instead, and they in turn referred the problem to the Directors in London.

The *Malacca Observer* of 30 June 1829 published a report saying that the Government intended to collect the 'tenth' from Naning. The situation was confused and the final match to set the whole affair on fire came when, in October 1830, Enche' Suring, a Malay property owner and farmer, who owned property adjacent to the Naning territory complained to the Resident Councillor that Abdul Said's men had trespassed on his lands and had removed fruit from his duku trees.[12] When challenged the Penghulu replied to the effect that the trees were within the border of the state of Naning. The Resident Councillor, instead of checking the facts of the case, commenced a tactic of trying to bully the Penghulu.

In the meantime the Directors in London considered that Naning was a part of the Malacca Territory but in order to avoid a conflict they considered it better not to impose this decision during Abdul Said's lifetime.[13]

The new Governor, Robert Ibbetson, considered that the Company should take a strong position otherwise the Penghulu's defiance might encourage the local population of the Malacca territory to also refuse to pay the 'tenth'. He also considered that he should teach the Penghulu a lesson and if he sent a token force of Company troops to Naning to frighten the Penghulu that such a force would be offered no resistance. The Supreme Government in India permitted the Governor to act as he thought advisable given local circumstances.[14]

This was the situation at the close of 1830. There have been a number of books and articles which discuss the rights and wrongs of the handling of the Naning problem but it does not fall within the scope of this book to add to that number and therefore this chapter deals with the actual military events in what later became known as the Naning War.

Fortunately one of the East India Company officers who was present during the two separate expeditions recorded in great detail his version of the events. He was Lieutenant

Peter James Begbie of the Madras Artillery. Peter James Begbie was born at Hendon, Middlesex on 30 August 1804, the son of Peter and Francis Begbie. He was educated at Edinburgh High School and was then a Cadet at Addiscombe during 1820 to 1822. He was accepted into the service of the Company and was commissioned as a Second Lieutenant into the Madras Artillery on 10 May 1822. He was promoted to Lieutenant on 11 May and arrived in India on 14 January 1823.[15] He served in the Burma campaign of 1824–6. On return from the campaign he married Miss Charlotte W. Morphett at St George's Church, Madras on 3 July 1826.[16] The first child, Frances Anne, was born in India during 1827 and the family moved to the Straits Settlements in March 1831.[17] Lieutenant Peter James Begbie moved to Malacca to take command of the Madras Artillery detachment in that garrison. The references to 'Lieutenant Begbie' require careful study as at this time his younger brother Lieutenant Alexander John Begbie was also serving in the Straits Settlements, he being stationed in the Singapore Garrison. Both were serving in the Madras Artillery and many of the orders and references merely refer to 'Lieutenant Begbie' and require careful analysis to ensure accurate interpretation.[18]

At the beginning of 1831, the forces in the Malacca garrison consisted of a company of the 25th Regiment, Madras Native Infantry whose headquarters were based in Singapore; a detachment of Madras Artillery under the command of Lieutenant Peter James Begbie, and detachments of the Madras Ordnance Department and the Madras Medical Establishment.

Begbie records in his account of the Nanning Campaign (the territory was spelt as either Nanning or Naning) that the expedition set forth from Malacca on 5 August and that the composition of the force that had been assembled at Malacca as '. . . being one Captain [this was Captain James Shaw Wyllie], one Lieutenant [Lieutenant John Milnes], one Ensign [Ensign Edward Holwell Short], one Assistant Surgeon [Edward Smith],[19] a proportion of Native Commissioned and Non-commissioned officers, and 150 rank and file of the 29th M.N.I.; accompanied by a brigade of six-pounders with their limbers and waggons, and one Lieutenant [Begbie], one Magazine Sergeant, and 24 Golondauze (inclusive of Non-Commissioned Officers); the whole commanded by Captain Wyllie, whilst W. T. Lewis Esq., Deputy Resident of Malacca, was appointed Commissioner with the force. There being no gun bullocks, buffaloes were, per force, substituted as cattle for the artillery . . .'[20] Begbie further commented that it was considered that the force was disproportionately large for the job that it was required to do. Some of the garrison at Malacca considered that a Havildar's Guard would be sufficient to bring the Penghulu under control and that '. . . a red jacket amongst the bushes . . .' would be sufficient. The whole expedition was looked upon as nothing more than a training exercise and a picnic. Lieutenant John Milnes, of the 29th Regiment, Madras Native Infantry, was placed in charge of the advance party and was formally appointed as Resident of Naning even before it had been conquered. The plans included setting up a strong point at Taboo and a Depot was to be formed at the Government bungalow at Soongei Pattye [*sic*].[21] The bungalow was some thirteen miles from Malacca town and had the advantage that supplies could be sent by means of the Malacca river which ran within half-a-mile of the bungalow.

The expedition set out on the road towards Malim, which was, according to Begbie, five miles from the 'bridge at Malacca' and then proceeded through a swampy area until passing through a pass between two hills, known as the Ching Hills and then on into the jungle. The progress of the force was being watched by one of the Naning chiefs who allowed the expedition to pass and was then able to give details of the strength of the force to the

Penghulu of Naning. The party arrived at the bungalow at Soongei Pattye and Captain James Shaw Wyllie[22] posted guards and the force remained encamped there until the following day. The border with Naning was due to be crossed on the next day and Mr Lewis sent a messenger with a flag of truce to the Penghulu with a request, in letter form, asking him to abide by the terms of the treaty. The messenger was stopped at the border '. . . at Kalama by the Panglima Dattoo who was, entrusted with the defence of the Nanning [*sic*] boundary, and threatened with death, if he attempted to proceed . . .'[23] After some discussion he was allowed to proceed but the mission failed to produce an agreement. The force continued on the following day and at the border met a party of Naningites lead by the Panglimah Dattoo. The numbers were never accurately known but on seeing the red coats of the Company's force two of the Naning sentries opened fire. The order was given for the force to halt and one of the guns was moved up to the fore and six rounds of grape-shot were fired which dispersed the Naning defenders who merely melted into the jungle only to regroup out of sight of the Company's force. Advancing once again the force arrived at a paddy field and as the track across it required repair the advance was again halted whilst these repairs were carried out. The Naningites appeared on the flanks at the edge of the jungle and once again the few rounds of grapeshot were fired. The troops then advanced through dense brushwood along a narrow path with the guns making their own road as they progressed. Begbie records that '. . . the low jungle was at length succeeded by trees and finally by the plain of Alu Gaja.[24] The distance between Sungei Pattye and Alu Gaja is just two miles . . .'

There was now some concern over rations as these had not yet arrived at the base camp but rice was obtained from a Chinese farmer living at Alu Gajah. The following day the force set out through the jungle and then made camp at 2.00 pm having taken seven and three-quarter hours to advance one and three-quarter miles through the jungle. The force had, during its advance fired the house of one of the Naning chiefs – Dattoo Embangin – and as the camp was set up the Naningites commenced firing into the camp at infrequent intervals. The next day the advance continued and brought the force to the foot of Bukit Seboosa where the road to Taboo diverged, one path leading over the hill and the other around the foot. Both routes were blocked and the officers at the head of the force saw signs of the enemy hiding in the jungle. Captain Wyllie ordered the coolies, who had been brought with the military force, to the front and ordered the felled trees to be cleared in order that the advance could continue. The Naning force had contructed a small stockade beyond the felled trees and as the coolies started to clear the obstruction the Penghulu's men opened fire and the first casualties of the conflict were inflicted upon the force in that one sepoy and one coolie were wounded. The advance in this direction was abandoned and the second track was cleared. The force continued until reaching the point where the two tracks rejoined and once again trees had been felled to cause an obstruction. These were also cleared with the odd sniping shot being fired from the jungle. A party under Lieutenant Milne, detached from the main force, visited the house of another Chief (this was Dattoo Malalu) but on finding it deserted Milne fired the house and rejoined the main force. Begbie comments on the action as follows:

> '. . . The unsparing manner in which house after house had been subjected to the flames was justifiable only by strong military necessity. Every one, which had been thus sacrificed, commanded either the road, or the encamping ground, and was thus injurious to the safety of the detachment . . .'[25]

In the meantime rations were beginning to become short and a request for supplies from Malacca brought about a message to the effect that none could be sent due to the lack of coolies. The townsfolk of Malacca, on hearing that the expedition had met constant opposition, had panicked so that '. . . not a cooly, for the transport of the rice, could be procured on any terms, and [it] was recommended to forage on the country . . .'[26] The force continued to be harassed with sniping from an unseen enemy and by the end of the day the provisions with the force had been completely expended and the sniping was increasing in ferocity. The camp was guarded and later the guards heard from the rear, at about 6.00 pm, firing which continued for about two hours. It was thought that the firing was probably the party bringing up the rations being attacked by the Naningites. Captain Wyllie despatched a party, under the command of a Havildar, to the rear expecting it to relieve the ration party. The shooting continued to be heard and at the time that the Havildar's party should have reached the ration party the noise of the shooting increased. At 2.00 am the Havildar's party returned as it had been unable to locate the ration party. At 5.00 am another Havildar's party, with seventy coolies, was despatched back to Soongei Pattye in the hopes of rescuing at least some of the supplies. At about 9.00 am the Havildar and the remnants of his party rejoined the force. He had lost one sepoy, had one cooly wounded and of the seventy coolies only seventeen returned.[26]

The coolies had during the attack deserted into the jungle or had been captured by the Penghulu's men. Whilst these attacks were proceeding the Penghulu of Naning had received unexpected help from the Ruler of Rumbau (also spelt Rumbow). The combined force now was more than a match for the small force sent out by the Company from Malacca. Not only were the numbers now in the favour of the Penghulu but also all his men, and the men from Rumbau, knew the jungle and were at home in its environment.

It was now clear that the only course of action open to the Malacca force was to retire to the bungalow at Soongei Pattye. The order was given and the baggage and less portable items such as tents, etc were destroyed and the column commenced to retire. With the loss of the baggage and stores the men were able to keep together as a body and more coolies were lost as there was the constant need to clear the path of trees that had been felled behind the force as it had advanced. Apart from sniping, the force met with little opposition until it reached Kalama, here it encountered a concealed stockade which opened fire on the column as it approached. The guns were once again brought to the fore and a few rounds of grapeshot were fired but to little or no effect. Begbie records the next move in the skirmish as follows:

> '. . . The advance guard, under Ensign Short, was accordingly ordered to detour to the rear of the enemy's position, whilst Captain Wyllie, accompanied by Mr Assistant Surgeon Smith, led another party round to the other flank. Ensign Short came suddenly on the rear of the stockade; poured a volley into it, and emptied it of its defenders . . .'[27]

The retreat to Soongei Pattye continued and the force reached there at 7.00 pm, bringing with them seventeen men who had been wounded and who had to be carried. It was then learned that the firing heard on the previous day had been the Havildar at Soongei Pattye and his party defending the camp against a strong force of the enemy. The attack had lasted about ten hours and the defenders had lost a third of the party to wounds. The Havildar, Peer Homed, was created a Subadar for this gallant defence but unfortunately before the news reached him he died.[28] The Company's force set up a stockade around the bungalow and cleared an area of jungle to give a clear field of fire. The Penghulu's men also erected stockades in the area and commenced to snipe at the troops within the Company's stockade.

The following day a small party of the 29th Regiment, Madras Native Infantry arrived at the camp from Malacca. It had encountered felled trees along the paths and had a grenadier shot dead and a sepoy and three Chinese coolies wounded. The remainder of the coolies abandoned their loads and the sepoys had to continue to Soongei Pattye without the supplies which, of course, fell into the hands of the enemy.

The defences at the bungalow were again strengthened but concern was felt over the vulnerability to fire as the roof of the bungalow was of atap.

In the meantime the panic that had occurred in Malacca Town when the force met its first reverses continued. Within the town some of the civilians had years before registered for service with a Militia which was to be called out in an emergency but its terms of service were such that it was merely confined to the defence of the town. The Militia was called out and such was the panic by the civil authorities that Captain Wyllie was requested to return to Malacca to defend the town against the combined force of the Naning and Rumbau states. Captain Wyllie however, considered that the best plan was for him to remain at Soongei Pattye and stop any advance at that point. Two days later Captain Wyllie and Mr Lewis, who had no function to fulfill now that an advance into the Naning territory was impossible, returned to Malacca mainly to placate the civil inhabitants. Captain Wyllie despatched some reinforcements to the advance base and exchanged ammunition parties. All movement was severely hampered by sniper fire from the jungle and Lieutenant Milne, who had been left in charge at Soongei Pattye was unable to do more than hold the stockade against numerous attacks. The force at this advance camp had now been reduced to one hundred rank and file, inclusive of wounded.[29]

Five days after the departure of Captain Wyllie to Malacca the defenders of the bungalow had been reduced to one barrel of spare ammunition. However at midday the sound of firing was heard from the direction of Roombiyah.[30] It was hoped that this relieving force would be the light company of the 29th Regiment Madras Native Infantry who were due to arrive from Singapore. At about 1.00 pm on 20 August Lieutenant Milnes despatched a party under the command of a Havildar to attack the enemy stockades from the rear and so help the reinforcement party to break through to the Soongei Pattye stockade. Shortly after, the defenders heard a light infantry call and the '. . . head of the column emerged into the paddy field on which the rear of the stockade rested . . . the company filed into the stockade, bearing the body of Lieutenant White who was numbered amongst the casualties, being mortally wounded through the lungs. The remaining officers, consisted of Captain Hibgame, Lieutenant Brodie, and Ensign Fothergill . . .'[31]

This party had in fact arrived from Singapore having departed from there on 15 August on the British brigantine *Leegoan* carrying Lieutenant James White, Ensign John William Fothergill and 120 Sepoys of the 29th Regiment, Madras Native Infantry.[32] The party had departed from Malacca not knowing whether the defenders of the stockade at Soongei Pattye were still alive or not. Lieutenant James White died at 2.45 pm during the afternoon of 20 August.

On 21 August Captain Hibgame, Lieutenant Brodie and seventy privates and nine of the most severely-wounded sepoys returned to Malacca without further incident. Ensign Fothergill remained at the advance camp. The 22nd, 23rd and 24th August passed with frequent attack and counterattack but the force from the stockade were at a disadvantage as when they came out from the stockade with fixed bayonets the enemy withdrew into the jungle leaving the sepoys no alternative but to return to the stockade. Negotiations were made with Syed Sabban, who was in charge of the force that was attacking Soongei Pattye and he received a promise from the Government that if he allowed the troops to retire to Malacca without further harassment he would be paid a sum of five hundred dollars.

At one o'clock in the morning of 24 August, Lieutenant Richard Hurlock of the 29th Regiment MNI arrived at the camp with a reinforcement party of forty men. He also carried orders that the whole force was to withdraw to Malacca town. It was directed that, if the guns could not be evacuated, they were to be destroyed. The limber, wagons and ordnance gunpowder were destroyed and the round shot buried. The guns were kept mounted until the last moment and Begbie then recounts the measures taken to destroy the guns as follows:

> '... a few spokes being knocked out of each wheel, and the cap-square keys &c, carried away, the carriages and wheels dragged on to the top of the heap of broken limbers, &c, and portfires strown profusely through the pile. The bags of rice were prepared for destruction in a similar manner ...'[33]

Almost all the coolies were employed in the carriage of the wounded and the officers and men each carried as many cannister grape as the weight of their arms and accoutrements would allow. The guns were lashed to platforms and conveyed by the remaining coollies. At 8.00 pm on 24 August the troops filed out of the stockade under the guidance of a bright moonlit night. Ensign Short and Assistant Surgeon Smith fired the piles of equipment and provisions and as the enemy rushed into the stockade made their escape and joined the rear-guard of the retreating Company's force. The enemy opened fire on the rearguard who returned the fire.

The retreat to Malacca was delayed due to the deterioration in the weather. The party managed to surmount the numerous trees that had been felled across the path. Finally after scrambling up a hill a large tree was found to be blocking the road. Begbie records that later the tree was measured (this was during the second expedition) and it was 'found to be thirteen feet in circumference. A few yards further on lay another, if I mistake not, was 21 feet in girth ...'[34] After much effort on behalf of all available men it was found to be impossible to carry the guns over the obstacle and finally Lieutenant Begbie spiked the guns and the party then moved on toward Malacca. The troops although unhindered by the guns made slow progress due to the weather and finally reached Malim at 4.00 am having covered the eight miles in eight hours. After a brief rest the remainder of the journey to Malacca town was made by 6.00 am.[35]

The seriousness of the situation is indicated by the fact that Captain H. Moore, Lieutenant Lewin and thirty-eight Royal Marines, who were in transit at Singapore[36] were diverted to Malacca on 17 August, departing from Singapore on the British barque *Freak*[37] and arriving at Malacca '... a few days previous to the return of the detachment [from Soongei Pattye] ...'[38]

The following month was devoted to political activity rather than military activity. The *Singapore Chronicle* on 15 September 1831 carried the obituary of Lieutenant James White:

> '*Death*
>
> At Rambeiya [also spelt Roombiyah], about 10 miles from Malacca whilst gallantly attacking the Stockade at that place, on the 20th August 1831, Lieutenant JAMES WHITE, of the 29th Regiment Madras Native Infantry, most sincerely and deservedly regretted by the Officers of his regiment, and all who had the pleasure of his acquaintance ...'[39]

The Naning Memorial at Fort Road Cemetery, Malacca.
Photograph by Mrs J. Harfield, 1977. The inscription on the obelisk reads:

Sacred
to the memory of
LIEUTENANTS WHITE and HARDING
of the 29th Regiment Madras N.I.
who fell in action with the Malays,
the former
on the 20th of August, 1831
and the latter
on the 29th of March, 1832.
This tomb was erected
as a mark of respect and esteem,
by their brother officers.

James White was born in July 1806, the son of Captain James White, Paymaster of the 89th Regiment and Charlotte, his wife. He was appointed as Ensign on 14 May 1824 and joined the 29th Regiment, Madras Native Infantry as a Lieutenant on 13 August 1827. He married Catherine, daughter of the late Captain George Lane, HM 69th Regiment, at Vepery, on 28 June 1830, and in the following year moved with his regiment to Singapore.[40]

Shortly after the return of the troops to Malacca Lieutenant Peter James Begbie wrote an account of the action in pamphlet form. This was reviewed in the *Singapore Chronicle Supplement* dated Friday 11 November 1831 which read:

> '. . . The pamphlet of which we have given the title [The Nanning Expedition[41]] and what may be called the introductory part, was written . . . with the benevolent object of applying the profits, if any, to the benefit of the Widow of the late Lieut WHITE who lost his life during the above unfortunate Expedition. The interest excited by the forlorn situation of this young lady, and the circumstances under which her gallant husband met with his untimely fate, have contributed materially to render the sale of the pamphlet, (here especially) more extensive than it would have been otherwise. But while we render due praise to the writer for his benevolent intentions, we think sufficient time has been allowed to elapse, to render a critique on the matter unlikely at this period, to affect the object of the pamphlet, in a pecuniary view . . .'[42]

The newspaper then proceeded to criticise Begbie's writing and to generally disagree with his comments and conclusions and even to challenge his account of the action of which he was a participant. He was probably being used as a scapegoat as he was the officer who had abandoned the guns and had spiked them.

By the beginning of 1832, it was clear that the question of the suppression of the Penghulu of Naning could only be resolved by force and in consequence the Government in India despatched reinforcements to the Malacca garrison. An extract from the *Madras Gazette* dated 20 December 1831 was published in the *Singapore Chronicle* of 26 January 1832. The report read:

> '. . . The Army. The Troops proceeding to Malacca embarked early on Saturday morning last, on board HM Ship COMET and the ship LADY McNAUGHTEN, and have left the road [Madras] since. The remaining companies of the 5th Regiment Native Infantry are at Palaveram, under command of Major P. Farquharson, but will embark as soon as shipping can be provided for them. Captain W. J. Bradford of the 35th Regiment Native Infantry, accompanies the force proceeding to Malacca. Captain J. S. Wyllie, of the 29th Regiment Native Infantry, has been nominated Brigade Major to the Troops serving at Malacca, as a temporary arrangement.[43] The 5th Regiment Native Infantry, the Artillery and Sappers and Miners ordered for Malacca embarked on board last Saturday on the Lady McNaughten, but it was soon discovered there was not sufficient room for the whole regiment, and one Company were obliged to be sent ashore again. The Sappers and Miners are on board the HM ship Comet and their tools on board the Lady McNaughten . . .'[44]

On 18 January 1832 the Honorable the Governor Robert Ibbetson, T. W. Lewis and J. B. Westerhout, accompanied by Captain Hibgame, commanding an escort of fifteen sepoys embarked on board the HC schooner *Zephyr*, with an additional escort of 38 sepoys on board a Chinese tope, (a small vessel with latteen rigged sails) named *Pakeen*, departed from Malacca to journey to Simpang via the Linggy River to meet with the Chiefs of Rumbau in order to negotiate a separate peace. Begbie, who had managed to obtain a passage on the Chinese vessel as a passenger recorded the events as follows:

'. . . In the evening [of 19 January], the whole party of gentlemen landed at Simpang, for the purpose of viewing the spot which the convicts were employed in clearing the jungle in order to have space wherein to pitch a tent . . .'[45]

The meeting was planned for the following day at 10.00 am but the Rumbow (Randau) Chiefs had not arrived. The tent had been pitched, forty sepoys had disembarked to form a guard of honour, but also to act as a defence detachment should it be needed. The *Zephyr* was lying an anchor with springs on her cable so that she may act should it be necessary and bring her guns into action. The party waited and it was not until 2.00 pm that the Chiefs arrived and even then they were not at all at ease seeing the detachment of sepoys. The sepoys were ordered to 'present arms' as a token of respect which alarmed the Rambau chiefs, but following an explanation their hands left their krisses and they proceeded into the Governor's tent. A peace was negotiated after a long discussion and presents were exchanged. A Royal Salute was fired in honour of the Penghulu of Rambau and the two parties departed to their own territories.

By the end of January, the force designed to reduce Naning had arrived. The force now consisted of the 5th Regiment, Madras Native Infantry; five companies of the 29th Regiment, Madras Native Infantry; two Companies of Sappers and Miners, and a '. . . proportion of European and Native Artillery, with a good park, and cattle for the Ordnance and Commissariat Departments . . .'[46] The staff were allocated as follows:

Commanding – Lieutenant Colonel C. Herbert
Brigade Major – Captain J. S. Wyllie, 29th Regt MNI
Superintending Engineer – Lieutenant J. H. Bell
Commanding Artillery – Captain F. Bond
Commissariat Officer – Lieutenant J. Milnes
Commanding 5th Regt MNI – Major P. Farquharson[47]

During the period between the end of the first expedition and the assembling of the second expedition the authorities in Malacca had employed coolies in cutting back the jungle on either side of the road between Malacca and Roombiyah to a width of about eighty yards. The party had been protected by a locally-enlisted force of soldiers which were given the title of Malay Contingent. The contingent was dressed in the sepoy uniform and was armed with muskets. Timber had been cut at Roombiyah and a stockade contructed as it was intended that this should form a base for stores and supplies.[48] The progress was reported in the *Singapore Chronicle* of 16 February as follows:

'. . . By the latest account from Malacca we learn that two companies of Sappers and Miners and one of Light Infantry had marched to Rumbier [*sic*] (nine miles from Malacca on the way to Nanning) where a Government bungalow or depot is erecting. The above detachment is accompanied by 150 Malays who for the consideration of 5 dollars each per month, besides having their provisions, firearms and ammunition found them, have volunteered to 'beat the bush' for the Panghuloo [*sic*] of Nanning, for whose apprehension a reward of one thousand Dollars is offered. . .'[49]

On 7 February 1832, the Light Company of the 5th Regiment, Madras Native Infantry under command of Captain William Justice, with Ensign George Wilford Walker, together with the Light Company of the 29th Regiment, Madras Native Infantry under command of Ensign Short, and a Company of Sappers and Miners under command of Lieutenant J. H. Bell, together with one hundred men of the Malay Continent set out for the stockade at Roombiyah. They arrived there without incident. The strength given by Begbie differs from that given in the report published in the *Singapore Chronicle*. It can be reasonably assumed that, as Begbie was on the spot and probably keeping notes, his figure is the more accurate.

On 9 February, the Grenadier Company of the 5th Regiment under command of Captain Charles Poulton moved out from Malacca and took up position on the road at Ching so as to secure a route should the force at Roombiyah have to retire to Malacca. The usual sniper fire broke out and kept the guards on stand-to and also caused some soldiers of the Malay Contingent to waste ammunition by firing into the jungle without any target at which to aim.

On 12 February, the Penghulu of Naning wrote to J. B. Westerhout asking for a meeting at Soongei Pattye and 'offering to give up the brigade of six-pounders, which had fallen into his hands at the termination of the first expedition . . .' and he also offered to vacate the Penghulu-ship in favour of his son or nephew if no further action were taken against him. The reply was to the effect that he should surrender himself and the guns to the Company's forces at Malacca. Needless to say he did not do so and it was in all probability a move on his part to gain more time to prepare for the advance of the new and enlarged force.

Further reinforcements arrived at Malacca on board the barque *Lady Munro*, these being the remaining company of the 5th Regiment, Madras Native Infantry, a Company of Sappers and Miners, and twenty gun-bullocks, out of thirty that had embarked. The following day the fifth Company of the Regiment under command of Lieutenant Matthew Poole and Ensign William Hertford moved out to Ching and relieved Captain Poulton who, as the senior Captain of the regiment,[50] moved forward to take command at Roombiyah.

On 21 February the Rifle Company under command of Captain John H. Winbolt and Ensign Joseph Wright moved up to the forward camp. Daily scouting parties were sent out under various officers whilst other detachments acted as guards to the Sappers and Miners who were working in the area. On the following day Captain Poulton advanced to Soongei Pattye and met with no opposition and on his return he fired the numerous stockades that the enemy had constructed alongside the track from Roombiyah to Soongei Pattye.

A further report was published in the *Singapore Chronicle* of Thursday 23 February 1832 which read:

> '. . . By recent accounts from Malacca we learn that the Lady Munro had lately arrived there with 20 bullocks sent from Madras to be employed in dragging the field pieces and mortars which are to be used in the Expedition against Nanning. The Lady Munro having at a previous period, transported a detachment of troops, from Madras sailed in return to convey the bullocks. We understand that some of those originally died on passage. As the troops only waited the arrival of the bullocks, previous to proceeding into the interior, we presume they have now entered on the campaign . . .'[51]

The artillery was moved to Roombiyah on 25 February under command of Lieutenant Peter James Begbie, this was a brigade of 6-pounders with the detachment of golondauze to serve the guns, the detachment was accompanied by a company of Sappers and Miners under command of Second Lieutenant Smythe. The party reached Roombiyah during the early afternoon with, as Begbie reports, '. . . the cattle being much distressed . . .' On 2 March, Lieutenant Colonel Charles Herbert moved to Roombiyah from Malacca with his staff, leaving Captain Edward Thurlow Hibgame, of the 29th Regiment MNI, in command at Malacca. The Sappers and Miners, guarded by the Malay Contingent continued to advance cutting trees on either side of the path (referred to as a road by Begbie). The task was undertaken by a seconded body of Chinese and Malay coolies and according to

Begbie's report the jungle was cleared to a distance of eighty yards on each side of the road. There were a number of casualties due to falling trees and on 6 March it was reported that three Sappers and Miners were severely injured by falling trees and had to be evacuated and took no further part in the campaign.

Ensign Short had to be relieved by Ensign Boulderson, also of the 29th Regiment MNI, when Short had to proceed to sea on a sick certificate. He in fact arrived back at Singapore on the HC schooner *Zephyr* on 25 March.[52] Lieutenant Colonel Herbert and his staff, accompanied by a covering party of grenadiers of the 5th Regiment MNI, advanced to Soongei Pattye found it still deserted and then returned to their base camp. This caused the Naningites to start to rebuild the stockades that had been destroyed during the previous reconnaissance and also to carry out attacks on the camp, by firing from the jungle and then retreating when troops moved out of the camp. An attack on the Naning stockades was carried out on 17 March which was completely successful and with only minor casualties to the Company's forces and when the stockades had been taken they were once again destroyed by fire. During the next three days various skirmishes were carried out and gradually the force advanced towards Soongei Pattye. Begbie reports on 22 March that:

> '. . . The ground at Soongei Pattye having been cleared sufficiently for an encampment, the flank companies of the 5th and 29th with the rifle company, and the two companies of sappers and miners, marched for that place at 10 a.m. having delayed until that hour by heavy rain. One of the 29th severely wounded by a ranjow, which, entering at the ball of the toe, re-appeared at the surface of the foot . . .'[53]

Colonel Herbert moved his headquarters to Soongei Pattye on the Sunday, 25 March, and in the meantime Captain Poulton and his men attacked and over-ran five Naning stockades at Kalama with the loss of two sepoys wounded by musket fire, one seriously and the man subsequently died. Two others were injured by the mantraps which the Naning forces had planted before withdrawing. The force under Colonel Herbert continued to press forward slowly and were gradually gaining ground and destroying the numerous stockades that the Naning defenders had erected so as to delay the advance of the expedition. Information was received on 29 March that a very strong stockade had been constructed at Ayer Mangis[54] which was located to the left of Soongei Pattye, and a party of the Malay Contingent was sent out under command of locally-commissioned officer Soon Kien, to reconnoitre the place. The party ran into trouble and the Light Company of the 5th Regiment MNI under command of Captain William Justice and Lieutenant Matthew Poole, together with the Grenadiers of the 29th Regiment MNI, under command of Lieutenant Edmund Vincent Harding set out to give assistance.

Captain Justice took command of the attack and ordered Lieutenant Poole to attack the right flank and Lieutenant Harding to attack the left. Captain Justice moved his force up the paddy field in a frontal attack. Fire from the stockade was brisk and the stockade being constructed of upright poles was unassailable. Justice then had to hold his men in a position in front of the stockade to await the flanking attacks. Firing from the left announced that Harding had reached his objective and with that signal Captain Justice ordered his men to advance. Begbie records what happened next:

> '. . . on wheeling round the face of the stockade, Lieutenant Harding was found lying on the ground mortally wounded. It appeared that his flank movement had taken the enemy completely by surprise, who fled, in the greatest confusion, after firing an ill-concerted volley, leaving two killed and one wounded, the latter of whom managed with great difficulty to effect his escape. The very last man that issued from the stockade

unexpectedly confronted Lieutenant Harding, there being a space of about twelve yards between them. This officer put his hand to his belt, but, unfortunately, the pistol was inextricably entangled, and, before he could draw it, the man fired from his hip. The ball passed through the throat, and injured the spine, down which it appears to have passed . . .'[55]

Other casualties included Subadar Shaik Byram, of the Grenadier Company who had his left arm shattered and a number of other sepoys were wounded, some seriously although no details of injuries appear to have been recorded. The stockade having been taken the main concern was then for the evacuation of the wounded. Lieutenant Harding whose nervous system was affected by his wound was lying on the ground 'suffering but little pain' however any attempt to move him caused him great agony and when he was eventually moved to camp it was found that the ball was lodged against his spine. The Surgeon was unable to remove it and Lieutenant Harding lingered until 9.45 pm on the following day. After he had died his body was sent back to Malacca to his widow and was subsequently buried in Fort Cemetery at Malacca. He was a well respected and friendly officer and his loss came as a sad blow not only to his wife but to his brother officers in his regiment.

Heavy sniping of the force continued and during one encounter Ensign Wright shot one of the enemy dead and the remainder fled. A large party of 100 sepoys under command of Captain Burgess and Ensign Hertford advanced from Roombiyah to a kampong called Loondoo but on arrival the kampong had been deserted and before moving on the sappers who had accompanied the force levelled the defences. Burgess and his party moved on to Ayer Itam,[56] but on arrival found that it had already been visited by a party under command of Captain Poulton. On 3 April the European artillery and eight of the Golondauze, with Captain F. Bond and Lieutenant P. J. Begbie moved forward to Soongie Pattye and brought with them a 12-pounder howitzer, one $4\frac{2}{5}$-inch howitzer, one 8-inch and one $5\frac{1}{2}$-inch mortar, and a brigade of 6-pounders. A Jemidar and twelve Golondauze remained at Roombiyah with a brigade of 6-pounders.[57]

With the arrival of the artillery at Soongei Pattye it was decided that a force would proceed to the house of Dattoo Membangin; this was a new house built on the site of the previous residence which had been destroyed during the first expedition. The force reached the objective at about 7.00 am and set up the artillery on the right flank. The remainder of the force arrived on 5 April and the house of the Dattoo was appropriated for use as a General Hospital. At this time heavy rain was falling and at Roombiyah it was found that there was three and half feet of water in the magazine and 10,500 rounds of musket ammunition and the whole of the rifle powder had been destroyed by the floods. It was a serious setback to the campaign that was, at last, beginning to show some progress.

The Sappers and Miners continued to cut the jungle back from the path and progress was now being made towards the enemy who were at Taboo. Covering parties were supplied daily on a roster basis, ie,

9th April –	Cover party –	C Company, 5th Regt MNI Ensign Wright
10th April –		F Company, 5th Regt MNI Ensign Thomson
11th April –		Light Company, 5th Regt MNI

On 12 April the covering party, which on this occasion was supplied by the Rifle Company, 5th Regt MNI came under fire from a Malay stockade on their right flank. With the first volley from the stockade the 5th Regiment suffered six casualties, one sepoy being killed and five wounded. Captain Winbolt ordered Ensign Wright to lead his section across the paddy field whilst he attacked the right flank of the stockade. Begbie records the next event as follows:

'. . . Ensign Wright sprang upon his feet calling upon his men to follow him, and dashed into the paddy field, without observing that his call had not been obeyed by any of the company except his orderly boy, the men being apparently panic struck by their losses. This gallant young officer had hardly reached the centre of the paddy field, exposed to the fire of the stockade, which was all concentrated on him, when a ball broke his right thigh, and brought him to the ground. The Malays rushed out, with the intention of cutting off his head, but were driven back by the orderly boy, Emaum Ally who, kneeling behind his master, fired seven or eights shots over the body. Whilst lying in this helpless state, Mr Wright received a second ball in the left shoulder . . .'[58]

The company rallied and the Penghulu's men were driven off, the casualties rescued and taken back to the camp.

On the same day Lieutenant Peter James Begbie left Malacca with 18-pounder carronades and at the same time another party of coolies with supplies set out from Malacca to the camp at Roombiyah and as a protection was given an escort of thirty-seven rank and file. Over the next few days escort duties continued with the duty company now supported by an artillery officer, a howitzer and a gun crew. The company changed daily but on occasions the same officer was despatched on escort duty, for example on 13 April Ensign Boulderson commanded the Grenadier Company, 29th Regt MNI and on the following day he had command of the Light Company to carry out the same duty.[59]

The routine continued with various skirmishes taking place, however on 17 April a party under command of Ensign Thomson, ran into an ambush. He was stunned by a musketball and on recovering his senses he found that his party had either been killed, wounded or had fled from the ambush. He, and his servant who had remained with him, dashed for the artillery position to his rear and reached this without further incident. Of the fifty-four sepoys in the party twenty-seven had been killed or wounded.[60] The combined infantry and artillery party, under command of Lieutenant Peter James Begbie, then retreated as far as Wright's stockade and set up a defensive position, having despatched the wounded to the rear and asked for reinforcements. At about 11.00 am Captain Justice and Lieutenant Poole of the 5th Regiment MNI arrived with two companies and a supply of ammunition for the artillery.

On 19 April the covering party was supplied from the flank companies of the 29th Regiment MNI, together with Lieutenant Lawford of the artillery and a mortar and a howitzer. On the following day Begbie records that as sickness and injuries had amounted to approximately three hundred the campaign was brought to a halt. Convoys moving out of Malacca were constantly attacked and sniping from the shelter of the jungle continued as before. The Naningites fired, at their camp at Taboo, one of the 6-pounder guns that they had captured during the previous year. Begbie comments that they fired this gun at about 8.20 pm '. . . in mockery of the British evening gun . . .'[61]

Ensign Wright and some of the more severely-wounded sepoys were evacuated back to Malacca. Ensign Joseph Wright recovered from his wounds but later, as Captain Wright, was transferred to the Madras Invalid Establishment on 15 February 1839.[62]

The remainder of April was spent with Sappers continuing to work on the widening of the roadway and clearing the jungle on either side and with the infantry and artillery acting as escorts. It was expected that reinforcements would arrive from Madras and Pinang and on their arrival the offensive would then commence which would bring to a close this costly and protracted campaign. On 3 May, just after 6.00 am Captain Justice was in the process of relieving the piquets on the right flank of the forward camp when the enemy attacked. It was perhaps fortunate that they attacked when they did as, although they were able to drive the piquets back, they did encounter twice the number of soldiers than they would normally have met at that particular point. The camp was roused by bugle calls and then stood-to. The enemy had also erected a stockade on the nearby hill known as Bukit Lanjoot, and from this position commenced firing into the camp. Captain Justice ordered his men to lie down and the artillery opened fire over their heads. Whilst this attack on the stockade was mounted from the camp another party, under command of Captain Poulton and with Ensigns Stoddart and Walker, moved along a path and made their way round the foot of Bukit Lanjoot. On reaching a junction in the track, Captain Poulton and Ensign Stoddart proceeded along one track with some of the sepoys whilst Ensign Walker moved along the second track. The account of the next sequence of events is given in Begbie's book:

> '. . . the enemy was fully prepared, at this stage of the war, for flanking movements, and a stockade had been erected on the path in anticipation. As the gallant boy came unexpectedly on this work, a ball took effect in his heart, and he fell dead to the ground . . . The Sepoy, next to his officer, and two others, were wounded by the volley . . .'[63]

The enemy, shortly after this, evacuated Bukit Lanjoot and the dead and wounded were recovered and it was recorded that

> '. . . the body of Ensign Walker was brought into camp and buried that evening. A handsome monument, designed by Lieutenant Smythe, Engineers, has been erected to the memory, and over the remains, of Mr Walker by his brother officers of the 5th . . .'[64]

An obituary was published in the *Singapore Chronicle* later in May 1832 which read:

> '. . . Killed in Action near Mambarngoin on the 3rd May 1832, Ensign George Wilford Walker, doing duty with the 5th Regt MNI. He was a most gallant young Officer, and died, beloved and respected by every Officer and Soldier in the Force . . .'[65]

The Company's force also had problems within its own ranks as in the evening of 5 May fighting broke out between the Malay Contingent and the convicts who were being employed as a labouring force. The riot was quelled by the officers and men of the infantry but not before there had been injuries on both sides. The expected reinforcements arrived at the advance base on Sunday 13 May when a company of the 46th Regiment Madras Native Infantry arrived under command of Captain John Wallace, of that regiment, together with Captain Charles Sinnock and Lieutenant John M'D. Minto both of the 5th Regiment MNI; Lieutenant Richard Henry Symes, 29th Regiment MNI and Lieutenant Charles Frederick Liardet of the 14th Regiment who was 'doing duty' with the 5th Regiment MNI. With the strengthing of the force and the construction of a stockade at Dattoo Membangin the position was renamed 'Bell's Stockade' as a compliment to Lieutenant Jasper Higginson Bell, Madras Engineers who designed and built the stockade.

The loss of active men due to sickness continued and on 17 May Begbie recorded that one half of his European artillerymen were in hospital due to sickness.[66] On 21 May an attack was mounted on the stockades that the Naningites had erected at Bukit Seboosa. The Malay Contingent, a Company of the 5th Regiment Native Infantry and a party of Sappers carried out the attack and the subsequent destruction of the eight stockades. The stockades were fired and the party then moved on to destroy a further seven stockades that had been built during the lull in the fighting. It was recorded that at Bukit Seboosa the troops found '. . . the skeletons, and severed skulls of the five men who had been left on the ground on the 17th . . .'[67]

The force was again strengthened on the 22nd by the arrival of a further company of the 46th Regiment Madras Native Infantry under command of Lieutenant Edward Boyd

Above: Tomb of Ensign George Holford Walker at Alor Gajah.

Right: Inscription plaque belonging to Ensign Walker's tomb. The plaque no longer exists.

Courtesy of The Malacca State Museum and Tan Sri Dato Mubin Sheppard, CMG, MBE.

Stevenson. As a result of this it was decided by Lieutenant Colonel Herbert that the force would now move to the attack and start to progress towards Taboo. The infantry continued to provide cover parties for the Sappers and the cover parties were also given the added protection of the artillery, generally under command of Lieutenant Peter James Begbie. On 25 May another attack was mounted on Bukit Seboosa as the enemy had again erected stockades on the hill. The attack opened with a barrage from the 12-pounder howitzer and this was followed by a charge by the infantry.

Captain Poulson was slightly injured with a ranjow and eighteen men were wounded, of which only eight were injured by musket fire the remainder were injured by ranjows. One of the European officers was also taken ill with jungle fever and had to be evacuated to Bell's Stockade.

At the end of the month, on 30 May, it was decided to take another hill during the advance on Taboo. Begbie reported the event:

> '. . . Captain Wallace, and Lieutenant Minto with a party of infantry, and Lieutenant Begbie with a 5½ inch mortar, started for Bukit Pur-Ling, in order to hoist the British Colors [*sic*] on its summit, whilst the sappers were employed in stripping one of the tallest trees of its branches in order to covert it into a flag-staff . . .'[68]

The force moved on past Tanjong Pur-Ling and after about a quarter of a mile came to an area devoted to paddy, with houses around the banks. The country now changed in favour of the troops as it was now a paddy and fruit-growing area and as the force approached Taboo the less enthusiastic of the Penghulu's Chiefs started to abandon him and came over to the side of the Company. Some of the former enemies were employed by the force to help the convicts and coolies in clearing the jungle and widening the road and were given the same rate of pay as that given to the hired cutters.

On Sunday 3 June, Mr Westerhout arrived at Bukit Seboosa having travelled from Malacca with the power to discuss peace terms with the Penghulu. On the following day Mr Westerhout travelled to Tanjong Pur-Ling with a guard of the Malay Contingent. The Penghulu met the Dutchman and they discussed the terms. During this talk the Penghulu admitted that he had the guns that had been abandoned during the campaign of the previous year. He commented that '. . . perceiving the Company's chaup on them, he had carried them to Taboo, as owing to his being a servant of the Company he was bound to take care of its property . . .'[69] An armistice was agreed in order that the Penghulu's terms could be studied at Malacca. However on the night of 5 June the Penghulu's men attacked Pur-Ling but were driven off. They retreated and planted the road with ranjows.

Bell's Stockade was strengthened by the arrival of a brigade of 6-pounders, and Lieutenant Stevenson, 46th Regiment MNI was given command of the post. On the following day the headquarters of the force moved on from Bell's Stockade to Tanjong Pur-Ling and on 11 June the detachment of artillery under command of Lieutenant Peter James Begbie also advanced from Bell's Stockade and joined the headquarters at its new location.

On 15 June two companies of infantry under command of Captain Justice and Lieutenant Minto, together with Lieutenant Begbie with the 12-pounder howitzer and 5½ inch mortar formed a covering party for the Sappers under Lieutenant Bell who was attempting to destroy the stockades at Bangkall Munji. These stockades were, in fact, the outer defences of Taboo.[70] The sappers were fired upon from one of the stockades by a jinjal and it was necessary to set up a battery in order to return the fire. During the setting up of the battery the force came under constant fire and as a result of the sound of firing Colonel Herbert and his staff moved forward to the newly set up artillery position. The Light

Company, under command of Captains Sinnock and Justice, accompanied by Lieutenant Mackenzie,[71] set off to reach the rear of the stockade by means of a concealed forced march through the jungle. The 12-pounder opened fire on the stockade, whilst the mortar fired in the direction of the Taboo lines. The fire silenced the enemy and Captain Bond, being in charge of the artillery, ordered the artillery to move forward but as they did so the Penghulu's men rallied and opened fire again, on this occasion they damaged a limber and wounded one of the convicts who had been employed with the artillery.

The attack was made more difficult as there had been heavy rain during the early morning and the paddy fields were knee-deep in water and all the nullahs were filled with water. Lieutenant Minto and his sepoys attempted to cross one of the streams but came under fire, although inaccurate, from one of the 6-pounders that had been captured by the Penghulu's men during the disasterous first campaign. Minto was able to press home his advance and was shortly in possession of the first of the enemy stockades. He was joined soon after by Captain Charles Sinnock, of the 5th Regiment, Madras Native Infantry, and his Company. This party, seeing that the stockade was secure, pressed on towards the Taboo lines. Captain John H. Winbolt with his rifle company joined the artillery, as did Lieutenants Liardet and Milnes, and, with the 12-pounder howitzer limbered up, the whole party advanced along the high road towards Taboo. Lieutenant Minto and his sepoys advanced from the stockades on the left flank but the artillery advance was checked by fallen trees. With the obvious delay in clearing the trees Lieutenants Begbie and Poole set off to cross the paddy fields with the mortar but on reaching the nullah which was flooded found that they were unable to cross. The artillery eventually reached the stockades and the section led by Lieutenant Liardet was able to advance and clear the enemy from the stockade on the hill which was known as Execution Hill.[72]

Whilst the troops were reforming after taking the stockades and the lines, the Malay Contingent, led by Syed Sabban, advanced on the Penghulu's house, and on seeing this contingent making the last attack the other companies joined in the general advance to the house. After a few shots from one of the captured 6-pounder guns the enemy fled into the jungle. Begbie records this final step in the long campaign as follows:

> '. . . the dinner of the chief and his people being left untouched on the mats. Everything eatable was immediately destroyed for fear lest poison should have been mingled with the food. The loss of the British, besides the convict, was only two sepoys wounded by musquetry, and three by ranjows. It was asserted, but I know not with what shew of truth that the enemy lost four wounded [all by artillery], one of whom was a son of Dool Syed [*sic* – the Penghulu], whose arm was said to have been broken by two fragments of a shell . . .'[73]

On the following day the Headquarters of the Malacca Field Force arrived and the British standard was hoisted at Taboo. Over the next few days negotiations were carried out with the various chiefs, with Lieutenant John Milnes acting as the spokesman for the Company as he was proficient in the Malay language. On 20 June the flank companies of the 5th Regiment Madras Native Infantry, with Captains Poulton and Sinnock, Lieutenant Minto and Ensign Stoddart occupied Sebang. The mopping-up continued and Mr Alexander Bewicke Anderson of the Pinang Civil Service arrived at Taboo as the Commissioner for Naning. He set about his task with zeal and determination and on 3 August set out with a party from Bell's Stockade, much against the advice of his friends and companions. He had hoped to capture one of the minor chiefs but was unsuccessful and arrived back on the evening of 4 August. He was taken ill and died '. . . of brain fever on the following day at 3 pm . . .'[74] Meanwhile, the gun carriages for the 6-pounders having been repaired permitted the guns to be transported back to Malacca on 12 July. The artillery were under the command of Lieutenant Peter James Begbie and they were escorted by a section of infantrymen commanded by Captain Justice. The main body retired from Taboo to Bell's Stockade and Taboo was made an advance post and a portion of a company stationed there under command of Captain Edward Thurlow Hibgame of the 29th Regiment, Madras Native Infantry. Taboo soon proved to be extremely unhealthy and was abandoned as a station so that the Company's forward post was then located at Bell's Stockade.[75] A report appeared in the Singapore newspaper which was dated 27 June 1832 and was merely signed 'An Officer of the Force' which substantiated the fact that Naning was unhealthy for both the European officers and the sepoy soldiers; the report read '. . . About two thirds of the Force were, during the month of May, in hospital . . .'[76]

Although the campaign was now officially over the Penghulu and a number of his followers were still at large. The Government in India sent further troops to Malacca so as to ensure the peace and prevent a further rising by the Naningites. The Singapore newspaper, published on 9 August, listed the troops that had arrived at Malacca from Madras:

'Arrived Malacca on HMS Alligator from Masulipatam (dep 25 June)
Captain J. B. Puget,
Lieut J. Kerr, Lieut F. Hamilton, Lieut F. Easor,
Ensign A. Barker
Asst Surg (J. G.) Malcolmson, plus 2 companies of (the)
Madras European Regt . . .

Arrived at Malacca on Royal William from Masulipatam (dep 25 Jun)
Captain Prior, Captain Ogilvie, 23rd NI
Lieut G. Balfour; Lieut S. W. Croft, Acting 2nd Lieut
R. Kinkead (both Artillery) and Ensign Hay,
23rd NI plus Det of Arty, 2 Coys of 23rd Regt NI
and a portion of the Ord Dept . . .

Arrived at Malacca on Resolution from India
Lieut T. W. Cooke and Lieut T. C. Hawkes, 23rd Regt NI.

Arrived at Malacca on HMS Imogene from Masulipatam
remaining companies of the Left Wing, ME Regiment

Arrived at Malacca on Brig Hamenshaw from Masulipatam
50 Dooley Bearers, Doolies [*sic*],[77]
and tents . . . all the baggage and Native Followers
of the ME Regiment . . .'

Arrived at Malacca, 8 July 1832, on Ganges from India
Headquarters and remaining companies of the
23rd Regt NI.

Arrived at Malacca on Sir Archibald Campbell from India.
A few Gun Bullocks and Fifty-nine bearers for
the use of the Troops at Malacca . . .'[78]

The Malacca Field Force consisted of the following units all of whom had joined by the middle of August 1832:

Headquarters of the Field Force, (commanded by Lieutenant Colonel C. Herbert).
Left Wing, Madras European Regiment
5th Regiment, Madras Native Infantry
23rd Regiment, Madras Native Infantry
29th Regiment, Madras Native Infantry (officially based at Singapore)
46th Regiment, Madras Native Infantry (officially based at Penang).

There were, of course, elements of the Madras Artillery, Madras Engineers, Ordnance Department and Medical Establishment all of which made it an expensive campaign as far of the Government in India was concerned.

Lieutenant Colonel Francis W. Wilson CB was appointed as Commander of the Force as Lieutenant Colonel C. Herbert had been granted sick leave to return to Europe in order to recover from the rigours of the campaign. Wilson, who was promoted to the rank of Colonel on 22 December 1832, departed from Madras on 27 July on HC ship *London* and arrived at Malacca on 26 August.[79]

It can be assumed that the situation in Malacca soon reverted to a peacetime routine and it is probable that Lieutenant Peter James Begbie was either stationed at Taboo or Bell's Stockade during August or visited the camps in that area as an obituary appeared in the *Singapore Chronicle* dated Thursday 30 August 1832 which read:

'. . . *Death* At Malacca, on the morning of the 23rd August, of dysentery, contracted in the Nanning district, FRANCES ANNE, eldest child and sole daughter of Lieutenant P. J. BEGBIE, Madras Artillery, aged 5 years and 4 days . . .'[80]

Frances Anne Begbie was buried in Fort Cemetery at Malacca and the inscription on her memorial reads:

'Sacred to the memory
of
FRANCES ANNE
Eldest child of
CAPTAIN P. J. BEGBIE
Madras Artillery
who entered upon life
August 23rd 1832
AETAT SUAE AN 5 DIE 4'[81]

The memorial plaque was probably added to the grave during 1833 as by that time Begbie had been promoted to the rank of Captain (from 2 January 1833) but, of course, at the time of the death of Frances Anne he was still a Lieutenant.

Bell's Stockade was demolished in the middle of 1833 as a defence work had been constructed nearby. This new work was named Fort Lismore as a compliment to the family of the Commander in Chief of the Madras Army, Lieutenant General The Hon Sir Robert William O'Callaghan, KCB.

Although the Malacca Field Force was now merely a defence force, (the actual fighting had ceased with the taking of Taboo) reinforcements continued to arrive and a report appeared in the *Singapore Chronicle* dated Thursday 30 August 1832 which gave details of further reinforcements:

'. . . The Madras Gazette of the 18th ultimo, notices the departure, on the previous day, of the ship Neptune, for Malacca. Two companies of Sappers and Miners, including

Tomb of Frances Anne Begbie, daughter of Captain Peter James Begbie, Madras Artillery, at Fort Road Cemetery, Malacca. Photograph by Mrs J. Harfield, 1977.

Inscription on the tomb of Frances Anne Begbie at Fort Road Cemetery, Malacca. Photograph by Mrs J. Harfield, 1977.

followers, about 147 persons and 90 bullocks, and the following Officers: Captain Lawe, commanding Sappers and Miners, Capt Kirby, 4th Regt NI, Lieut Roberts, 29th Regt, Lieuts E. Lawford, T. Ditmas and W. H. Horseley of Sappers and Miners, Ensign Boddington, 23rd LI and Assistant Surgeon Curling. These two companies of S and M are, we suppose, as the relief to the companies whose labors [*sic*] were so efficient during the late campaign . . .'[82]

The whole situation in Malacca remained quiet with the major problem being the accommodation of such a large force. Prior to the two campaigns Malacca had permanent accommodation for one company of Sepoys, a detachment of artillery and a small ordnance store and a hospital. It now had five regiments to accommodate and feed. Moves were made to arrange for the withdrawal of some of the troops and this fact was hinted at in a report that was published in November in the *Singapore Chronicle*. It was proposed to '. . . remove all troops at the close of the year, with the exception of one Regiment, of which two companies are to remain at Bell's Stockade, in Nanning. Colonel Herbert had left for Madras, in the Neptune, through ill-health, and had been succeeded in command by Colonel Wilson . . .'[83]

There is evidence that certainly some members of the Malacca Field Force had been able to bring their families with them to Malacca. This is shown by a notice published in the *Singapore Chronicle* on 8 November which read:

'*Births* At Malacca on the 1st November, the Lady of Capt J. S. WYLLIE, Major of Brigade, Malacca Field Force, of a son . . .'[84]

The withdrawal from Malacca started in December 1832 when 'Lieutenants Ditmas and Boulderson and 1 company of the 29th Regt MNI' arrived on the *Ahmad Shaw* at Singapore, from Malacca.[85] On 14 March 1833 a report was published which read: '. . . We learn from Malacca that the 23rd Regt MLI [*sic*] will shortly be removed from Malacca . . .'[86] The Headquarters of the Field Force and troops of the 29th Regiment MNI returned to Singapore on 13 May on the ship *Edward*. The removal of the surplus troops from that settlement continued throughout 1833.

The Penghulu of Naning surrendered in March 1834 and was permitted to live in Malacca in comfort on a government allowance and where, Buckley records, his house became '. . . the daily resort of health seeking followers of Mohamed, as he set up as an unqualified medical practitioner and was believed to have a miraculous power in the cure of diseases . . .'[87]

The two Naning campaigns were recorded in detail in the book *The Malayan Peninsula, embracing its history, Manners, and Customs of the Inhabitants, Politics, Natural History &c., from its earliest Records* written by Captain P. J. Begbie, Madras Artillery. This work was 'Printed for the Author at the Vepery Mission Press' in India in 1834. Begbie's previous work which dealt with the first Naning campaign received a very hostile reception from the Singapore press and this second work of his also brought severe criticism from the Editor of the *Singapore Chronicle*. Viewing the comments after nearly a century and a half it is difficult to see why the press were so hostile. His book gives a good account of the campaign and he did not try to hide his faults, nor, of course, did he try to be modest over his achievements. Perhaps the press disliked the idea of a military man becoming an author. The *Singapore Chronicle* of Saturday evening 11 July 1835 carried a report on Begbie's book and also a poem which was to say the least unkind, but is reproduced here as, with the passage of time, it is not without interest:

Who was't in August thirty one,
Was sent in charge of shot and gun,
To help to crush the growing pride
Of Nanning's Chief, the bold Dool Syed –
Why Peter,

Who was it, in the woods of Ching
So much admired every thing,
Whose martial soul was quite distressed
To find he passed there unmolested –
T'was Peter.

Who when Panglima like an ape
Stood dancing, at him rattled grape
And took him (t'was – Panglima's fault)
The third round, in a demi volte.
Why Peter.

Who spoilt his powder, buried shot,
Left his gun carriage to rot
And lastly gave ungunner like
Without a cause, the word to spike
Poor Peter.

Who not withstanding toil and pain
Got to Malacca safe again
And six months after, spite rebut
Starts off again for Soongei *Putt*
Brave Peter.

Who on the 10th of April starts
To bring up Powder, one seven carts,
And Halting when attacked, who,
Drove off the enemy, wounding two
T'was Peter.

Who When he's ordered takes command
Of Ensign Thomson's little band?
To whom instead of true direction
Does Jack pass on a 'misconception'?
To Peter.

Whose wise permission then was granted
To separate as Thomson wanted –
Whose tactics once again defeated.
Whose force to 'Wrights' Stockade retreated.
Our Peter's.

Who is it first (the lucky wight)
Hoists British Flag on Purling height
And when flag is hoisted, who,
Throws several shells towards Taboo!
Bright Peter.

Who in the middle of the action
Near Taboo, heard with satisfaction
Those guns discharged by Nanning's hosts
Which he so foolishly had lost.
Why Peter.

Who in this war with the Malay,
Was always hero of the day
And who, when all was over took
The wise resolve to write a book!
Shrewd Peter[88]

The final comment was published on the following Saturday when the *Singapore Chronicle* copied remarks that had previously appeared in the *Madras Courier* of 5 June 1835. The comments were made, again, of Begbie.

'. . . in conclusion, we cannot but feel that much praise is due to Capt Begbie for the use he has made of the time and opportunities he has enjoyed, and we can only wish that others more capable of forming and expressing ideas would follow his example. For himself we would strongly advise him to attempt less in future – to eschew that silly attempt at flowery description, which is so gross an encroachment on good taste – to attend *more* to arrangement; and lastly – to make himself less the Hero of any Military operations he may have to recount . . .[89]

Malacca settled back into its normal routine and during 1835 Lieutenant Colonel John Henry, 48th Regt MNI joined the garrison to take over the post of commander. Little else is recorded of the events within the garrison. One final notice appeared in the press in 1836 that concerned Malacca and that was an obituary notifying the sad death of Nancy Henry on 14th December 1836 at the age of 45 years. She was buried in the Christian cemetery at Malacca and her grave carries the following inscription:

Herein are deposited
the remains of
NANCY HENRY
the beloved wife of
LT COL HENRY Madras Army
Commanding the Troops at Malacca
She died 14th December 1836
Aged 45.[90]

On the other side of the main path of the old Christian cemetery at Malacca is an obelisk which dominates the whole of the cemetery. It is the memorial which was erected to two of the European officers who lost their lives during the Naning campaign and has the following inscription:

Sacred
to the memory of
LIEUTENANTS WHITE AND HARDING
of the 29th Regiment Madras N.I.,
who fell in action with the Malays,
the former
on the 20th of August, 1831:
and the latter
on the 29th of March, 1832.
This tomb was erected
as a mark of respect and esteem
by their brother officers.[91]

The other European officer who was killed during the campaign was Ensign George Wilford Walker of the 5th Regiment, Madras Native Infantry who was killed at Bukit Lanjoot and who was buried at 'camp'. His grave has been located at the back of the Government Secondary School in the centre of Alor Gajah. The actual tomb is near the playing field and at the back of one of the school buildings. Unfortunately the inscribed plaque has now disappeared. There is a second tomb which still remains unidentified. The tombs are enclosed by a metal railing and it has been suggested that the area may also be the burial site of some of the sepoy soldiers who were killed in that area.

NOTES TO CHAPTER SEVEN

1. The 1829 census in the Straits Settlements Records 169, 8 August 1829 gives the population of Naning as 4,875 but Begbie gives the figure as 6,000.
P. J. Begbie, *The Malayan Peninsula*, Oxford University Press: Oxford in Asia Historical Reprints, 1967), p149.
2. T. J. Newbold, *Political and Statistical Account of the British Settlements in the Straits of Malacca*, (Oxford University Press, 1971), vol1, pp 198–222.
3. Begbie, op cit, p62. The commutation is also contained in Bengal Secret and Political Consultations, vol 363, 25 November 1831, nos 69 and 70.
4. Bahara – a weight (3 pikul = 400lb).
5. Begbie, op cit, pp 3 and 4.
6. The East-India Register and Directory for 1831. The Establishment of Singapore, Prince of Wales Island and Malacca (p110) lists William Thomas Lewis as the assistant to the resident at Malacca.

7. *Journal of the Malayan Branch, Royal Asiatic Society,* no 111, p 118 ff.
8. Begbie, op cit, p 148. Sukus, or Sookoos, is the title given to a Chief of a village or district.
9. Straits Settlements Records 169, January 1829.
10. Ibid, 11 March and 20 March 1829.
11. Ibid, 18 July 1829.
12. Duku. John Cameron in his book *Our Tropical Possessions in Malayan India,* (by Smith, Elder & Co, London, 1865), describes the fruit as follows: '... Dookoo. This delightful fruit is the produce of a large tree. It grows in clusters: each is about the size of a cricket-ball. The brownish skin being broken, displays the pulp in six cloves, of a pleasantly acid taste, enclosing a greenish kidney-shaped seed. It is by many reckoned the finest fruit in the peninsula. The month of July is the season at Malacca, in which it is had in greatest perfection ...'
13. Public Letters to Bengal, 1830–1831, vol 14, 2 June 1830, no 18.
14. Bengal Public Consultations, Range 126, vol 127, 2 April 1831.
15. The Hodson Index, National Army Museum, London.
16. Ibid.
17. Singapore Chronicle, no 14, dated 7 April 1831. Entry under 'Arrival of Passengers' shows '... By the Lady Munro. Mrs Begbie [and] Lt Begbie. Arrived on 20 March 1831. Left Madras 22 February 1831 ...'
18. The Hodson Index, National Army Museum, London, Card shows: Begbie, Alexander John. Born 24 May 1807. 4th son of Peter and Francis Begbie. Addiscombe Cadet 1822–1824. 2nd Lieut 16 December 1824. Lieutenant 10 June 1825. Married to Miss Eliza Ann Wilson on 17 September 1827. Promoted to Captain 30 July 1842. Major 11 November 1851. Invalided 31 July 1852. Promoted to Hon Lieutenant Colonel. Retired 28 February 1855. Died near Southampton 31 March 1881. East-India Register and Directory, 1833, 1st Edition Entry shows: Jan 5, 1832. At Singapore, the Lady of Lieut A. J. Begbie, of Artillery, a daughter.
19. East-India Register and Directory for 1831, 2nd Edition, p 58, (Fort St George) shows: Captain J. S. Wyllie, Lieut J. Milnes and Ensign E. H. Short as serving with 29th Regiment, Madras Native Infantry.
20. Begbie, op cit, pp 163 and 164.
21. Soongei Pattye. Begbie gives the interpretation in his book at p 164 which reads: '... Compounded of Soongie, a river, and Pattye, a tree which produces a fetid bean, eaten by the Malays and which grows here in great abundance ...'
22. East India Register and Directory, for 1831, 2nd Edition, p 58, (Fort St George).
23. Begbie, op cit, p 169.
24. The translation in English means 'the place of the elephant hunt'.
25. Begbie, op cit, pp 174 and 175.
26. Ibid, pp 178 and 179.
27. Ibid, p 180.
28. Ibid, p 181.
29. Ibid, p 183.
30. Translation of 'Roombiyah' into English is 'Sago Palm'.
31. Begbie, op cit, p 185.
32. *Singapore Chronicle,* no 33, Thursday 18 August 1831. Under Departures from Singapore: '... 15 Aug. British Brig Leegoan. for Malacca. Lieuts White and Fothergill, 120 sepoys, 29th Regt MNI ...'
33. Begbie, op cit, p 187.
34. Ibid, p 188.
35. Ibid, p 189.
36. *Singapore Chronicle,* no 31, 4 August 1831. Under Arrivals at Singapore: '... 1st Aug 1831. British ship Eliza from Hobart Town on 15th June Captain H. Moore, Lieut Lewin and 38 Marines ...'
37. *Singapore Chronicle,* no 33, 18 August 1831.
38. Begbie, op cit, p 190.
39. *Singapore Chronicle,* no 37, Thursday 15 September 1831.
40. The Hodson Index, National Army Museum.
41. *The Index Malaysiana* published in 1970 shows Naning as being spelt as 'Naning' whereas Begbie consistently uses the spelling 'Nanning'.
42. *Singapore Chronicle* 45, Friday 11 November 1831, Supplement.

43. *Singapore Chronicle,* vol2, no4, Thursday 26 January 1832.
44. Ibid.
45. Begbie, op cit, p195.
46. Ibid, p200.
47. Ibid.
48. Ibid.
49. *Singapore Chronicle,* vol2, no7, Thursday 16 February 1832.
50. East-India Register and Directory for 1832, p34.
51. *Singapore Chronicle,* vol2, no8, 23 February 1832.
52. Ibid, vol2, no13, 29 March 1832.
53. Begbie, op cit, p213. A 'ranju', 'ranjow' or 'ranjou' is a man-trap.
54. Ayer Mangis – the 'Mango stream'.
55. Begbie, op cit, p216.
56. The translation of Ayer Itam is, literally, Black water.
57. Begbie, op cit, p219.
58. Ibid, p222. Begbie notes that Emaum Aly was later executed for the murder of Lieutenant Colonel John M. Coombs at Palaveram on 10 October 1833. The East-India Register for 1834, p159, (Fort St George) does carry the following entry under Casualties: 'Deaths' '... Infantry. Lieut Colonel John M. Coombs. 10th Oct 1833. [at] Palaveram ...'
59. Ibid, p223.
60. Ibid, p227.
61. Ibid, p228.
62. East-India Register and Directory for 1855, p107.
63. Begbie, op cit, p233.
64. Ibid, p234.
65. *Singapore Chronicle,* vol2, no20, 17 May 1832. The plaque that was placed on the grave of Ensign Walker, which is now missing read:

To the memory of
Ensign George Holford Walker
doing duty
with the 5th Regiment M.N.I.,
who fell
while gallantly leading on his division
to storm a stockade at Allegata
on the 3rd of May 1832.
This tablet
was erected by his brother officers
as a mark of their esteem
for one so universally beloved.

66. Begbie, op cit, p235.
67. Ibid, p237.
68. Ibid, p246.
69. Ibid, p250.
70. Taboo, a drum, the name was given to the village (kampong) as it held a '... great drum ... made of a trunk of a tree, the internal diameter of the cylinder being about 2 feet, the length 5½ feet, and both ends covered with parchment, similar to drum heads. It was suspended between two posts near the mosque, and was heard distinctly at Pur-Ling ...'
71. East-India Register and Directory for 1832, p34. Lieutenant Alexander Mackenzie, 5th Regt MNI.
72. Execution Hill was so named as this where the Penghulu had offenders put to death by 'krissing'. That is execution by the use of the Malay hand weapon the kris.
73. Begbie, op cit, p256.
74. East-India Register and Directory for 1833, p110 (Singapore, &c) and Begbie, op cit, p258.
75. Begbie, op cit, p259.
76. *Singapore Chronicle,* vol2, no28, 12 July 1832.
77. Colonel Henry Yule and A. C. Burnell, *Hobson-Jobson,* (John Murray, London, 1903), p313, Doolies – also Dhooly, a covered litter.

78. *Singapore Chronicle,* vol2, no32, 9 August 1832.
79. Ibid, vol2, no33, 16 August 1832.
80. Ibid, vol2, no35, 30 August 1832.
81. A. G. Harfield, *Christian Cemeteries in Malacca, (British Association for Cemeteries in South Asia,* 1984), pp 26 and 27.
82. *Singapore Chronicle,* vol2, no35, 30 August 1832.
83. Ibid, vol2, no44, 1 November 1832.
84. Ibid, vol2, no45, 8 November 1832.
85. Ibid, vol2, no51, 20 December 1832.
86. Ibid, vol3, no11, 14 March 1833.
87. D. and J. Moore, *The First 150 years of Singapore,* (Donald Moore Press, Singapore, 1969), p159.
88. *Singapore Chronicle,* vol5, no28, 11 July 1835.
89. *Singapore Chronicle,* vol5, no29, 18 July 1835.
90. Harfield, op cit, p16.
91. Ibid, p4.

CHAPTER EIGHT

The Straits Settlements Garrison, 1830–58

On 1 May 1830 Singapore, Prince of Wales Island and Malacca combined to become a single Residency under the control of Fort William. The Governor Robert Fullerton suggested that the headquarters of the new Residency should be established at Malacca but this was not implemented and the Headquarters for the administration of the Straits Settlements remained at Prince of Wales Island until 1832 when it moved to Singapore. The Governing Council of the Presidency of Fort William imposed severe cuts on the civil administration at this time but there does not appear to have been any immediate change to the military garrison.

During the remaining months of 1830 there is very little recorded of note. Movements of officers continued to be published in the *Singapore Chronicle and Commercial Advertiser* and the same newspaper, which was Singapore-based, continued to publish details of births, marriages and deaths of the European members of the garrison force. For example, on 17 June the following notice appeared:

'*BIRTH* On the 25th of May, 1830, The Lady of Lieutenant R.N. Campbell, 4th Regt M.N.I., Cantonment Adjutant, Commissariat Officer &c., of a daughter . . .'[1]

This type of item is of interest as the East-India Register and Directory for 1830 only shows 'Lieutenant Robert Nutter Campbell, 4th Regiment, Madras Native Infantry' as serving in 'P. W. Island'[2] but the newspaper qualifies his actual job as being the Cantonment Adjutant, etc. In September of that year two Assistant Surgeons died whilst serving in Singapore, their deaths were recorded in the *Singapore Chronicle* as follows:

'*Death* On the 7th Inst Dr John Black, Assistant Surgeon 13th Regt, Bombay NI . . .'[3]

'. . . *Death.* On the 10th Instant J. G. Sim Esq MD., civil Surgeon at this Station, aged 26 years, generally regretted . . .'[4]

In October, 1830 a further notification of death was published and on this occasion the obituary was followed, in the next edition of the newspaper, with a Notice giving details of the settlement of the officer's effects. The two entries are reproduced to show the type of entry that was published on such occasions:

'*Death.* On Thursday Morning, the 30th ult., Captain CULLEY 2nd Regt Bengal NI . . .'[5]

'*Notice* Captain George STOTT 11th Regiment MNI Commanding the troops at Singapore, having taken charge and disposed of the Effects of the late Captain Thomas CULLEY, 2nd Regiment Bengal Native Infantry, by Public Auction has, after deducting Funeral Expenses, Servants Wages, and other authorised charges, this day remitted the balance amounting to Sicca Rupees three hundred and thirty six, two annas and 11 pie (Sa Rs 336 2 11) to the Sub-Treasurer to Government, Fort William.

GEORGE STOTT Capt
11th Regiment MNI
Commanding Troops
at Singapore.

Singapore, 20th Oct 1830 . . .'[6]

In the same edition of the *Singapore Chronicle* an interesting report appeared concerning a recruiting campaign that was being held in the East Indies for local soldiers. The report read:

'Thursday, October 21st, 1830. On the 16th instant HM Colonial barque Anne, anchored in the Harbour, from Ceylon on the 9th September. The purport of her voyage to this part of the world is to raise recruits for HM Ceylon Rifle brigade, which, from the time that Island was in the possession of the Dutch, has been composed of natives of the Eastern Archipelago, principally Javanese, we believe. As the Anna has not been very successful in the Straits in accomplishing her object, she will sail for Java in a few days. We learn there are a number of discharged native soldiers on board, with a detachment of HM 58th Regt, to protect the vessel . . .'[7]

In the 'Movements' section of the same newspaper the arrival of the vessel is recorded as follows:

'Arrivals
16th Oct 1830. Anne. from Ceylon 9th Sep.
Lieut Skinner, DAQM General, Lieut Mann, 58th Regt; Dr Kemmis, Staff Assist Surgeon, and a detachment of HM 58th Regt and Ceylon Rifle Regiment . . .'[8]

At the close of the year a further death of a European officer was recorded at Singapore. This was Ensign Alexander Dennistoun, of the 11th Regiment, Bengal Native Infantry, who died on Saturday 11 December, 1830, aged nineteen years. The East-India Register and Directory for 1830 shows 'Ensign Arch Dennistoun' of the 11th Regiment BNI as serving at Barrackpore. It is therefore probable that Dennistoun was on furlough in Singapore at the time of demise. The East-India Register for 1832, shows under the section 'Casualties – Bengal':

'. . . 11th Nat Inf. Ensign Alexander Dennistoun 11 Dec 1830. Singapore . . .'[9][10][11]

An examination of the remaining memorials in Fort Canning Cemetery in Singapore (opened 1823 and closed 1865) has failed to reveal any trace of the graves of the two officers who died during the latter half of 1830, nor are they mentioned in the report on the cemetery compiled by H. A. Stallwood in 1912.

At the close of the year 1830 the total population of Singapore Island was in excess of 18,000. At the time of the setting up of the Settlement by Sir Stamford Raffles the estimated number of inhabitants was 150; but by 1823 the rapid development of the Settlement had increased the number to 10,683, of which only seventy-four were European. By 1826 the population had risen to 13,725, with eighty-seven Europeans and the population as at 1 January 1829 was recorded as being 17,664 with an increase in the European population, which by then numbered 122.[12] With the Settlement at Singapore growing in importance conversely Penang, or Prince of Wales Island declined in importance.

In January 1831 the newspaper the *Singapore Chronicle* commenced a new series of papers, which were, from 6 January, sold weekly, and two items appeared in the first two issues which dealt with the European Officers serving in the Garrison. In the edition of 6 January, a birth was notified, this was shown as:

'*Birth* At Singapore, on the 2nd instant, the Lady of Captain R. D. Odell, Commanding the 25th Regt MNI, of a son . . .'[13]

At the time of the publication of the birth notice, Captain Robert Deane Odell was the senior Captain of the 25th Regiment Madras Native Infantry, and the senior officer of that unit on station, and in command. In this connection notices that have appeared in the newspaper have given important clues which cannot be obtained from documents such as the East-India Registers.

The newspaper published on Thursday 13 January contained details of the estate of the late Ensign Alexander Dennistoun, who had died on 11 December 1830. The notice reads:

> '*Notice* Captain George Stott, 11th Regt MNI Commanding Troops at Singapore, having taken charge and disposed of the Effects of the late Ensign Alexander Dennistoun, 11th Regt, Bengal NI by public auction, has, after deducting Funeral Expences, Servants Wages, and other authorised Charges, this day remitted the balance amounting to Sicca Rupee four hundred and fifty four, annas twelve and pie eight (454 12 8) to the Sub Treasurer to Government Fort William . . .'[14]

From this entry it has been established that Captain George Stott, of the 11th Regiment, Madras Native Infantry was the Officer Commanding the Troops on Singapore Island. His unit was however serving at Bellary, therefore Captain Stott must have been serving on an attachment at Singapore. The entry in the East-India Registers for this period fails to show him as being on detached duty. Although Stott was the Officer Commanding the garrison, Captain R. D. Odell was at that time the senior officer in charge of the Sepoys of the 25th Regiment Madras Native Infantry.

Another notice which appeared in the local newspaper in March 1831 shows that there were some European NCOs serving with the 25th Regiment MNI in Singapore. The notice was, once again, a notification of a birth and read:

> '*Birth* At Singapore, on the night of the 11th instant, the wife of Mr G. Speechley Quarter Master Sergeant of the 25th Regt MNI, of a son . . .'[15]

During 1831 and 1832 there was considerable movement within the three garrisons due to the Naning War of 1831–1832 but this detail has been dealt with in the preceding chapter.

On 1 August 1831 the British ship *Eliza* arrived from Hobart Town. The journey had taken over six weeks; the vessel had commenced her voyage on 15 June and had Captain H. Moore, Lieutenant Lewin and thirty-eight Royal Marines on board. This party was accommodated in transit accommodation at Singapore and departed bound for Malacca and Calcutta on the British barque *Freak* on 17 August 1831.

With the activities of Malacca and Naning dominating the scene there is very little recorded for the remainder of the year relative to Singapore and Penang. In August 1832 a Royal Salute was fired from the Battery at Singapore and the event was recorded in the *Singapore Chronicle* as follows:

> 'On Tuesday, last, being His Britannic Majesty's Birthday a salute of 67 guns, corresponding with the age of our gracious sovereign, was fired from the Battery at Sunrise. A Royal Salute was like-wise fired at noon . . .'[16]

It is not clear as to which battery actually fired the salute. It was probably the battery at 'Scandal Point' which was retained for the sole purpose of firing salutes. Fort Fullerton was built during the period 1830–32 and whilst it had a battery firing seaward, the former battery was retained for ceremonial purposes. In fact the battery at Scandal Point was retained in that form until 1845 when it was recommissioned as a combat battery and remained as an active service battery until the guns were removed in 1851 when the sea wall along the Esplanade was rebuilt.

On 1 January 1833, the population in Singapore had again increased and was now recorded as being 20,978, with 119 Europeans, of whom ninety-one were male and twenty-eight female.[17] With the completion of Fort Fullerton and the move of the infantry regiment lines the garrison at Singapore was experiencing a fairly quiet period, with the usual routine of parades, drills and guard duties. During the early years of the Settlement being established the military Cantonment had been built adjacent to what is now Stamford Road. It became necessary for the Cantonment to be moved as the area was required for development and on 26 November 1828 a committee of officers assembled and chose a new site in the area of Rochore. The ground was later found to be too low and subject to flooding and the infantry sepoy regiment was then moved to an area adjacent to Pearls Hill. The infantry camp was known as the 'Sepoy Lines' and in more recent years can still be identified in the name of the school which occupied the old site, the school known as 'The Sepoy Lines Malay School'.[18]

Towards the end of 1833 another notification of a death was published in the *Singapore Chronicle* which again confirms that European NCOs were serving with the Madras units in the Straits Settlements garrisons. The notice reads:

> '*Death* On the 1st inst: after a lingering illness of two years, which she bore with Christian resignation Mrs McSWINEY, aged 33 years, wife of Staff Sergeant McSWINEY . . .'[19]

A check on the remaining memorials in Fort Canning Cemetery revealed that Mrs McSwiney and her infant daughter had been buried there. The memorial inscription read:

Sacred
To the Memory of
MRS ANNE McSWINEY
Wife of D. L. McSWINEY ESQr.
who departed this life
on the 1st of November 1833
Also in Memory of
ELLEN
her infant daughter
who died on the 25th June 1829
Aged 8 days[20]

The 29th Regiment Madras Native Infantry, which moved to Singapore to take up garrison duties in 1832, had a Regimental Band and on the last day of 1833 the wife of the Band Master W. White, gave birth to a son.[21] On 14 January 1834 a further birth was notified which read:

> '*Birth* On the 14th instant, the Lady of Lieut Alexander Begbie, Madras Artillery, of a son . . .'

Lieutenant Alexander Begbie was the younger brother of Peter James Begbie, also of the Madras Artillery who had command of the artillery detachment in the Nanning Campaigns of 1831 and 1832. P. J. Begbie was also the author of the book *The Malayan Peninsula* which was published in 1834. Peter James Begbie eventually became a Major General, but Alexander John Begbie retired on 28 February 1855 in the rank of Lieutenant Colonel.[22]

In January 1834, the 29th Regiment Madras Native Infantry departed from Singapore and were not immediately replaced by another unit. The Settlement was garrisoned, for the next year, by a detachment of the 15th Regiment, Madras Native Infantry which was stationed at Penang Island. The movement of the 29th Regiment, Madras Native Infantry was published in the *Singapore Chronicle*

'*Departures*
27 Jan 1834. Dutch ship Pascoa, (bound) for Malacca and Madras.
Lieut Col King, Capt Hibgame, Capt and Mrs Servante, Lieut and Mrs Hurlock, Lieut Symes, Lieut Wood, Ensign Boulderson, Ensign Free, and six companies 29th Regt MNI . . .'[23]

At this time the 23rd Regiment, Madras Native Infantry were stationed at Malacca, but the senior Captain of that regiment Captain Henry Prior was acting as the Officer Commanding the Troops at Singapore. The appointment is confirmed by a notification which appeared in the 6 February edition of the local newspaper. This read:

'*Birth* At Campong Glam on the Morning of the 24th ult the Lady of Capt H. Prior, 23rd Light Infantry, Commanding Troops at Singapore, of a daughter . . .'

The three stations in the Straits Settlements attracted a number of officers to spend their furlough in the area, it was considered to be a 'change of air' leave but in a number of cases the officer, or a member of his family fell ill and died in the Settlement that they had selected to visit. One such case occurred in 1834. Captain David Lester Richardson of the Bengal Invalid Establishment visited Singapore, with his family. Unfortunately his young daughter became ill and died. The obituary notice was published in the *Singapore Chronicle* – '. . . On the 21st April, JESSY HAY, daughter of Captain D. L. Richardson, of the Bengal Army, aged two years and seven months . . .'[24] A check of the Fort Canning cemetery revealed the following memorial inscription which is now bricked into the south wall of the old cemetery. The inscription is of interest and reads:

Sacred
to the memory
of
JESSY HAY RICHARDSON
Eldest daughter of
Captain D. L. Richardson
Bengal Army
Born Sept 15th 1831
Died April 21st 1834.
Calcutta. O.P. Lawrence Sc-[25]

The name of the stonecutter indicates that the memorial stone had been inscribed in Calcutta and shipped to Singapore. This frequently happened as there was little or no suitable material for memorials available on the island.

Very little is recorded for the year 1834 that is of interest to the military historian. The activities of the garrison troops were routine and it is not until October of that year that anything of significance happened and then it is only of marginal interest. The Christian Cemetery, which was on the eastern slope of Government Hill, and which had been opened in 1823, was finally consecrated by the Bishop of Calcutta, Bishop Daniel Wilson, on the morning of 6 October 1834. The area was normally used for Protestant burials and the cemetery was extended on the land side to cater for Catholic burials, but this segregation was not strictly carried out until 1845.[26]

In July 1835 the new infantry unit for the garrisons of Singapore and Malacca commenced arriving by detachments. The first detachment arrived on 4 July on the *Swallow* and the second on 7 July on the *Sir A. Campbell*, both vessels having left Madras on the 13th of the previous month.[27] The senior officer of the new unit, 48th Regiment, Madras Native Infantry was at that time Captain Richard Rodney Ricketts.[28] This regi-

View from Government Hill, Singapore, *c*1835, later to become Fort Canning Hill. By an unknown artist. Photograph from a Christmas card used by HQ Singapore Base District. The regiment can be identified as either the 23rd Regiment Madras Native Infantry, who were stationed in the Straits Settlements 1832–1835, or the 15th Regiment Madras Native Infantry who relieved them during August 1835.

ment provided the garrison troops for Singapore and Malacca whilst the 15th Regiment, Madras Native Infantry provided the garrison troops for Penang (Prince of Wales Island). It was during this year that the 15th Regiment lost their Commanding Officer, Lieutenant Colonel Gregory Jackson.

During his first tour of duty in Penang, Lieutenant Colonel Jackson purchased a clove plantation known as Mount Elvira. The name was given to the hill by the Reverend Hutchings who purchased the hill during the early 1820s and having cleared it and set up a clove plantation named it after his wife. In 1827, whilst he was supervising his clove plantation he was stricken with fever and died some nine days later. It was at this time that Jackson purchased the plantation. A substantial bungalow was erected on the highest point of the hill and the Jackson family used this residence as a retreat to escape the heat of the plain on which the officers' quarters had been built. Towards the end of March 1835, whilst on their second tour of duty on the island, the family, consisting of Colonel Jackson, his wife and son went to the bungalow on Mount Elvira to spend a few days holiday. On 31 March they were all attacked by fever and '. . . died within twenty-four hours, and were buried in one grave in the church yard. Several of their native servants were also carried off . . .'[29] The Jacksons were in fact buried in Northam Road Cemetery, at Georgetown. A report published some years later in the *Overland Singapore Free Press* carried an interesting sequel to the story. The report read:

> '. . . When we visited Mount Jackson (formerly Mount Elvira) about 20 years ago the bungalow was still standing but partially delapidated. On a recent visit (1860) we found it had long fallen down from further decay. Very little remained of it except the low granite pillars on which it rested . . . the road to Jackson's Hill is quite overgrown with jungle . . .'

The report also commented that during the time that the Jackson's contracted 'jungle fever' few of those '. . . who were seized by it ever recovered from the attack, and few indeed lived many days after it . . .'

Interest had obviously been raised by the account in the newspaper as in the next edition of the *Overland Singapore Free Press* the following comment was published: '. . . The Melancholy event referred to in our last issue which has caused Jackson's Hill to be avoided for 25 years, even by Chinese squatters shunning close proximity to it . . .'[30] From other contemporary accounts it is known that the local Malay population also avoided Mount Jackson, or Jackson's Hill as it was also known, believing that it was visited by a hantu (ghost).

During the latter part of 1835 there is again a rare reference to a senior NCO when a notification of a birth was published in the *Singapore Chronicle* on 28 November 1835.[31] The notice showed that Sergeant R. W. Stonehewer was a member of the Commissariat Staff at the Malacca Garrison. At this time there were detachments of both the 15th Regiment and the 48th Regiment Madras Native Infantry serving in the Malacca Garrison.

It was during the year 1835 that a new publication appeared in Singapore, this was the *Singapore Free Press* which commenced publication in October 1835 and proved to be a more attractive publication than the *Singapore Chronicle*, which had previously been the only newspaper published in Singapore. The *Singapore Chronicle* suffered as a result of the new publication and finally closed with the edition of Saturday 30 September 1837. Unfortunately copies of the editions covering the final two years of the publication of the *Singapore Chronicle* are not available so that the movements of officers to and from

Singapore cannot be recorded. The *Singapore Free Press* had a completely different format and did not publish this type of information.

A belated report appeared in the *Singapore Free Press* in April 1836 which recorded the partial destruction of Fort Victoria at Amboyna. The report is interesting in itself and reads:

'. . . During three weeks in the month of October a thick sulphury mist lay over the islands (Molucca Islands) and on the 1st November at about 3 o'clock in the morning an uncommonly strong earthquake was felt at Amboina and several of the surrounding islands, and this was followed on the 4th of the same month by several new shocks . . . it is but too probable that there must have been an eruption of the Volcano of Banda . . . The earthquake has been dreadful in its consequences at Amboina [also spelt as Amboyna]: fifty eight men, women and children lost their lives by the falling of the Barracks in Fort Victoria. Sixty-six persons in all have been killed . . .'[32]

There is no record of East India Company troops being stationed on Amboyna at this time and although the Fort still retained the name of 'Fort Victoria' it is considered that the troops stationed there would probably have been locally-recruited Dutch troops.

During the month of April one of the officers stationed at Malacca married. The event was published in the *Singapore Free Press*:

'*Marriage* On the 9th instant, at Malacca, Lieutenant G. M. Gumm of the Madras Artillery to Agnes Maria Sri Singapoorah, eldest daughter of F. J. Barnard Esq . . .'[33]

F. J. Barnard first arrived at Singapore with Sir T. Stamford Raffles and was appointed Acting Master Attendant and Marine Storekeeper, later he became chief of the Malay Police at Singapore and married Colonel William Farquhar's daughter. In January 1824 it was Frederick James Bernard that first established a newspaper in the Settlement, which was the *Singapore Chronicle.* He and his wife obviously had become very attached to the settlement to have named their daughter Sri Singapoorah.

In December, the 48th Regiment, Madras Native Infantry suffered the loss of one of their officers. Lieutenant John P. Germon[34] died on the afternoon of 20 December 1836 and in the same newspaper that notified his death a notice was published concerning the disposal of his effects. The notice read:

'To be sold at Public Auction.

By Order of Major Robert Alexander, Commanding the Troops on Monday the 26th instant at 11 o'clock in the forenoon.

The Personal Effects of the late Lieut John P. GERMON.

NB. Terms of Sale - CASH.

Singapore 21st December 1836. RAPPA & Co . . .'[35]

Major Robert Alexander was the senior officer of the 48th Regiment Madras Native Infantry in the Straits Settlements and spent time in both Singapore and Malacca. His wife gave birth to a daughter whilst at Malacca during July 1837.[36] Very little is recorded for that year except that the *Singapore Free Press* noted that on Sunday 17 September 1837 '. . . minute guns to the number of 72 corresponding with the age of our lamented Sovereign [William IV] were fired both from the battery on shore and from Her Majesty's Sloop-of-War Zebra in the Roads: and on the following day a Royal Salute was fired in honour of the accession of the Princess Victoria to the throne . . .'[37]

The newspaper *The Penang Gazette and Straits Chronicle* was started on 7 April 1838, and this was published weekly as was the other paper on sale in that Settlement which was the *Prince of Wales Island Gazette.* The *Penang Gazette and Straits Chronicle* covered news items from all three settlements and during 1838 three items were published which referred to the military, these were:

'*Marriage.* Pinang [*sic*]
At St George's Church on Thursday the 22nd instant by the Revd Mr Jones, Lieutenant T. A. C. Godfrey, Staff Officer of the Madras Artillery stationed at Fort Cornwallis to Eliza Bruce, fourth and youngest dauthter of Charles Buchan Esq., of Edinburgh . . .'[38]

'On Friday evening the 5th inst the gentlemen bachelors of Pinang entertained their friends of the town and neighbourhood . . . the Band of the 12th Regt MNI attended and played with great correctness and spirit . . .'[39]

The 12th Regiment, Madras Native Infantry had replaced the 15th Regiment as garrison troops during March 1837 and had by 1838 become firmly established in the community. The garrison at Malacca was now provided solely by a detachment of the 48th Regiment Madras Native Infantry which had its headquarters on Singapore Island. The third notice was again a marriage notice but does give confirmation that members of the 48th Regiment were serving in the Malacca Garrison. The notice read:

'*Marriage.* At Malacca on the 6th instant, at Christs Church, by the Rev J. Hughes, Lieut I. Ferrier, 48th Regt MNI, to Miss Catherine Maria, eldest daughter of J. B. DeWind Esq . . .'[40]

This wedding was also published in the *Singapore Free Press and Mercantile Advertiser.*[41] At the end of 1838 the 48th Regiment Madras Native Infantry returned to India and were replaced by another Madras Infantry regiment, this time the 8th Regiment. In February 1839 a notice appeared in the newspaper complimenting that regiment as follows: '. . . One of the most elegant entertainments . . . we have witnessed here was that given last week by Colonel Cooke and the officers of the 8th [Regt MNI] to Commodore Read and the officers of the American Squadron now at this station . . .'[42]

An obituary appeared in the Singapore newspaper in respect of William Farquhar, the first Resident. It referred to him in his rank of Major General. The notice read:

'. . . On the 13th May, at Early Bank, near Perth, in his 69th year, Major General Farquhar, of the Madras Engineers Corps, many years Resident and commandant of Malacca, and subsequently employed in the formation of the new settlement of Singapore in the East Indies, the Chief local authority (both civil and military) of which he exercised for nearly the first four years of its infancy . . .'[43]

During 1840 Singapore became a transit station for troops proceeding to China and during September 1840 two items were published in the newspaper which concerned the movement of these troops:

'. . . On the 8th instant the Golconda came in from Madras, from which she sailed on the 27th ultimo, with the Headquarters of the 37th Regt MNI, on board, en route to China for which they get underweigh this morning. The Sophia was to sail on the following day and the Minerva a few days later with other detachments of the Regiment on board . . .'[44]

'. . . The Thetis, Sophia and Minerva have arrived during the week from Madras with detachments of the 37th Regiment on board for China. The Thetis took the route from this [settlement] direct to Chusan, the others proceeding to the mouth of the Canton River . . .'[45]

Where it was necessary for the troops to disembark they were accommodated in tented lines which had been erected on the Singapore Padang so as to create a temporary transit camp.

During 1840, the garrison regiment at Penang again changed, the 12th Regiment Madras Native Infantry departed to Palaveram and 24th Regiment, Madras Native Infantry took over the garrison duties. In the following year, 1841, the 8th Regiment, Madras Native Infantry were relieved from garrison duties at Singapore and replaced by the 39th Regiment, Madras Native Infantry. The 39th Regiment was only to stay in the Settlement for a short period as in June 1842 it was transferred to join the China expedition. The notification of its departure appeared in the local press:

> '. . . Since our last the transport SHAH ALLUM has arrived from Penang and she will leave on Saturday next, with Lieut Colonel Taylor and a portion of the Officers and men of the 39th Regt MNI, to join the China expedition . . .'[46]

The next unit to provide garrison duties was the 4th Regiment, Madras Native Infantry. As the 39th Regiment MNI was withdrawn hurriedly the responsibility for providing the garrison troops for Malacca was passed to 24th Regiment Madras Native Infantry stationed at Penang Island.

On Monday 19 September 1842 a 'Royal Salute of 21 guns was fired [during the] forenoon on receipt of the intelligence of Peace being concluded with China.'[47]

During the following year, 1843, one of the Madras Artillery officers stationed at Singapore died. He was 2nd Lieutenant Thomas Bernard Cox who died on 16 February 1843.[48] The memorial plaque belonging to this officer's tomb is still in existence and is bricked into the south wall of the Fort Canning Cemetery. The plaque reads:

Sacred
to the
Memory
of
THOMAS BERNARD COX
2nd Lieut. Madras Artillery
died at Singapore
16th February 1843
Aged 19 years.

The garrison troops at Penang and Malacca changed during 1843 and on this occasion the handover was not as easy as usual as the relieving force was delayed due to severe bad weather in the Bay of Bengal. The incident was reported in the local press:

> '29th June 1843. The Champion left on the 27th instant with two companies of the 24th Regt for Madras. The departure of the Rifle Companies is delayed indefinitely in consequence of the non appearance of the Lord Elphinstone and Lord Lyndoch . . .'[49]

and in the subsequent edition:

> '. . . 6th July 1843. We are glad to learn from the Penang Gazette that the Lord Lyndoch which it was feared had foundered in the late gale has arrived at Penang with a further portion of the 27th Regiment from Vizagapatam . . .'[50]

Whilst the 27th Regiment Madras Native Infantry were settling in at Penang and Malacca the 4th Regiment had become well established in Singapore, and were fulfilling their civic duties. The *Singapore Free Press* recorded that:

> '. . . On Friday 15th instant the officers of the 4th Regt MNI entertained the Hon'ble the Governor at Dinner at the Mess House when, as is usual on such occasions, the heads of departments Civil and Military were invited to meet his Honour . . .'[51]

The use of the term 'Mess House' is interesting as prior to this date reference to social entertainment has indicated that such events took place in a room in the town, or in a private residence.

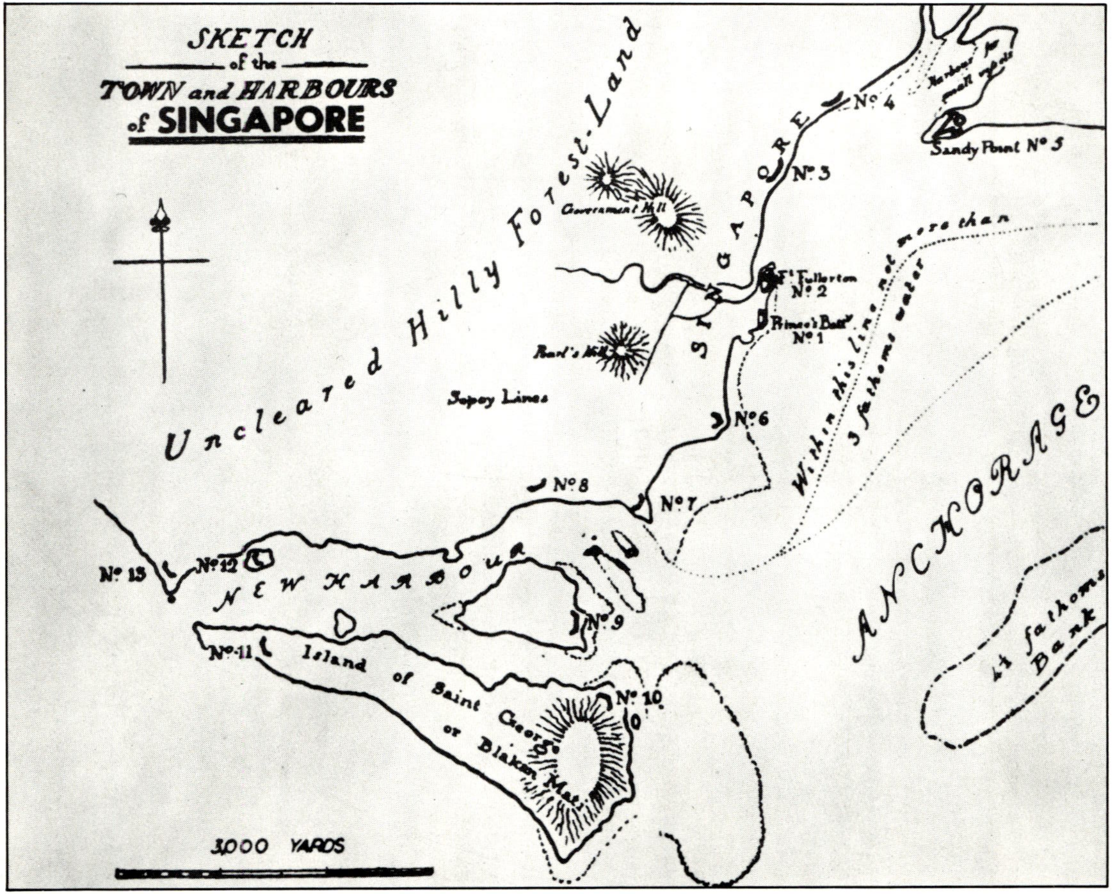

Sketch of the town and harbour of Singapore by Captain Samuel Best, Madras Engineers dated 1843. Courtesy of the Director, National Museum, Singapore.

During the month of October the wife of Conductor W. Deare gave birth to a son and the notification was published:

'*Birth.* At Singapore, on Thursday morning, the 19th inst, Mrs Deare, wife of Mr Conductor Deare, of a son . . .'[52]

but the event was quickly followed by the tragic loss of Mrs Deare, whose obituary was published on the 9th November 1843:

'*Death.* At Singapore, on Thursday morning, the 2nd instant, Mrs Diana Deare, wife of Mr Conductor Deare, aged 34 years, leaving a disconsolate husband, six children, and a large circle of friends to bemoan their irrepairable loss . . .'[53]

The son, who had been born on 19 October 1843 died on 10 January 1844.[54] The plaque commemorating the sad loss of the wife and son can be still seen on the south-west wall of Fort Canning Cemetery.[55]

Early in 1844 the Troopship *Rattlesnake* arrived in Singapore Harbour from Hong Kong carrying wounded and fever-stricken troops from China. The local report read:

'. . . HM Troopship Rattlesnake sailed for England on Thursday last having on board a party of invalids from Hong Kong under the care of Dr Allan . . . We are sorry to learn that great mortality prevailed on board the Rattlesnake, both on the passage from Hong Kong, and during her stay in Harbour here. In the former period between 10th and 21st ulto, nine men, and in the roads 12 men died, and 13 more were so dangerously ill that they were not expected to recover . . .'[56]

In a Supplement published to the weekly *Singapore Free Press* newspaper on 4 July 1844 a further report concerning the sickness among troops in China was published this stated that

'. . . the sickness among the Madras troops is unquestionably partially caused by the improper way in which they are lodged, the huts not being fit for any human being to occupy in this climate . . .'[57]

The report was part of a military despatch received by the newspaper from Hong Kong. As late as December of that year troops were still passing through Singapore having been invalided from the China command. One vessel, the *Defiance*, was recorded as having arrived from Hong Kong '. . . with invalid troops of the 4th MNI . . .' the journey having taken from 16 November to 20 November during which time the vessel had encountered '. . . very bad weather . . . by which she suffered considerable damage . . .'

During the period 1843–1844 the defences of Singapore were again reviewed. The proposal that had been submitted by Major Edward Lake for the fortification of the town and harbour had been shelved in 1829. Captain Samuel Best, of the Madras Engineers, submitted a proposal for the fortification of the town and the 'new harbour' area. His work was based largely upon the recommendations made by Lake. Best submitted a proposal allocating a chain of twelve batteries between Tanjong Berlayer to Kampong Gelam, with a redoubt at Tanjong Rhu. Seven of the batteries were sited to defend the New Harbour and the remaining five to cover the waterfront from Telok Ayer to the Kalang Basin. Captain Best also suggested that part of the Sepoy garrison should be moved from the Sepoy Lines to a new barracks that were to be erected on Blakang Mati Island. Best referred to the island in his plan as the 'Island of St George'. Like Lake's plan the plan submitted by Captain Best was also shelved. The batteries were not built, the plan for a barracks on Blakang Mati was discounted as it was generally believed at that time that the island was far too unhealthy for habitation. Now, in 1984, Blakang Mati Island is called Sentosa Island and is a holiday and health resort. The only part of the 1843 plan to be implemented was the rehabilitation

Fort Canning Cemetery: memorial tablet to Diane Deare, wife of Conductor Deare, in the south-east wall of the cemetery. This photograph was taken in 1971 prior to the memorials in this section of the wall becoming covered with creepers. Photograph by Mrs J. Harfield.

albeit only temporarily, of the battery at Scandal Point and for the improvement of Fort Fullerton at the mouth of the Singapore River. This had a seawall built around it during the years 1844 and 1845.[58] Two maps are held in the Singapore Museum collection, one by J. T. Thomson dated 1843 which has annotations signed by Captain Samuel Best and the other is a manuscript map drawn by Best.

In August 1844 Captain Charles Edward Faber of the Madras Engineers was appointed Superintending Engineer of the Public Works of the Straits Settlements.[59] At this time the cemetery located on the side of Government Hill was still in use and in 1846 a brick wall was built which enclosed the whole cemetery. At the time of the construction of the wall two arches were designed by Captain Faber and built one on the south side and the other on the land side. The arches and the exterior walls still stand and as the tombstones and memorials have gradually been cleared from the centre of the ground the walls on the south, west and north sides have been used for the preservation of the memorial plaques. A wall was constructed to divide the Protestant and Catholic sections of the cemetery but this was removed sometime prior to 1949.[60] Captain Faber's duties included supervising work in the other two settlements and the *Singapore Free Press* notified his departure in October 1845 on one such visit:

> '. . . The HC Steamer Diana left this Settlement on Friday morning for Malacca, conveying Capt Faber, Superintending Engineer for the Straits Settlement . . . proceeding on his tour of inspection in the Diana . . .'[61]

News of the change-over of garrison regiments was notified in the local press in April 1846 when a report was published saying:

> 'We observe the 21st Regt Madras NI is to relieve the Regiment at present stationed in the Straits. The exchange will probably be effected by means of HM Steamer Spiteful and HC Steamer Tenasserim . . .'[62]

By 1846 the garrison force for the three stations had been reduced from two Native Infantry regiments to one and the unit covering the three stations in 1845 was the 27th Regiment Madras Native Infantry. On the last day of April 1846 the following report on the changeover of regiments was published:

> '. . . We understand that the HC Steamer Tennesserim [*sic*] has arrived in Pinang with part of the 21st Madras NI. The detachment destined for Singapore left Madras in the Argyle on the 20th and we suppose may be expected in about a month hence . . .'[63]

During May 1846 it was recorded that '. . . the Fort fired a salute at one o'clock (on Sunday last) being Her Majesty's birthday . . .', the fort on this occasion being Fort Fullerton. It was during that month that the 21st Regiment Madras Native Infantry arrived in the garrison. In June of that year it was reported that:

> '. . . The Band of the 21st Regt MNI performed at the Lines on Monday evening – the Band will perform twice a week, Monday and Friday evenings at the lines wind and weather permitting – the Band would be permitted by the officers of the 21st to perform once a week on the Esplanade . . .'[64]

In the same newspaper the account appeared giving details of the departure of the officers and men of the 27th Regiment. The report read:

> '. . . Early on Wednesday morning Lieut Col Watson, the Officers and Privates of 17th Regt MNI embarked on board the Duke of Argyle bound for Madras – and on Sunday morning Lieut J. D. Scott and the Artillerymen on Board the Indian Queen. During the residence of Lieut Scott in this settlement he won the respect and esteem of the community . . .'

Fort Canning Cemetery, the North Gate, *c*1912. Courtesy of the Director, India Office Library and Records.

Fort Canning Cemetery: the North Gate, 1972. Gateway designed by Captain Charles Edward Faber, Superintending Engineer, Straits Settlements, 1846. Photograph by Mrs J. Harfield.

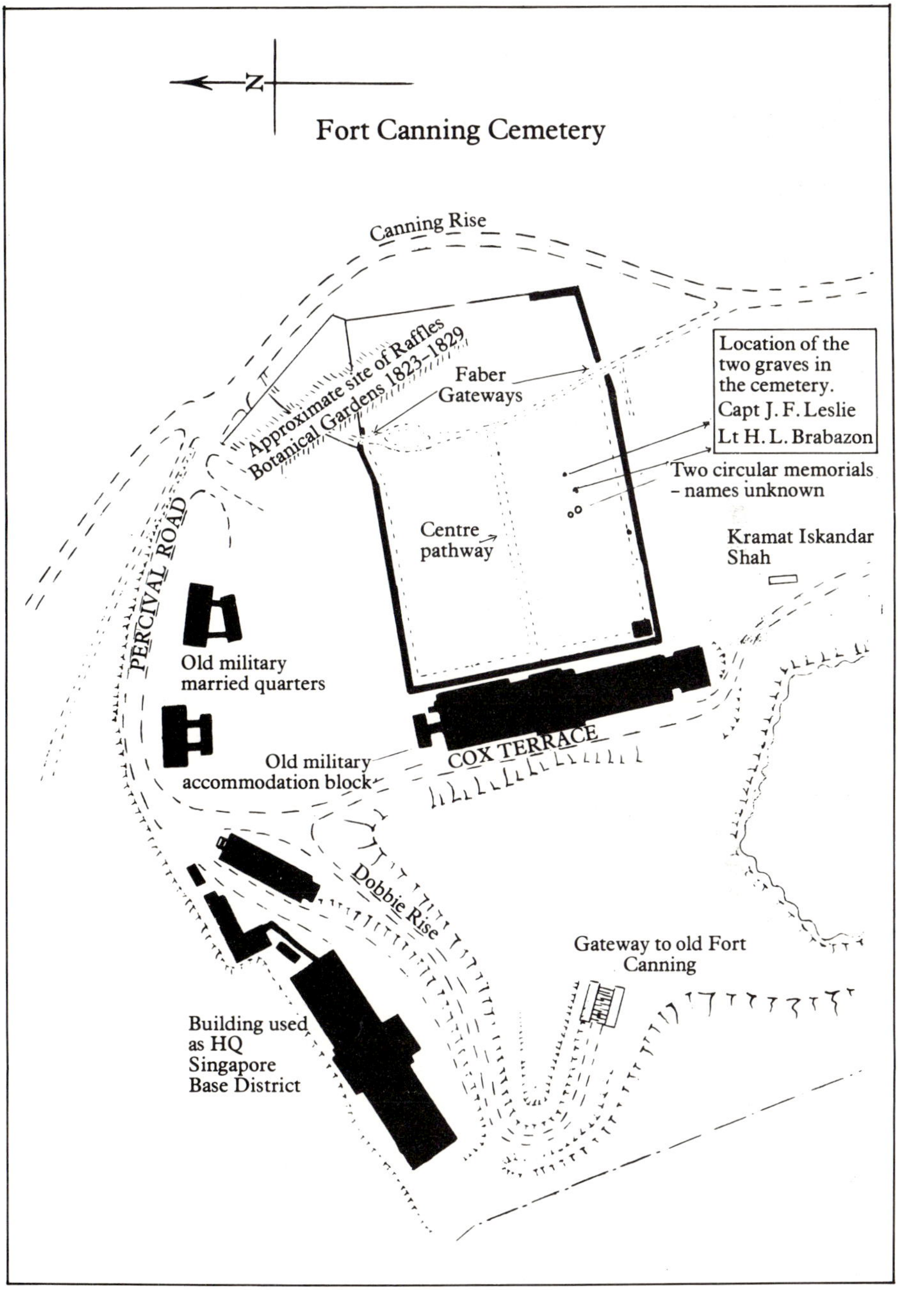

Plan of Fort Canning buildings as at 1971 and the cemetery.

Fort Canning Cemetery: south-west corner, *c*1912. Courtesy of the Director, India Office Library and Records.

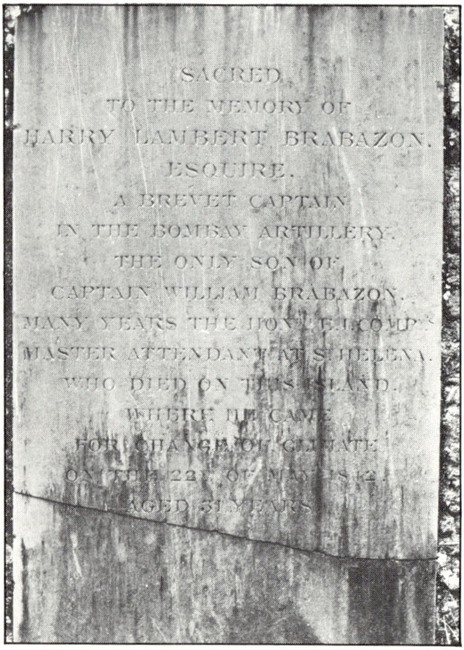

Fort Canning Cemetery: the Brabazon Memorial tablet in the south wall. Photograph by Mrs J. Harfield, 1977.

An important occasion took place in August 1846 which was commemorated in a painting by J. T. Thomson and was later published as a lithograph in 1856. Charles Burton Buckley gives a detailed account of the occasion in his work *An Anecdotal History of Singapore*:

> '... On Tuesday, the 31st August, the Sword of Honour, which the late Sultan of Johore carried with him on state occasions, and which was often seen at Government House on the Queen's birthday, was presented to the Tumoongong. The following inscription is on the sword – Presented in the year 1846 to the Tumoongong of Johore Sree Maharajah, by Lt-Col Butterworth, CB, Governor of Prince of Wales' Island, Singapore and Malacca, as a testimony of the high estimation in which the services of the Sree Maharajah in the suppression of piracy are held by the Government of India ...'[65]

The Governor invited the whole of the community to see the presentation which was held on Government Hill. A number of tents were pitched on the hill for the convenience of the public and '... preparations made on a most liberal scale ...' It was considered to be a holiday and from early in the morning Chinese, Malays, Javanese, Hindoos, etc all gathered on the slopes of the hill. Guns were removed from the battery and sited near to the Government House and two companies of the Madras Native Infantry, together with the Officers and the Band were all in attendance or 'on parade'. Buckley records that the unit was the '27th M.N.I.' but this would appear to be an error as that regiment had departed in June of that year and the garrison duties had been taken over by the 21st Regiment, Madras Native Infantry. It has been established that the 21st MNI had a Band which was already performing at functions in the Settlement and it is unlikely that two companies and the Band from the 27th MNI would have remained on station for this function. Buckley also stated that 'all the Officers' were in attendance, such a statement is far more likely to apply to the 21st Regiment who had its full complement on station.

At two o'clock '... His Highness left the Court House, accompanied by the Resident Councillor, the Sultan of Johore, the Sultan of Lingin, Tuanku Jaffar, Major Carthew &c and on arriving at the foot of the hill proceeded in the Governor's carriage until he reached the guard of honour, when he alighted and walked to the Government House, a salute being fired ...'[66]

On the presentation of the sword by the Governor a further salute was fired and after a number of speeches the Tumoongong departed with the firing of a further salute. At about 3 o'clock the guests, who amounted to about ninety, sat down to a '... splendid tiffin at the hospitable table of the Governor ...'

The officer mentioned in Buckley's book was in fact Major Morden Carthew, of the 21st Regiment, Madras Native Infantry, who was in command of the regiment at that time as the Commanding officer, Lieutenant Colonel G. Hutchinson, was on furlough. A close examination of the picture by J. T. Thomson also confirms that the unit present was that of the 21st Regiment MNI. The officer, accompanied by his family, in the centre of the picture is wearing buff facings with gold lace, which was worn by the 21st Regiment, whereas the 27th Regiment had black facings with silver lace.

During the following year, 1847, it was recorded in the *Singapore Free Press* '... At Singapore, on the 27th April, the Lady of Brevet Major Edward Faber, Madras Engineers, of a daughter ...'[67] and at the close of the year the press was very critical of the temporary loss of two officers from the garrison at Penang Island. The report read:

> '... The Mail Steamer started this morning at six o'clock. Captain Ferrier, Suptd of Province Wellesley, and Lieut Macpherson, Madras Artillery, in charge of the Fort and

Executive Officer, here, proceeded with her, passengers to Madras. We believe each of these officers have been summoned to appear and give evidence at a Court Martial to be held on Lieut McLeod of the 24th MNI, for an affair which happened four years ago when that regiment was stationed in the Straits . . . Thus for a private matter, between certain officers of the 24th Regiment, the public in the Province is deprived of our Suptd without there being any person in the place competant to discharge his duties in the interim . . .'[68]

At this time the military were getting bad press coverage and during the year 1848 several reports were published in the press which illustrated the deep rift between the military community and the editor of the *Free Press.*

The first of the articles appeared in January when it was reported that half a Company (*sic*) of European Artillery were being sent from Penang to provide reinforcements for the Hong Kong garrison. The report commented:

'. . .The small force stationed in the Straits has prevented the Governor from detaching any but the half Company of Artillery in answer to the requisition for troops which it is understood Sir John Davis has . . . requested . . . (the Governor considers it) imprudent to have diminished the troops at Singapore by a single man, both on account of the smallness of their number, and the large body of convicts (upward of 1,400) and the turbulent Chinese population whom they are required to overawe. Many of the Chinese are ready and prone to outrage, and they might have considered the small force left, had one or two Companies been despatched, as affording a favourable opportunity for plunder . . .'[69]

In the same paper a comment was published in the 'Monthly Summary' which again criticised the military and the system, the editor reporting on the departure of one of the military staff officers from the settlement:

'. . . Captain Cuppage of the Madras Army, lately Assistant Resident Superintendent of Police &c at Singapore embarked on board HMS Vernon for Madras on the 13 instant having previously resigned his appointments here . . . [Cuppage was being replaced by civilian personnel] . . . We trust that this is the beginning of a better order of things in regard to such appointments . . . and that men are to be chosen in future, rather from their ability to perform their duties of the office, than from motives of private friendship, or the possession of a showy uniform fitting the wearer to act on occasions as a gratuitous aide-de-camp . . .'

In the newspaper published on 27 January 1848 the editor continued his attack on the military by commenting on the need to provide reinforcements for Hong Kong. He comments:

'. . . We do not wish to draw invidious comparisons between the two stations, when we say that, in our opinion, the necessities of Pinang, could as little spare any reduction of the military force at the Station. Here the force is already sufficiently small for the usual duties of the place in a time of quietness. The troops do not number more than 300 men, and those in Hospital and on various guards give employment to fully the one half. Our population is large and mixed, our territory of considerable extent – a long line of which borders on the Kedah country, and we have no mounted police to afford protection. The events of July 1847 should not be forgotten. The Chinese forming a large portion of our inhabitants – then showed a dangerous spirit, collected in thousands in the Province to disturb the peace, and became so riotous that the Superintendent – Capt Ferrier, deemed it necessary to apply for a party of Sepoys to put them down . . .'[70]

With the setting up of the new settlement on Labuan there was an increase in the overall establishment for the Straits Settlements as the troops for Labuan were to be found from the resident unit stationed at Singapore. In May 1848 it was reported in the *Singapore Free Press* that '. . . the 51st Regiment MNI is expected here soon, and will be permanently stationed at Singapore furnishing a garrison of 200 men to the Settlement at Labuan. The present Regiment in the Straits will be divided between Pinang and Malacca . . .'[71] In fact the information was incorrect as the 21st Regiment Madras Native Infantry continued to serve at Singapore until it was relieved in April 1849. The 51st Regiment Madras Native Infantry was earmarked to serve at Penang.[72] It was supposed to provide the garrison troops for Penang Island and Malacca but in the event came to Singapore in April 1849.[73] The note of their arrival was published as follows:

> '. . . The HC Steamer Hugh Lindsay arrived here on the 5th from Madras having brought a detachment of Officers and men of the 51st MNI, which regiment will relieve the 21st MNI at present doing duty in the Straits. The Hugh Lindsay landed part of the troops at Malacca . . .'[74]

The decade closed quietly. The plans for the new defences of the town of Singapore and the New Harbour had, once again, been shelved, the work of the troops was a routine of drill and guard duties with the officers attending the social functions that were frequently held in what was now Victorian Singapore, Malacca and Penang. Nothing of great import was recorded in the local newspapers for 1849 and the main social event of the year was the marriage, in Singapore, of Lieutenant Walter Stuart Mann, of the Madras Infantry to Harriet Georgina, eldest daughter of the Hon'ble T. Church, the Resident Councillor at Singapore. The marriage took place on 26 April at St Andrew's Church which is opposite the site of the former gun battery which was known locally as Scandal Point.

The 1850s opened with a visit of the Governor General from India. Buckley records the event as follows:

> '. . . On Sunday night, the 17th February, shortly before midnight the EI. Company's man-of-war Feroze anchored in the roads, having on board the Marquis of Dalhousie, Governor-General, and the Marchioness of Dalhousie, attended by a numerous suite, which included Sir Henry Elliot, K.C.B., the Foreign Secretary to the Government of India . . . At an early hour on Monday morning the Governor, Colonel Butterworth, C.B., repaired on board the Feroze, when it was arranged that the landing should take place at half-past nine o'clock. Long before that hour arrived, the roads leading to the landing place were thronged with natives, all in their gala dresses, hastening towards the scene of disembarkation, where the 51st Regiment M.N.I. was drawn up. Two lines of sampans, manned chiefly by the Tumonggong's followers, in bright bajus and sarongs, formed a lane from the entrance of the river to the shipping, through which the procession of boats bearing His Lordship and suite passed to the landing place, where the great body of the European residents, H. H. the Tumonggong and sword bearers, the Heads of the Chinese tribes, and other principal native inhabitants were drawn up to receive him . . .'[75]

The local press reported that the visit had been a complete success and the Governor General departed on Wednesday 20 February at '. . . half past two o'clock pm., under a salute of 19 guns, the attendance at the landing place being similar to that which had assembled to honour his arrival . . .' The salute would have been fired from the battery at Fort Fullerton. A subscription was raised to commemorate the visit and a monument was erected on the Esplanade. This was later moved when the Esplanade was widened but in fact the 'Dalhousie Monument' can still be seen adjacent to Cavenagh Bridge. The obelisk was designed by John Turnbull Thomson and was erected during the same year as Lord Dalhousie's visit.

In the following month, March 1850, Colonel James Low a retired officer from the Madras Native Infantry left Penang Island. Low had first arrived at Penang, then referred to as Prince of Wales Island, during 1818 and over the years held important military and civil posts within the Straits Settlements and was a very controversial figure. James Low was born at Kettle, Co Fife on 4 April 1791 and received his education at Edinburgh College.[76] He was commissioned as an Ensign in the 25th Regiment, Madras Native Infantry on 11 June 1812, and was appointed Adjutant of the Regiment in May 1817. He was promoted to Lieutenant and joined the 23rd Regiment, Madras Native Infantry in August 1817. He moved to Penang and came under orders of the Government of Prince of Wales Island on 1 January 1818. Low soon settled into the social life of the Settlement and in June 1818 he was a witness at the double wedding of the elder daughter and niece of the Governor of the Settlement, Colonel Alexander Bannerman.[77] During 1820 Lieutenant Low was placed in charge of the newly-formed Penang Local Corps which had been raised by the East India Company to save expense of having a Regiment of the Madras Army permanently stationed on the island. The unit had been composed of men mainly recruited from '. . . the Upper Provinces of Bengal . . .' and was used to deal with the pirate menace and for the defence of the portion of the settlement on the mainland – Province Wellesley. Low remained in command of this local unit until mid-1827 when it was disbanded,[78] and the majority of the soldiers returned to their homes in Bengal.[79]

Low used his early years in Penang to good advantage and studied both the Malay and Thai languages and in 1824 was awarded a special bonus for his proficiency in the latter.[80] Due to his undoubted ability with languages he was given additional employment with several political missions, the first of which was in mid-1824 when he acted as envoy to the Raja of Ligor. With the outbreak of the Anglo-Burmese War it was hoped to obtain the loan of local boats for the British expedition up the Irrawaddy River to Ava.[81] The mission failed but Low's report showed his flair for this type of work and in September 1824 he was again selected and on this occasion served as assistant to Lieutenant Colonel Snow on a mission to Tennasserim. During the next twelve months, and in fact until October 1825 he spent the greater part of his time conducting surveys in Siam, Cambodia and Laos. On 6 May 1826 he was promoted to Captain in the 46th Regiment, Madras Native Infantry but continued to be employed in the Settlements. In September of that year he was given command of forty sepoys and a cruiser and despatched to Perak as the '. . . political and military agent of the Government . . .' His mission was to warn the Siamese, who had invaded Perak, to leave the state. The Siamese withdrew before Low arrived, nevertheless he remained at the capital of the state for a month and on 18 October 1826, grossly exceeded his authority in that he signed a Treaty with the Sultan of Perak which gave the Sultan the right to expel any Siamese from the State and committed the Company to come to the aid of the Sultan should he require assistance.[82] The Penang Council were not empowered to approve such a Treaty and it was then referred to the Government in India but never appears to have been formally ratified.

In the following year Low was again censured over another incident, which occurred when dealing with a local Chief, Nakhoda (Captain) Udin. Udin was in alliance with the Raja of Ligor and had been carrying out raids on shipping and during 1826 had even raided shipping anchored in Penang Harbour. Captain Low was ordered, early in 1827, to deal with the problem and set out with half a company of his Local Corps to attack and destroy Udin's stronghold. The stockade was located and attacked and without too much opposition destroyed. The followers of Udin scattered although Udin and some of his men were captured and brought back to Penang. The Penang court had no jurisdiction over the Chief and he was returned to the Raja of Ligor to be punished.[83] The Raja complained about Low's handling of the situation and declared that Udin was one of this own officials. The complaint was passed on by the Secretary – Captain Henry Burney (at whose wedding Low had been as witness) – to the Supreme Government and the outcome was that the Governor, Robert Fullerton and Captain Low were both censured. Low was suspended from all political employment. Fullerton prepared a good defence which convinced the Supreme Government that the situation had been handled correctly and the suspension on Low was lifted.[84] It is interesting to ponder on the situation at that time when Burney, the Secretary, had sent papers to India and had succeeded in having his superior and a fellow officer censured. It is no doubt significant that Captain Burney ceased to be Secretary to the Penang Government in 1828.

Low's next appointment was Superintendent of Police, Magistrate, Commissioner of the Court of Requests, Coroner, Supervisor of the Excise Farms and Superintendent of the Land Office for Province Wellesley, the whole being summed up in a short title of 'Superintendent of Province Wellesley'.[85] He was later to be criticised for the handling of the '... grants of leases and other revenue affairs ...' and he was instructed to submit '... more regular and precise reports ...'[86] J. T. Thomson, the Government Surveyor and author, alleged that Low had used his official position for personal gain and that he had obtained land within the Province by evicting the rightful native owners. This allegation was unjust as it was not publicly made until after Low's death and therefore he was unable to refute the allegations made in Thomson's book.[87] Low most certainly did acquire land but this was in all probability due to his being on the spot at the right time when land was being vacated by the local owners. T. J. Newbold, the author and also a fellow officer, refers to Province Wellesley as being '... under the able superintendence of Captain Low, of the Madras Service ...'[88] In December 1837 Low was appointed temporary Acting Resident Councillor at Prince of Wales Island[89] and on 23rd November 1839 he was promoted to the rank of major.[90] In 1840 he was called upon to relinquish his position as Administrator of the Province or dispose of his sizeable holdings of land in the Province. He objected to these alternatives and refused to surrender his properties. He had by this time been living in the Settlement for twenty-two years and, as he understandably did not wish to dispose of his lands he was given a new appointment as Assistant Resident for Singapore.[91] He remained in Singapore until March 1844 and Charles Burton Buckley lists, in his history of Singapore, the principal residents for 1841–1842 and shows Major Low as residing with '... Miss Low ... in the Pavilion in North Bridge Road ...'[92] An account of a farewell dinner in his honour was given in the *Singapore Free Press* and it would appear that he was well respected in the community. The report reads:

> '... it was not the intention of the Community to allow Major Low to leave the Settlement, without carrying with him some suitable manifestation of their sentiments and feeling in his favour and it was decided to give him a public dinner on his retirement from his present duties, in consideration of his long and useful career in the Straits and of the good feelings entertained towards him by the country. The dinner took place on Saturday last, at the house of the Recorder, a numerous assemblage, including nearly everyone of the principal gentlemen of the Settlement and many military officers of the station, sat down to table ...'[93]

He returned to Penang and on 31 December 1844 was promoted to Lieutenant Colonel in the 16th Regiment, Madras Native Infantry. He spent the next six months on leave in Penang and retired from the Madras Army on 21 November 1845. He remained in the Settlement, probably on his estate in Province Wellesley, until he returned to Europe in March 1850.[94]

Little of note is recorded for the years 1851, 1852 and 1853 and the routine of the three garrisons remained unchanged. The resident regiment was changed in April 1852 when the 51st Regiment Madras Native Infantry departed to Vellore and the 43rd Regiment Madras Native Infantry arrived on station. During 1853 the Deputy Assistant Commissary of Ordnance at the Arsenal at Penang died and his deputy was appointed to take charge. The notice read:

> '. . . Acting Deputy Assistant Commissary of Ordnance William Owen, to be Deputy Assistant Commissary of Ordnance from the 29th October 1853, vice Platt deceased, and to continue in charge of the Arsenal at Penang.'

The Garrison at Penang were, by this time, merely a subordinate unit to Singapore. The regiment providing the force for the defence of the Straits Settlements had its headquarters on Singapore Island and outstations at Penang Island and Malacca. A note in the Governor's Diary for 4 June 1853 records that the sum of Rupees 806-1-6 had been sanctioned for the renewing of nineteen huts which had been used by the 21st Regiment, MNI whilst they were stationed at Penang.[95]

During 1854 the question of the defence of Singapore was again raised and yet another report was submitted for consideration by the authorities. This report was submitted by Captain Henry Yule of the Bengal Engineers and in it he recommended the construction of two batteries. The Governor also ordered the construction of a powder magazine which had been sanctioned and at that time, April 1854, had not been built.[96] During the same year a deputation '. . . waited on the Governor to offer the services of a local Volunteer Rifle Corps to the Government and requested his Honour to propose a set of Rules for the Guidance of the Corps . . .'[97] This was, of course, the first of the Singapore volunteer units being formed and the history of this and the subsequent volunteer units is given in Chapter 12. The volunteer service of the inhabitants was accepted and the Governor forwarded a list of the volunteers to the Government in India.

Concern was felt over the lack of defence of the Settlement which consisted, at this time, of one regiment which was required to cover the three stations, detachments of artillery at each station and supporting services. The actual defences had been reduced to Fort Fullerton at Singapore, plus the recommendation of two additional batteries as yet unbuilt; at Malacca a small battery but no fort; and at Penang Fort Cornwallis, which had been allowed to deteriorate and was by this time little more than a guarded store-house. The Governor requested that reinforcements be provided for the Straits Settlements and recommended one additional regiment of Native Infantry, a company of Golandauze and two small howitzers for each of the stations, ie Penang, Malacca and Singapore.[98] To add to the problems Captain I. Ferrier, of the 48th Regiment Madras Native Infantry, who was serving as Resident Councillor at Malacca, died at that station on 24 July 1854.[99] In August 1854 a reply was received from India to the effect that a Company of Golundauze and a 'Wing of the 38th Regt NI' will be embarked for the Straits '. . . as soon as transport may be provided . . .'[100]

The defence position began to improve. Two small batteries were constructed, one at Mount Palmer, which became known as Fort Palmer and the other on Mount Faber, which then was known as the Mount Faber Battery. There is a record of practice firing having taken place from these two batteries during March 1855.[101]

Early in 1855 the 29th Regiment Madras Native Infantry arrived in the Straits Settlements and became the garrison regiment for Penang Island. The 38th Regiment Madras Native Infantry were due to arrive in April 1855 and the 43rd Regiment commenced embarkation in March of that year. The entry in the Governor's Diary read:

'. . . Colonel Cameron, 43rd Regiment [reported] regarding the embarkation of the Head Quarters and a further portion of the Regiment for Vizagapatam on the ship 'Burlington' . . . Enclosing a subsequent communication from the above officer reporting the embarkation of himself, Lieut C. James, Ensign Dobree, Asst Surgeon Mackenzie, 2 Native Officers, 91 non-commissioned rank and file and 49 followers on the above vessel in addition to the Malacca Detachment, it leaving the remainder of the Regiment to be forwarded by some other opportunity . . .'[102]

As a result of the death of Captain Ferrier at Malacca and later the death of Conductor James Craig, of the Ordnance Department, serving at Singapore a number of posting changes were made and these were published in the *Fort St George Gazette* and are quoted to give an example of the type of order that was published to authorise such moves during the mid-nineteenth century:

'The following removals are ordered:

Conductor R. Taylor	*From* Malacca	*To* Singapore – to remain at Malacca until relieved.
Conductor T. McLoughlin	*From* Penang	*To* Malacca to join forthwith
Conductor J. Crocker	*From* Arsenal of Fort St George	*To* Penang'[103]

'*Ordnance Department*

Acting Conductor William James to be Conductor and Acting Sub Conductor James Burke to be Sub Conductor from 20th December 1854, vice James Craig deceased . . .'[104]

'Captain H. Man Superintendent of Convicts and Roads at Singapore to be Resident Councillor at Malacca.

Captain R. Macpherson, Executive Officer and Superintendent of Convicts and Roads at Penang to be Superintendent of Convicts and Roads at Singapore . . .'[105]

The main source of labour for the building of roads and any fortification works, including the new batteries on Singapore Island, would have been from the convicts sent to the Straits Settlements and therefore of necessity the appointment of Superintendent of Convicts and Roads was created; one such appointment was held on the establishment of each of the three garrisons.

In May of that year a further posting order was issued and this confirmed the fact that each of the three garrison stations was provided with a military hospital. The posting order moved Assistant Apothecary J. Hargreaves, of the Madras Establishment from the '. . . Station Hospital, Singapore . . .' to the '. . . Station Hospital, Malacca . . .'[106]

An outbreak of diphtheria struck the Malacca garrison early in 1856 and four members of an officer's family died. Lieutenant Edward Lloyd Malcolm Evans of the 51st Regiment Madras Native Infantry served in Malacca during that regiment's tour of duty in the

Straits Settlements. On 16 January 1849 Lieutenant Evans married, at Malacca, Anna Maria Velge, the daughter of John Henry Velge of Malacca. The 51st Regiment left the station in April 1849 and either Lieutenant Evans family remained behind at Malacca, Evans visiting whenever he was granted leave, or the family may have returned to the Velge family home at Malacca for the birth of their third child. The tragic events commenced with the death of Robert Malcolm aged 2 months and 19 days on 16 January. On 24 January the mother, Anna Maria, and Edward Nagel, aged 3 years, 5 months and 12 days, both died. The eldest child, Louisa Augusta died shortly after on 2 February 1856, she was aged 4 years 9 months and 17 days. A tomb was erected in the St Paul's Cemetery at Malacca and a plaque was placed in Christ Church at Malacca. This was probably placed there after 29 May 1857 the date on which he was promoted to the rank of Captain as the inscription shows him as being 'Captain Evans'.

By 1856 the 29th Regiment Madras Native Infantry had settled down at Penang and was taking part in numerous social events. The band of the regiment was extremely popular and gave band performances in the Lines. The performance was given advance notification in the newspaper *Penang Gazette* in the following format: '... The Band of the 29th Regt MNI will perform the following at the Lines, on Monday and Tuesday next...' there then followed a list of the items that were to be played.[107]

The 29th Regiment were also to be remembered as within their ranks was the very accomplished artist Captain Charles Henry Cazalet. Cazalet was born on 12 May 1818, the son of Peter Reade Cazalet of the Madras Civil Service and his wife Caroline, who was the fifth daughter of Major General George Wahab. Charles was baptised at Vizagapatam on 28 November 1818 and was accepted as an Ensign from 31 January 1837. He joined for service on 6 July 1837 and was promoted to Lieutenant on 3 October 1840. He reached the rank of Captain on 12 August 1848 and in the following year proceeded on leave to the Cape of Good Hope from 1849 until 1851 and then on leave to Europe during 1852–1855. He returned to his regiment in time to proceed to Penang. The paintings by Cazalet are all dated 1856 and 1857 and give a superb insight into Georgetown in the mid-nineteenth century. His painting entitled *Penang from the Harbour 1856* is probably the one of greatest interest to military historians as it shows clearly Fort Cornwallis and Georgetown and shipping in the harbour. The paintings were sold as Lot 229 at Sotheby's auctions on 28 and 29 June 1971 and were at that time obtained by Messrs Francis Edwards Ltd of London. The portfolio was purchased by Mr Eric Nonweiler and Mr George Caldwell of Singapore and later six of the paintings were sold as a limited set of prints by the Times Publishing Berhad, Singapore in 1977. The six titles issued were:

Flagstaff, Penang Hill	Dated July 1856
Government House, Penang Hill	Dated August 1856
The English Church, Penang	Dated August 1856
Penang from the Harbour	Dated August 1856
Malay Huts MacPherson Bridge	Dated 15th September 1856
Glugor Residence of Forbes Brown Esq.	Dated 1st January 1857.

All the paintings were inscribed as having been presented to Mrs Forbes Brown. Cazalet left Penang Island with his regiment when it moved to Masulipatam. The regiment subsequently moved to Kamptee early in 1860 and whilst stationed there Captain Cazalet contracted cholera and died on 5 March 1860.[108]

The *Penang Gazette* recorded in June 1856 details of an accident which occurred during artillery practice from Fort Cornwallis. The report read:

'. . . We regret to mention that while the artillerymen were practicing on the ground near the Fort on Wednesday a ball struck a Kuag in a boat that was passing at the time . . .'[109]

and the edition published on the following Saturday carried another report:

'. . . The Coroner's Inquest in the case to which we referred to in our last issue closed on Saturday last with a verdict of "Accidental death" accompanied by a recommendation that more stict precautions should be used in future firing . . .'[110]

The Band of the 29th Regiment Madras Native Infantry continued to perform regularly and notices were published in the *Penang Gazette* giving details of the programme to be played, such notices were signed by 'John Brahm, Bandmaster'. The regiment lost its Commanding Officer, Lieutenant Colonel J. Richardson, who died suddenly at Penang on 15 September 1856. He had only taken over command of the regiment a short time before his death.[111]

Towards the end of the year 1856 the defences of Singapore once again received attention and authorisation was given for the construction of 'explosive magazines' to be erected at Fort Fullerton, Mount Faber and also for the new battery at Sandy Point.[112]

During the following year the Resident Councillor at Penang requested guidance on the question of the control of Sepoy military guards over the convicts housed in Jail and the convict Lines. The Officer Commanding Troops at Penang pointed out that '. . . a Military Guard over a Convict Jail is placed there to support the Civil Power in the exercise of its authority over its inmates, and in no way intended to act as the immediate Guardian of those inmates . . .'[113] With this statement the Governor concluded that situation was quite clear and that he could see no reason for any steps to be taken to modify the existing military regulation. There was, however, obviously some unrest brewing and in April 1857 the following commendation appeared in the Governor's Diary:

'Officer Comanding Troops Straits commends the conduct of the Garrison of Penang during the riots which lately occurred at that station and hopes that the men of the Guard who behaved so nobly on being attacked may be found deserving of approbation . . .'[114]

Buckley records in his *Anecdotal History of Singapore* that in May 1857 rumours began to arise '. . . in a vague uneasy way, about threatenings of coming trouble in India . . .' These were only suspicions which were founded on nothing more than bazaar gossip. It seemed to Buckley that it was possible that the convicts in Singapore may have had, as many natives in India had, some news of what was about to happen. On Sunday 31 May the opium steamer *Fiery Cross* arrived from Calcutta and with it arrived the first news of the mutiny in India. Some troops that had been destined for China were to be diverted to India and on 23 July 1857 a note was given to the Officer Commanding the Troops Straits requesting that '. . . he will be good enough to give directions to the Commissariat Depot at Singapore to furnish whatsoever supplies may in his opinion be considered necessary for the comfort of the European Troops shortly expected during their residence at Singapore . . .'[115] The troops stationed in the Straits Settlements remained firm and there was no doubt that those in command had every confidence in their own regiments staying loyal.

It was during the year 1857 that two troopships ran into trouble near to Singapore. The first was the P&O troopship *Himalaya* which was bringing the 90th Regiment of Foot to the Far East and was diverted from China to India. The vessel ran aground on a shoal in the Banka Straits but was able to get off again. She eventually arrived in Singapore and then whilst in port the regimental band of over fifty men, played on the Saturday evening on shore at New Harbour.[116]

The second incident occurred on 10 July when the troopship HMS *Transit* ran onto a sunken reef in the Straits of Banka. Buckley puts the location as off Cape Oelar on the Island of Banca [Banka]. Within ten minutes of the accident the ship had reached a sinking state and water was already covering her engines. The troops on board were listed as being from the 59th Regiment, 119 men; 90th Regiment, 286 men; Royal Engineers, 30 men and 193 men of the Medical Staff Corps. The ship sank so rapidly that only part of the arms were saved and many men had to leave the ship without boots and stockings and all the officers and men lost their kit. The *Transit* had had a most unfortunate voyage as she had departed from Portsmouth on 8 April for China and the following day she had to return to port as she had holed herself on the anchor. On 15 April she left dock but in doing so damaged her stern, however, this was not apparent until she ran into a storm in the Bay of Biscay and took in water and had to make for Corunna to effect repairs to the stern. She made good time to the Cape but then ran into more heavy weather and once again took in water from the stern.

On nearing her first port of call, Singapore, she was steaming at her best speed of eight knots when the vessel ran onto a sunken rock which was not shown on the chart. Buckley records that:

> '. . . Perfect discipline was maintained, the boats were got out, but they only held 200 men, and the ship seemed likely to go down before half could be landed. Captain Chambers ordered them to be landed on a reef two miles away, which was uncovered as it was low water. So all went first to the reef, and then to the shore; and the last trip was accomplished just in time, as the tide on the reef was rising, and was up to the knees of those who remained to the last. The discipline was compared at the time to that memorable incident of the Birkenhead . . .'[117]

Buckley made no mention of any casualties but in H.C.B. Rogers' book *Troopships and Their History* he quotes from a diary of a soldier of the 90th Regiment which records the following: '. . . we were all taken off on boats and landed all safe with the exception of one man . . .'[118]

The survivors all arrived at Singapore on 25 July and the European inhabitants seeing the deplorable state in which the men were clothed raised a subscription which, in two days, amounted to a sum which supplied each man with a pair of boots, socks, a flannel shirt and cloth sufficient to supply each soldier with a smock and a pair of trousers. The men were accommodated in a temporary barracks that had been erected for their use.[119]

In October 1857 the Governor wrote to India stating that: '. . . under existing circumstances I do not consider it advisable to send away the 38th Regt from the Straits thereby leaving only one Regiment to guard [the] garrison of Singapore, Malacca and Labuan . . .' He also commented on the high rate of sickness in each regiment and that in numbers one complete regiment would in fact produce the same number of men as were available for duty in the two regiments that were stationed in the Settlement at that time.[120] A reply was received from India and passed to the Officer Commanding Troops to the effect that the '22 Regt NI is under orders to relieve the 29th and 38th NI in the Straits . . .' the report also gave the strength of the 22nd Regiment MNI as being 'about 750 men'.[121] The total number of men at Singapore, Malacca and Labuan at the time was estimated to be 933 and it was suggested that a communication should be sent to Madras recommending that the 22nd Regiment Madras Native Infantry should be brought up to a strength of 1,200 men before embarking for duty in the Straits Settlements.

Notification was received in March 1858 that Madras had agreed to increase the strength of the 22nd Regiment to 1,000 privates. The 14th Regiment Madras Native Infantry had, in the meantime, arrived and had commenced to act as the garrison troops for Singapore. It was recommended at the time that the 22nd Regiment's strength should be not less than 1,000, that the strength of the 14th Regiment 'may not be allowed to fall below the original strength at which it arrived here [Singapore]'. Apart from the three garrisons within the Straits Settlements the combined force now had to man the newly-acquired Island of Labuan.

In March 1858, Captain Tranchell of the Ceylon Rifle Regiment[122] arrived at Singapore and in a letter to the Governor dated 9 March, he asked if the Governor thought that if he, Tranchell, went to the Island of Bally [Bali] for the purpose of enlisting Malays for the Ceylon Rifle Regiment he would be opposed by the Dutch Resident at Rhio under whose jurisdiction the island of Bali was for purposes of administration. He was advised that he should apply, in person, to the Resident at Rhio and that, if he required it, the Resident Councillor at Singapore would supply him with a written introduction.[123]

It would appear that Captain Tranchell failed to get permission to recruit on the island of Bali as on 23 March 1858 he asked the Governor for a letter to His Majesty the King of Siam with a view to obtaining permission to recruit in the state of Calantan and requested that a '. . . gunboat might be placed at his disposal to convey him to Bankok, Calantan, Tringano and Pahang . . .'[124] However a gunboat was not available and Tranchell had to be content with a letter which was addressed through HBM's Consul in Siam to His Majesty the King of Siam. A communication was received from the British Consul in Siam dated 14 July 1858 in which it stated that:

> '. . . Captain Tranchell of the Ceylon Rifle Regt was most graciously received by His Majesty the 1st King of Siam who gave him letters to the Governors of Kelantan and Tringano in order to assist him in his duties of recruiting . . .'[125]

Unfortunately the records do not show whether Captain Tranchell was successful in his recruiting campaign in Siam and Malaya.

In April 1858, Brevet Major James Jackson of the 14th Regiment Madras Native Infantry was appointed Secretary and ADC to the Governor of the Straits.[126] The 14th Regiment were at this time the garrison regiment for Singapore.

In the following month the question of local recruiting again appears in the Governor's Diary when Lieutenant Colonel David Pott, of the 47th Regiment Bengal Native Infantry, arrived in a Company vessel and wrote to the Governor saying that he had received instructions to enlist Malays into his Regiment (the 47th Regiment) for the purpose of filling existing vacancies and asked '. . . what kind of Malay is best suited for the purpose . . .' and '. . . the localities they are most likely to be found . . .' and concluded his letter stating that he intended to land a recruiting party.[127] He received a short sharp reply from the Governor stating that '. . . there is no chance of getting any Malays for enlistment in these Settlements . . .' and advising that a '. . . recruiting party would be useless . . .'[128] That appeared to be the finish of that particular recruiting episode.

At the end of May 1858 it became necessary for Captain Ronald McPherson, Madras Artillery, who was serving as the Resident Councillor at Malacca, to be relieved. Major James Jackson was appointed to that post and this set up the usual chain reaction of postings. Lieutenant Frederick Lyon Playfair was appointed Secretary and ADC to the Governor, in place of Major Jackson, and Lieutenant Charles Alexander McMahon who was stationed on Labuan Island was recalled and posted to Malacca, to relieve Lieutenant Playfair and take command of the artillery detachment at that station. Major Jackson assumed charge of the Malacca Residency on 26 May.[129] Captain R. Macpherson (*sic*) reported to the Acting Secretary (Lieutenant Playfair) prior to his departure from the Station on 3 June 1858. He was proceeding to Europe on a Sick Certificate but was later to return to the Straits Settlements and eventually died there in 1869. During the months of June and July other moves were notified as Lieutenant George Towers Hilliard, of the 50th Regiment Madras Native Infantry, who was serving as Superintendent of Convicts and Executive Engineer at Penang was recalled to India to rejoin his regiment who were engaged in 'Field Service' in the supression of the mutiny.[130] The *Government Gazette* of 23 July 1858 authorised his replacement as follows:

'... With the permission of the Colonel Commanding the Troops in the Straits, Lieutenant W. S. Mitchell of the 22nd Regiment Native Infantry has been appointed to act from the 12th instant, as Superintendent of Convicts and Executive Engineer at Pinang [*sic*] during the absence of Captain G. T. Hilliard or until further notice ...'

Mitchell was later to die in Penang on 22 February 1865 and it can be assumed that he was still filling a staff post as his regiment had departed from the Settlement in 1861. The entry in the Indian Army List for 1865 records his death as follows: '... 22nd NI Brevet Capt W. S. Mitchell. 22 Feb 1865. Penang ...'

The actual inscription on his tombstone reads: 'Sacred to the Memory of Major William Somerville Mitchell, 22nd Regiment. Aged 38 ...'[131]

Captain G. T. Hilliard departed from Penang on 17 July on board the P&O Company steamer *Pekin* which was bound for Galle and Madras.[132]

A Notice was published during June and July calling for the provision of ponies. The Notice read:

'... Wanted to purchase for the Artillery of this station [this was Singapore] 8 ponies suitable for draught, and from 12 to 13 hands in height. For further particulars apply at the Office of the Executive Engineer ...'[133]

At this time there were various moves taking place in preparation for the construction of the new defence fort that was to be built on what was then called Government, or Flagstaff Hill. In August, Captain George Chancellor Collyer, of the Madras Engineers was posted to the Straits Settlements from his previous post of Civil Engineer at North Arcot in India. He was appointed on 13 August 1858:

'... Captain G. C. Collyer of the Madras Engineers has this day assumed the charge of the whole of the Public Works Department in the Straits, as Chief Engineer, under the Bengal Code of Regulations ...'[134]

One other change was promulgated in the *Government Gazette* and that was in respect of the Public Works Department post in the garrison of Penang. The order read:

'... Corporal Michael Sweeney of the C Company 2nd Battalion Artillery, at present extra Overseer of Convicts at Penang, may be placed at the Governor's disposal to act as Assistant Overseer in the Public Works Department at that station ...'[135]

This order is one of the few that have been traced that actually nominated a soldier to an appointment.

Buckley, in his history of Singapore, records the events leading up to the construction of what was to become the major fortification of the island of Singapore as follows:

'... The whole labour of the convict body, both skilled and ordinary, was placed at his [Collyer's] disposal. Some of the military works comprised in the proposed fortification were at once commenced, and the convicts were placed on Government Hill to form a battery; and on Fort Fullerton with a view of rendering that Battery more serviceable by extending and widening it. The work executed by convicts in those two Batteries was considered to be of excellent quality, as good, if not probably better, than could have been obtained from free Chinese labour; and the convict body proved most useful in the new scheme of covering the hills and shores of Singapore with Batteries, Redoubts, Barracks, Magazines . . .'[136]

An 'Extraordinary' *Government Gazette* notice was published on Saturday 13 November 1858 which notified that:

> '. . . the Proclamation of Her Majesty the Queen in Council, of the assumption by Her Majesty, of the Government of Her Indian Territories, will be publicly read on the Maidan on the morning of Friday 19th instant, at seven o'clock. All officers of Government, of every class are hereby ordered to attend . . .'

It was with this notice that the Straits Settlements ceased to come under Company rule and then became a Colony.

With the reconstruction work being carried out at Fort Fullerton the final *Government Gazette* order for 1858 notified a change of procedure for the firing of salutes which read:

> '. . . To prevent accidents and for general information it is hereby notified that until further orders all Salutes will be fired by the Artillery from the Esplanade near the large tree . . .'[137]

This site was adjacent to the old battery site that had been known as Scandal Point. The year 1858 closed with the prospects of new and enlarged defences being provided for Singapore in the following year.

NOTES TO CHAPTER EIGHT

1. *Singapore Chronicle,* No 163, Thursday 17 June 1830.
2. East-India Register and Directory 1830, p211.
3. *Singapore Chronicle,* No 169, Thursday 9 September 1830.
4. East-India Register and Directory, 1830, p108. Medical Establishment 'Assist Surg James George Sim.'
5. *Singapore Chronicle,* No 171, Thursday 7 October 1830, Supplement.
6. Ibid, No 172, Thursday 21 October 1830.
7. Ibid.
8. Ibid.
9. *Singapore Chronicle,* No 177, Thursday 30 December 1830.
10. East-India Register and Directory 1830, p56.
11. East-India Register and Directory 1832, p269.
12. R. Montgomery Martin, *History of the British Colonies,* vol 1, (Cochrane and McCrone, London, 1834), pp430–431.
13. *Singapore Chronicle,* No 1 New Series, Thursday 6 January 1831.
14. Ibid, No 2, Thursday 13 January 1831.
15. Ibid, No 11, Thursday 17 March 1831.
16. Ibid, No 31, Thursday 4 August 1831.
17. Montgomery Martin, op cit, p431.
18. *Singapore Guide and Street Directory,* 10th Edition, 1975 p9.
19. *Singapore Chronicle,* No 45, Thursday 7 November 1833.
20. *Fort Canning Cemetery, Singapore,* (published by BACSA, 1981), p57.
21. *Singapore Chronicle,* vol4, No 1, Thursday 2 January 1834, 'QM Sgt and Band Master W. White.'
22. The Hodson Index, National Army Museum, London.

23. *Singapore Chronicle,* vol4, No5, 30 January 1834, and East-India Register and Directory 1834, p58 (Fort St George).
'Lieutenant Colonel Thomas King
Captain Edward T. Hibgame
Captain Edward Servante
Lieutenant Richard Hurlock
Lieutenant Richard Henry Symes
Lieutenant Alexander Wood
Ensign William Lovell Boulderson
Ensign Henry George Free.'
24. *Singapore Chronicle,* vol4, No20, 15 May 1834. East-India Register and Directory 1834, p52 (Madras).
25. *Fort Canning Cemetery Singapore,* op cit, p74.
26. Charles Burton Buckley, *An Anecdotal History of Old Times in Singapore 1819–1867,* (University of Malaya Press, 1969), p288.
27. *Singapore Chronicle,* vol5, No28, Saturday evening 11 July 1835. '4 July Captain Ricketts and Lieutenant Watt' arrived and on '7 July Lieutenant Taylor' arrived (see Note 28).
28. East-India Register and Directory 1836, p77 (Fort St George).
Captain Richard Rodney Ricketts
Captain Robert Watts
Lieutenant Charles Taylor
29. *Singapore Free Press,* Tuesday 6 March 1860.
30. Ibid, Thursday 8 March 1860. '. . . The melancholy event referred to in our last issue which has caused Jackson's Hill to be avoided for 25 years, . . . is thus recorded:

This Monument
is erected
By his Brother Officers and Friends
To the Memory of
Lieutenant Colonel Gregory Jackson
15th Regiment MNI
Aged 55 years
Also of
His wife Matilda Jackson
Aged 39 years
And their Eldest son Gregory Jackson
Aged 21 years
Both parents expired on the morning of
The 1st April 1835, and their Son
The previous afternoon, from Jungle Fever'

31. *Singapore Chronicle,* vol5, No46, Saturday evening 28 November 1835.
32. *Singapore Free Press,* No27, Thursday morning 7 April 1836.
33. Ibid, vol1, No30, 28 April 1836.
34. East India Register and Directory 1836, p77 (Fort St George) Lieutenant John Pinsent Germon.
35. *Singapore Free Press,* vol1, No64, 22 December 1836.
36. Ibid, vol2, No31, 3 August 1837.
37. Ibid, vol2, No38, 21 September 1837.
38. *Pinang Gazette and Straits Chronicle,* vol1, no34, Saturday 24 November 1838, (National Library, Singapore, Microfilm No5147).
39. Ibid, vol1, No37, 15 December 1838.
40. Ibid, vol1, No39, 29 December 1838.
41. *Singapore Free Press and Mercantile Advertiser,* vol3, No50, 13 December 1838.
42. Ibid, vol4, No8, 21 February 1839.
43. *Singapore Free Press,* vol4, No36, 5 September 1839.
44. Ibid, vol 5, No37, 10 September 1840.
45. Ibid, vol5, No38, 17 September 1840.
46. Ibid, vol7, No23, 9 June 1842.
47. Ibid, vol7, No38, 22 September 1842.

48. Ibid, vol8, No8, 23 February 1843.
49. Ibid, vol8, No26, 29 June 1843.
50. Ibid, vol8, No27, 6 July 1843.
51. Ibid, vol8, No38, 21 September 1843.
52. Ibid, vol8, No43, 26 October 1843.
53. Ibid, vol8, No45, 9 November 1843.
54. Ibid, vol9, No4, 25 January 1844.
55. Memorial Inscription at Fort Canning Cemetery:

Sacred
to the memory of
DIANA,
The beloved Wife of Mr Conductor W. Deare,
of the Ordnance Department,
Who Departed this life 2nd Novr 1845,
Aged 33 years & 4 months.
ALSO
Of their infant Son
WALTER MILLER,
Who departed this Life,
On the 10th of January 1844,
Aged 2 Months & 22 Days.
Yeathard & Co. Scts.
Calcutta.

56. *Singapore Free Press,* vol9, No1, 4 January 1844.
57. Supplement to *Singapore Free Press* dated 4 July 1844.
58. G. A. Gibson-Hill, MA, *Singapore Old Strait & New Harbour 1300–1870,* (Government Printing Office, Singapore, 1956), pp 83 and 106.
59. *Singapore Free Press,* vol9, No35, 29 August 1844.
60. Fort Canning Cemetery. The author first visited the cemetery in 1949 and although there were, at that time, a reasonable number of tombstones and memorials still standing there was no trace of the wall between the two sections.
61. *Singapore Free Press,* vol10, No44, 30 October 1845.
62. Ibid, vol11, No17, 23 April 1846.
63. Ibid, vol11, No18, 30 April 1846.
64. Ibid, vol11, No25, 18 June 1846.
65. Buckley, op cit, pp 449 and 450.
66. Ibid, p450.
67. *Singapore Free Press,* vol 12, No17, 29 April 1847.
68. Ibid, vol12, Supplement dated 30 December 1847.
69. Ibid, vol13, No1, 6 January 1848.
70. Ibid, vol13, No4, 27 January 1848.
71. Ibid, vol13, No18, 11 May 1848.
72. East-India Register and Army List for 1849.
73. East-India Register and Army List for 1850.
74. *Singapore Free Press,* vol14, No10, 8 March 1849.
75. Buckley, op cit, p527.
76. The Hodson Index, National Army Museum, London.
77. *Prince of Wales Island Gazette,* vol4, No164, 4 July 1818.
78. James Low, 'An Account of the Origin and Progress of the British Colonies in the Straits of Malacca', published in the *Journal of the Indian Archipelago,* vol4, 1850.
79. Penang Orders and Notices, Military. Entry for 20 February reads '. . . The party of the late Local Corps sent to Bengal on the ship Flora . . .'
80. L. A. Mills, *British Malaya 1824–67,* (Oxford University Press – Oxford in Asia Historical Reprints, Kuala Lumpur, 1966). Mills states '. . . In some cases a much larger bonus was granted (for language qualifications). Lieutenant Low for example was given £450 in 1824 for proficiency in Siamese . . .'
81. Mills, op cit, p137.

82. Ibid, pp 157–158.
83. Ibid, p160.
84. Ibid, p161.
85. Penang Orders and Notices, 1828. Appointments. Dated 10th September 1828. '... Captain James Low – Suptd External Police and Superintendent Lands on a salary of 500 Rupees p. month...'
86. James Low, *The British Settlement of Penang*, (originally published Singapore 1836, republished in Oxford in Asia Historical Reprint series in Kuala Lumpur, 1973), pxi (introduction).
87. J. Turnbull Thompson, *Some Glimpses into Life in the Far East*, (London, 1864).
88. T. J. Newbold, *British Settlements in the Straits of Malacca*, vol1, (originally published in London 1839, republished in the Oxford in Asia Historical Reprint series in Kuala Lumpur, 1971), p105.
89. Madras Army Service List, India Office Library, L/MIL/11/39, folio 472.
90. The Hodson Index, National Army Museum, London.
91. Madras Army Service List, L/MIL/11/39, folio 473.
92. Buckley, op cit, pp 376–377.
93. *Singapore Free Press* dated 4 July 1844.
94. James Low returned to his native Scotland and died within two years at Portobello on 2 May 1852. He wrote some twenty-five books and articles including *A Grammar on the Thai or Siamese Language*. His main work was published in Singapore, by the *Singapore Free Press* in 1836 under the title: *A Dissertation on the Soil and Agriculture of the British Settlement of Penang or Prince of Wales Island in the Straits of Malacca: including Province Wellesley on the Malayan Peninsula, with brief references to the Settlements of Singapore and Malacca, and accompanied by incidental observations on various subjects of local interest in these Straits.*
95. Governor's Diary – Military 1852–1857, National Library, Singapore, film No3139, entry No 1 of 4 June 1853.
96. Ibid, Entries 13 and 14 of 19 April 1854.
97. Ibid, Entry No4 of 10 July 1854.
98. Ibid, Entry No9 of 15 July 1854.
99. Ibid, Entry No19 of 28 July 1854.
100. Ibid, Entry No3 dated 2 August 1854.
101. Ibid, Entries 26 and 28 of 21 March 1855.
102. Ibid, Entry No10 of 8 March 1855. Officers mentioned are: Lieutenant Colonel M. Cameron, Lieutenant Charles James, Ensign F. W. Dobree, Assistant Surgeon D. MacKenzie.
103. *Fort St George Gazette*, No2861, 23 January 1855. GO dated 18 January 1855.
104. Ibid, GO dated 23 January 1855.
105. Ibid, No2868, 16 February 1855. GO dated 31 January 1855.
106. Ibid, No2906, 18 May 1855. AGs Office order dated 16 May 1855.
107. *Penang Gazette*, volXIV, No16, 6 March 1856 and No18, 3 May 1856.
108. Captain Charles Henry Cazalet died of cholera at Kamptee, India on 5 March 1860. His will which was written in May 1859 left most of his estate to Charlotte Elizabeth Jellico. She had, on 11 May 1859, married his brother Captain Peter Grenville Cazalet who also served in the 29th Regiment, Madras Native Infantry. To 'his brother officer and friend, Lieutenant William Moore Dickenson' also serving in the 29th Regiment, he left five hundred rupees and a further five hundred rupees to his 'bearer' Shaik Adonis. There is a memorial plaque in the church at Kamptee which reads:

'Sacred to the memory of
Charles Henry Cazalet
29th Regiment M.N.I.
who departed this life at Kamptee
5th March, 1860. Aged 41.
This tablet is erected to him by his
brother officers as a mark of their
respect and esteem of his memory.'

109. *Penang Gazette*, volXIV, No23, Saturday 7 June 1856.
110. Ibid, volXIV, No24, Saturday 14 June 1856.
111. Ibid, volXIV, No39, 27 September 1856.
112. Governors Diary, Military 1852–1857. Entry No4 dated 5 November 1856.
113. Ibid, Military. Entry No19, 9 March 1857.
114. Ibid, Military. Entry No9, 26 April 1857.

115. Ibid, Military. Entry No20, 23 July 1857.
116. Buckley, op cit, p655.
117. Ibid, p656.
118. Colonel H. C. B. Rogers, *Troopships and their History,* (Seeley Service, London, 1963), p126.
119. Governors Diary, Military. Entry No21, 25 July 1857.
120. Ibid, Military. Entry No5, 12 October 1857.
121. Ibid, Military. Entry No11, 25 December 1857.
122. Army List 1858. Officer is Captain Edward Frederick Trenchall, Ceylon Rifle Regiment, p323.
123. Governors Diary, Military. Entry No7, 9 March 1858.
124. The two States referred to are now part of Malaysia and are now known as Kelantan and Trengganu.
125. Governors Diary, Military. Entries dated 3 April and 2 August 1858.
126. Ibid, Military. Entry No5, 5 April 1858 and East-India Register and Army List for 1858.
127. Governors Diary, Military. Entry No1, 5 May 1858.
128. Ibid, Military, Entry No2, 5 May 1858.
129. Ibid, Military. Entry No9 of 31 May and entries No1 of 1 June and No7 of 29 June 1858, also *Straits Settlements Government Gazette* No22 of 28 May 1858.
130. *Straits Settlements Government Gazette,* No30 of 23 July 1858. (National Library, Singapore, Film No994).
131. Major William Somerville Mitchell is buried in the Northam Road Cemetery at Georgetown, Penang Island. The inscription is from notes taken by the author during a visit to Penang Island in 1977. During this visit details of all the inscriptions on the military tombstones were recorded, both in note form and by photographic record.
132. *Straits Settlements Government Gazette,* No33 of 30 July 1858.
133. Ibid, No24 of 11 June 1858.
134. Ibid, No33 of 13 August 1858.
135. Governors Diary, Military. Entry dated 10 July 1858.
136. Buckley, op cit, pp 665 and 666.
137. *Straits Settlements Government Gazette,* No53 of 31 December 1858.

CHAPTER NINE

Labuan Island 1846 to 1935

The main problem when writing the military history of Labuan Island is that, apart from a few general references, very little has been written dealing solely with the island.

A book entitled *Labuan Story*, written in 1958,[1] records that nearly all the books dealing with the history of the island were destroyed during the bombardment and battle which ensued in June 1945 when the Allied forces re-occupied the island after defeating the Japanese forces stationed there. The author states that '. . . the town of Victoria was in fact blown off the face of the island, and the only books that survived were found later in the Hospital, secure under the Red Cross. These few books were sent to the Agricultural Station on Hamilton Road in the middle of the island to be sorted out. They were law books. There was not a history amongst them . . .'[2]

Labuan Island, or Pulo Labuhan, is from the Malay and Javanese word *labuh*, meaning to drop or let fall, in this context to drop or cast an anchor, thus referring to an anchorage or harbour.[3] The island lies to the north-east of the estuary of the Brunei River, and is triangular in shape with the base being in the south. It is twelve miles in length and is about six miles across at its widest point. Assessment of its actual area varies but generally it is accepted as being either thirty-four or thirty-five geographical square miles.

The general reason for the occupation of the island by the British was given as the suppression of piracy and the need to establish a naval station in Borneo, although it was decided that Labuan Island provided a better site for the naval anchorage than did the coast of Borneo. The decision to use Labuan Island as a base was debated as early as 1844 when it was stated that the island was not to be a new colony but '. . . only a naval and military station . . .'[4] it was also stated at that time that the island was not to be under the control of the East India Company. There were arguments for the abandonment of the whole project but these were set aside as there was the distinct danger that if the British failed to occupy Labuan Island then the Dutch, or the French or even the Americans, all of whom were seeking a foothold in that area, might take over the option of the occupation of the island. The whole question of the occupation or non-occupation was complicated by the negotiations that were being carried out by James Brooke with the Sultan of Brunei. In the midst of the negotiations various members of the Brunei Royal family were assassinated. The Government reacted to the situation by agreeing that Labuan should be occupied and a 'Naval Station and Harbour of Refuge for British vessels' established, as it was feared that Sultan Omar Ali, to whom the attrocities were attributed, might renew his association with the pirates that abounded in the area and might then place '. . . British Subjects and property in greater jeopardy than before . . .'[5] The Navy were instructed that they were not to take part in the conflict within Brunei and that their objective was to merely maintain a force sufficiently large enough to safeguard British interests.

The forces available were to be employed in the suppression of piracy and in the setting up of the naval station. The instruction stated that '. . . The Admiral should occupy Labuan

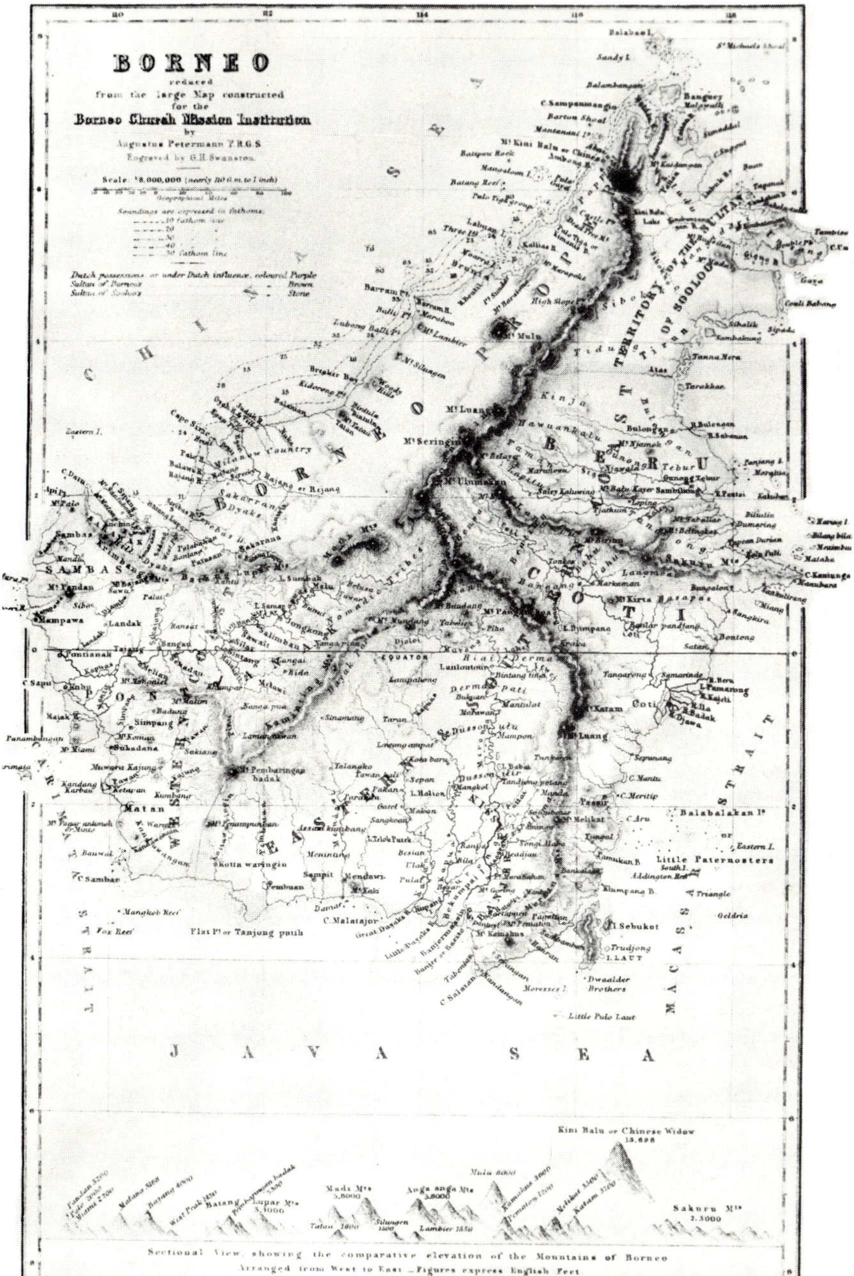

Map of Borneo, *c*1870.

and place some ships there pending the formation of a settlement . . .'[6]

In the meantime the situation was complicated as in 1845 HMS *Driver*, under command of Captain Bethune, visited Labuan and a report published in the *Singapore Free Press*[7] records that he brought back specimens of coal which was said to be of an excellent quality. This was therefore a further reason for the establishment of a settlement on the Labuan island.

After consultations with James Brooke, Admiral Sir Thomas Cochrane decided to visit Brunei Town but as he approached, the ships in his squadron were fired upon by the Brunei batteries. Cochrane ordered fire to be returned and the Brunei batteries were quickly put out of action and the town captured. The Sultan Omar Ali fled but was later reinstated.

The actual occupation of the island took place at the end of 1846 and for those readers who wish to follow the long and complicated story of the political negotiations leading to the occupation these are recorded fully in two main works, these being *Britain, the Brookes and Brunei* by Nicholas Tarling[8] and his other work dealing with the area entitled *British Policy in the Malay Peninsula and Archipelago 1824–1871.*[9]

The expedition to establish the settlement on Labuan Island formed at Singapore. Captain G. R. Mundy RN, commanding HMS *Iris* waited at Singapore for the arrival of HMS *Wolfe* from Hong Kong and on arrival had the two ships loaded with '. . . an ample supply of tents, awnings, tools, and other gear for clearing the jungle and setting up the first camp in the new colony . . .'[10] Prior to the departure of the two ships a Mrs Church, of Singapore, had a final gift of flowers, plants and seeds sent on board. The expedition set sail from Singapore at daylight on 1 December 1846 calling first at Kuching in order that Captain Mundy could meet with James Brooke prior to his proceeding along the coast of Borneo and negotiating the final treaty with the Sultan of Brunei. The month of December is generally considered to be the wettest month of the year on most of the west coast of Borneo and the time when the north-east monsoons can be expected, consequently the warships had a rough passage and sheltered in the Sarawak River whilst Captain Mundy visited James Brooke. After a stormy passage along the coast the two warships anchored off Baram Point, and Mundy recorded in his diary:

> '. . . Suddenly we observed the summit of the great mountain of Kinabalu towering above the clouds, its fantastic and rugged peaks clearly pencilled on the blue sky beyond, the distance from the ship at the time being upwards of 120 miles . . .'

At noon on that day he observed Labuan and wrote:

> '. . . Ten days from this date Labuan will be British territory. I am proud to think that it will be my lot to hoist the British flag thereon for who can predict the results which may arise from this insular possession, commanding as it does the whole north-west coast of Borneo? If followed up, our influence will be firmly established in the eastern archipelago . . .'[11]

By this time HMS *Iris* had been separated from HMS *Wolfe* but Mundy decided not to wait for his support ship to catch up with him and despatched Lieutenant Heath in the *Iris*' boats with written orders which were translated to the Pengirans of Brunei. The order set out to the Sultan and his advisers the outrage felt by Britain over the last visit of a British Naval vessel, which had been fired upon, and the need for Captain Mundy to formally take possession of Labuan Island from the Sultan of Brunei. The alternative was stated to be another visit by Admiral Sir Thomas Cochrane, James Brooke and the whole British Navy's China Squadron. Mundy had to wait all day for the return of Lieutenant Heath who, on his return confirmed that the Sultan would meet Captain Mundy and that he

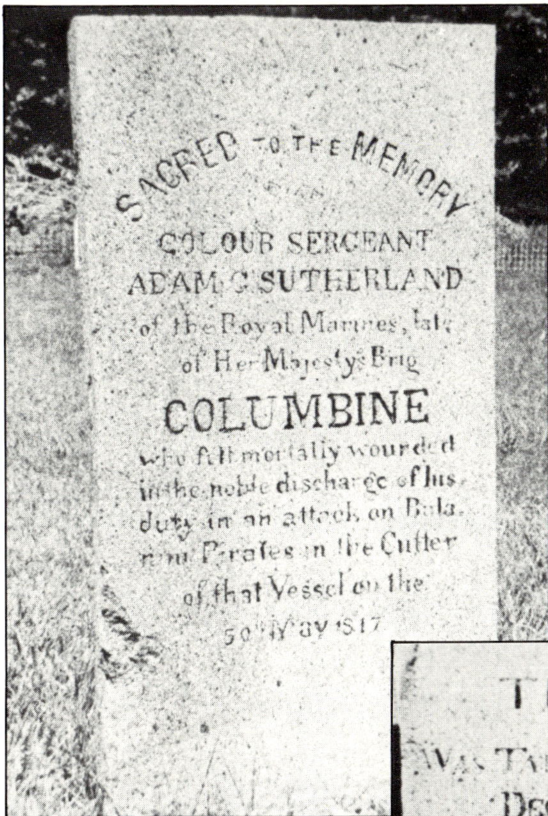

Labuan: gravestone of Colour Sergeant Adam G. Sutherland, Royal Marines, this stone was originally in the Old Naval Cemetery on the foreshore of Labuan Island. It and other memorial stones are now relocated in the Botanical Gardens.

See note 17.

Photograph by Mrs J. Harfield.

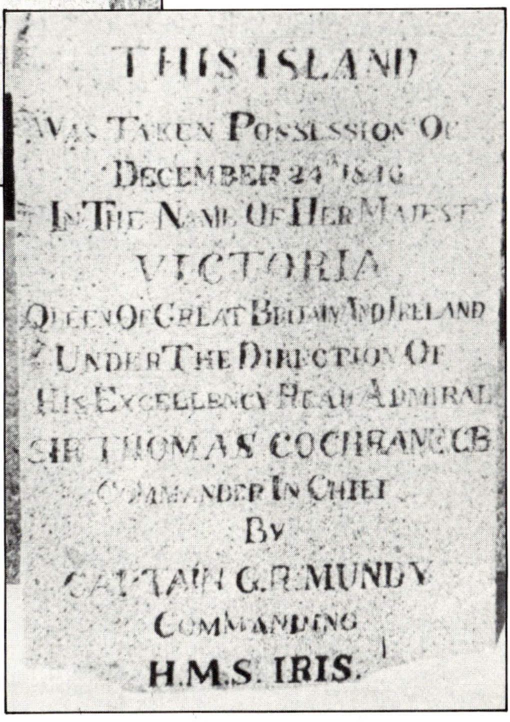

Labuan: commemorative stone at Bandar Labuan.

See note 14.

Photograph by Mrs J. Harfield.

would sign the treaty. On the following day Lieutenant Little and a party of twenty Royal Marines went by ship's boat to Brunei Town and awaited the arrival of Captain Mundy. After the long row up the Brunei River Captain Mundy arrived and reviewed the twenty marines who had lined the shore with bayonets fixed. The treaty was signed by Sultan Omar Ali and by Captain G. Rodney Mundy, on 18 December 1846 and read:

'1. Peace, friendship, and good understanding shall subsist for ever between Her Majesty the Queen of Great Britain and Ireland and His Highness the Sultan of Borneo, and their respective heirs and successors.

2. His Highness the Sultan hereby cedes in full sovereignty and property to Her Majesty the Queen of Great Britain and Ireland, her heirs and successors for ever, the island of Labuan and its dependencies, and islets adjacent.

3. The Government of Her Majesty the Queen of Great Britain and Ireland hereby engage, in consideration of the cession above specified, to use its best endeavours to suppress piracy and to protect lawful commerce, and the Sultan of Borneo and his ministers promise to afford every assistance to the British authorities.

Done and concluded at Brune' the 18th day of December, 1846 . . .'[12]

At the time of the treaty, the State of Brunei, which then covered the areas later known as Sarawak and North Borneo, as well as the present-day State of Brunei, was often referred to as the State of Borneo. On the conclusion of the signing of the treaty the Brunei batteries fired a salute of fifteen guns.

Captain Mundy and Lieutenant Heath landed on Labuan Island and fixed the spot for the flagstaff and for the tents for the initial landing parties. The area chosen was flat, clear of jungle and covered with a coarse grass although the area was swampy in places. In the afternoon a working party was landed, clearing was started and the tents were pitched. Good water was found only about three or four feet below the sand near to the seashore.[13] HMS *Wolfe*, which had been delayed in the Brunei River effecting repairs after it had been damaged by severe storms during its voyage from Kuching, arrived and further working parties were put ashore. The flagstaff was erected twenty paces from the shoreline and guns were landed and placed in a battery thus setting up the first British defence position on Labuan Island.

On 24 December a granite slab, bearing an inscription, was erected close to the flagstaff. The slab had been brought from Hong Kong by the warship HMS *Wolfe* and the inscription had been made by the sailors on board ship.[14] Pengiran Mumin representing the Sultan of Brunei arrived to take part in the ceremonies. One hundred and eighty sailors and marines landed first and at 2.00 pm, on a blazing hot day, Captain Mundy and Commander Gordon each left his ship and proceeded ashore where they arrived just ahead of the Pengiran and so were able to meet him and escort him through the files of marines to an elevated platform that had been erected for the occasion. Captain Mundy made a short speech in which he declared that '. . . I make known to all present that on this day I take possession of the island of Labuan and of the small islands around it in the name of Her Majesty the Queen of England . . .'[15] Midshipman Henry Morgan then hoisted the Union Jack. The ceremony was closed with refreshments being provided for Pengiran Mumin and the party of nobles.

During the last few days of the year working parties continued to clear the ground and a guard house was constructed as well as a pier. There were a number of cases of fever and dysentery and Commander Gordon of HMS *Wolfe* was taken ill and eventually died on 6 January 1847.

Captain Mundy selected a site for the cemetery, which was off to the right of the main settlement area. At five o'clock in the evening the funeral party landed and after a brief service which was read by Captain Mundy, Commander Gordon was buried. Some officers of HMS *Wolfe* remained ashore to plant a few shrubs. A memorial stone was erected which read

Sacred to the memory of
James A. Gordon Esquire
Commander of H.M.S. Wolfe
Son of Rear Admiral Sir James A. Gordon, K.C.B.
who departed this life
at Labuan, 6th January 1847
Aged 28 years[16]

The British having established the settlement had initially, until troops could be sent, to defend the new settlement with part of the force that had been used to take possession of the island. Consequently when Captain Mundy finally left Labuan on board HMS *Iris* on 12 January 1847 he left Lieutenant Heath in charge of the island and its defence. Heath was promoted to Commander and given command of the warship HMS *Wolfe.*

Having established themselves on the island the policy of breaking the pirate rule along the coast continued and a gravestone in the cemetery in memory of Colour Sergeant Adam C. Sutherland of the Royal Marines records that he was mortally wounded whilst attacking pirates on 30 May 1847.[17]

Initially the defence of Labuan appears to have remained the responsibility of the Royal Navy and the Royal Marines of whichever naval ship was stationed at the island. HMS *Meander* was on duty at Labuan Island during the latter part of the year 1848 and a report appeared in a Singapore paper on 21 December of that year giving a very adverse account of the conditions of the forces stationed on the island reading:

> '. . . Amongst the Marines of the Meander who were left as a guard at Labuan the fever has been extremely fatal. Out of 30 men disembarked for duty on the island one Sergeant, two Corporals and seven privates have been cut off, and it appears that 15 were on the sick list at one time . . . we learn that within a short distance of the quarters of the Marines, a grog shop keeper from Singapore planted his store and had a most active demand for his liquor . . . can it be wondered at the persons who, after indulging themselves to excess in the stuff, expose themselves to the climate, lying down and sleeping upon the damp ground, should be speedily cut-off . . .'[18]

This report is a prime example of bad press coverage and stems from the fact that the editor of the newspaper concerned had not been invited to a reception in Singapore given by the future administrators of the Labuan settlement and feeling himself slighted gave bad press coverage to items concerning Labuan. It is unlikely that the Marines from HMS *Meander* would have been undisciplined, as the article implies, as the Captain of the 44-gun warship Captain The Hon Henry Keppel RN, who was later to become Admiral Sir Henry Keppel, was a strict but fair disciplinarian. The British Government had appointed James Brooke as Governor of Labuan and William Napier as the Lieutenant Governor. The Governor's party had sailed from Portsmouth on 24 January 1847 and arrived at Singapore on 20 May. On 9 June the Hon'ble Company's Ship *Phlegethon* arrived at Singapore with the Civil Engineer, John Scott, who was destined for the new settlement. On 7 August, news reached Singapore that Her Majesty Queen Victoria had been pleased to confer the Order of the Bath on James Brooke. On 29 August HMS *Meander* departed from Singapore for Labuan with Sir James Brooke KCB on board. The Lieutenant Governor William Napier followed on the HCS *Phlegethon.* Sir James Brooke landed at Victoria, wearing the uniform of the first Governor of Labuan on 29 September 1847. Other officials followed including Hugh Low as Government Secretary and Spencer St John as Private Secretary to the Governor. A description of the settlement paints a very dismal picture and reads:

> '. . . temporary residences erected near the port consisted of small square cottages on posts about four feet above the ground. The walls and partitions were of matting and the roofs of atap. They did not look inviting. The flat selected for the settlement was below the level of the sea and protected by a self-formed sandbank. It had a swampy unhealthy look. The row of thatched house[s] stood on the sandy shore between the sea and the swamps behind. The air buzzed with millions of mosquitoes. Sandy flats in front may have appeared attractive, but the swamps of black mud behind abounded with mosquitoes, and crabs breeding in thousands ruined any paths built across them . . .'[19]

At the time of occupation the order had been given by the Government that no fortification was to be built and that the island was to be a naval station and nothing more. With the arrival of the new Governor it was declared that the settlement would now be opened up for settlers from 1 August 1848.

The fact that the area in which the settlement had been established was unhealthy was quickly proved as, in October 1847, after a particularly heavy period of rain the whole area turned into 'a huge pool of fetid water'.

The Governor, the Lieutenant Governor, the Settlement Doctor, Mr and Mrs Low, Captain Hoskins, Mr Grant and Captain Brooke and the Marines were all struck with the fever at the same time. Fortunately higher ground had been selected for the barrack site which was being prepared for the arrival of a detachment of the Madras Native Infantry. HMS *Meander* having been at Labuan sailed for Singapore and left a detachment of Royal Marines as the military guard for the settlement, pending the arrival of the troops from Singapore. Whilst on the short voyage from Labuan to Singapore malarial fever broke out amongst those Royal Marines on board who had previously been on shore duty at Labuan. Two died during the short voyage and concern was felt for those remaining at Labuan. The HCS *Auckland* was despatched from Singapore to Labuan to check on the situation and it returned with the news that the majority of the detachment were sick with fever, etc. HMS *Meander* embarked a company of the 21st Regiment, Madras Native Infantry and set sail for Labuan. On arrival Captain Keppel found that some of the Marines had died and that the majority of the Europeans at the settlement were ill. A newspaper report in mid-1848 gave details of the reinforcements due to be despatched to Labuan saying '. . . We learn that the 51st Regiment MNI is expected here soon, and will be permanently stationed at Singapore furnishing a garrison of 200 to the Settlement at Labuan. The present Regiment in the Straits will be divided between Pinang and Malacca . . .'[20] During July 1848 a further report showed that progress was being made towards the construction of the barracks for the Sepoy troops which read '. . . Labuan. A road is in process of formation from Fort Point to the Cantonments and the erection of the necessary buildings for the reception of the troops will by this time have begun . . .'[21]

After the construction of the Barracks on higher ground the buildings of the settlement were also moved from the original area to adjacent higher ground and in consequence the prevelance of sickness appears to have diminished. Another report in the *Singapore Free Press* in March 1849 gives a more optimistic view saying '. . . At Labuan there had been no deaths amongst Europeans since last accounts and the Sepoy troops kept their health . . .'[22]

Militarily the settlement gave little cause for concern. By 1853 the troops had settled into the barracks and apart from the detachment of Madras Native Infantry, sent from the resident battalion at Singapore, there was also a battery of artillery. There were, at that time, still no plans for the construction of a fort. Hall comments in his book *Labuan Story* that the garrison was '... strong enough to drive off any marauders from the mainland opposite and to protect the island against a sudden raid by pirates from further east...'[23] In July 1853 the Governor at Labuan wrote to Singapore and suggested that the '... Officer Commanding in the Straits should periodically visit and inspect the Troops garrisoning Labuan in the same manner as detachments within the Straits [Settlements]...'[24] He also requested that he be informed if there was any reason why this could not be implemented. The reply from the Governor of Singapore was, to say the least, unhelpful in that it contained the following '... the absence of the Officer [Commanding] from the Straits Settlements without due provision being made for the performance of his duties during such absence would be detrimental to the service...'[25] This appears to have been the end of the attempt to have the troops regularly inspected. However at this time there appears to have been problems over the provision of a European officer to take command of the troops stationed on Labuan Island. On 30 June 1853 the Governor wrote to the Officer Commanding the troops in the Straits Settlements asking why more than three months had elapsed since Captain R. P. E. Watt, of the 43rd Regiment, MNI, who had been appointed to command the Garrison had not yet taken over and it was suggested that '... some inquiry be made with the cause of his incapacity...'[26] A report was sent to the Governor some three months later in which the Officer Commanding the Troops informed the Governor that Captain R. P. E. Watt '... has been recommended by a Medical Board to proceed to Europe for 3 years for the benefit of his health...'[27] in other words he had been granted Furlough on a Sick Certificate which was a normal procedure in the early and mid-nineteenth century. The situation of a relief for the officer on Labuan seems to have been unresolved until November 1853 when the Officer Commanding the Troops asked formal permission from the Governor for Ensign F. W. Dobree of the 43rd Regiment, MNI to relieve Lieutenant A. G. W. Burn of the same regiment who was at that time in charge of the Sepoy contingent on Labuan Island.[28] The change-over was agreed by the Governor on the same date.[29] The garrison on Labuan Island remained unaltered until mid-1854 when the European artillery detachment was replaced by a '... detachment of 5 native artillery-men under a European Officer...'[30]

The life of the garrison on Labuan Island appears to be quiet with little of note happening except the daily round of guard duties and drills. There was a periodic change-over of detachments and those employed in specific appointments were subject to normal military posting procedures. An example of this is seen in Madras Orders, published at Fort St George as the *Fort St George Gazette* in which Assistant Apothecary C. Marrow was posted to 'Detachment, Native Infantry, Labuan'.[31]

By May 1858 the problem of the high percentage of soldiers being unfit for duty on the island once again came to the fore and the Officer Commanding Troops in the Straits wrote to the Governor on 21 May of that year and brought to his notice '... the very sickly state of the Troops on Labuan...'[32] and mentions that there was an urgent need for their immediate relief and suggested that a steamer be engaged for the purpose of evacuating the troops from the island. He also forwarded to the Governor the suggestion from the Officer Commanding the Garrison, which was supported by the Assistant Surgeon stationed on Labuan Island, that the Garrison be withdrawn completely and that one or two gunboats be stationed there in place of the troops. It was also suggested that if the troops were to remain a programme of jungle clearance and a scheme of 'thorough drainage' of the area occupied by the troops be implemented. The Governor wrote to the Secretary to the Government of India and informed him that he had sanctioned the engagement of a steamer at the cost of $5,000 for the evacuation of the troops who were unfit but also drew attention to the fact that there was '. . . heavy expenditure falling on the Government of India consequent on the deaths, invaliding and Pensioning of men who have served in Labuan . . . [and suggested] to the Honourable President in Council the propriety of either sanctioning a large expenditure for sanitary measures, or withdrawing the Garrison and substituting one or two Gun Boats . . .'[33] The Secretary to the Government of India replied on 24 July 1859 to the effect that the policy of furnishing troops for the Labuan garrison had been referred back to the United Kingdom but that authority was given for the relief of troops stationed on the island at frequent intervals.[34] The unhappy situation of continued sickness amongst the troops on the island prevailed and nothing further was done until 25 February 1859 when the Officer Commanding the Troops in the Straits again wrote to the Governor and submitted with his letter correspondence from the Officer in command of the troops stationed on Labuan who requested that because of the '. . . sickly state of the troops on the Island and the necessity of strengthing the Detachment of Native Infantry to bring up the number to the original complement . . .' reinforcements should be sent as soon as possible. The Officer Commanding Artillery in the Straits also made a similar request. The report continued by saying '. . . Both detachments have been there a little more than six months and nearly a month has elapsed since the date of Captain Dent's letter and it will be at least six weeks or two months before any additional men could reach the place. It is impossible to say what may be the result[ing] state of the detachment by that time from sickness, death or removals of men requiring change of air to the coast . . .'[35] The report suggested that the whole party of Artillery and Native Infantry should be relieved as quickly as possible and that a vessel should be chartered for that purpose. The Governor replied to the effect that the attempts to charter a vessel for the relief of the Labuan detachment had proved to be fruitless and that the Resident Councillor at 'Pinang'[36] was requested to provide a vessel. He chartered the ship *Juliet* for the sum of $3,150 to take the '. . . Troops to LABUAN and back . . . [and] . . . to take a Detachment of 25 Artillery and one officer with military and commissariate stores . . .'[37] The medical situation on the island continued to give cause for concern and in May 1859 it was requested that Assistant Apothecary McCloskie, who was serving with the 22nd Regiment, MNI, be attached for temporary duty. The request was approved and the Officer in charge of the Penang Garrison was requested to despatch McCloskie. It was also recommended that a 'dresser' be attached to the Labuan garrision to accompany the sick who were frequently sent to Singapore.[38] The Governor forwarded the request for the provision of a 'dresser' to the Secretary to the Government in Madras with the recommendation that the request should, if practicable, be carried out.[39]

The situation on Labuan Island appears to have improved, for a while at least, as a report from the Governor of the Straits Settlements to the Government at Madras dated 30 April 1860 recorded that '. . . the detachment of the 22nd Regt NI at Labuan is now extremely healthy . . .'[40], however if that problem was temporarily resolved then other problems were occurring on the island. His Excellency the Governor of Labuan wrote, on 9 April, that there was an '. . . almost total absence of provisions for the Troops in the Commissariate Store . . . [and that it was necessary to purchase] . . . such few articles as were procurable from the local market . . .'[41] The health problem on Labuan Island for the troops appears to have improved as there were no further requests for change of detachments to be made for medical reasons and the Madras Golundauze were relieved during 1861 as a mere routine[42] and in the following year the detachment of 40th Regiment, MNI were also relieved as a matter of course.[43] However by mid-1863 the situation had again deteriorated and on 1 June of that year the Officer Commanding the Troops wrote to the Governor pointing out that '. . . of 98 privates at that station [Labuan] 13 men [were] sick and leaving only 85 for duty of whom 28 are on Guard thus giving only two nights in bed out of three instead of 4 out of 5 as ordered in India . . .' the letter then strongly recommended '. . . the increase of the Detachment by 70 privates with officers and non-commissioned officers which would give the proper number of reliefs for the Guards and consider that the general health of the detachment will be greatly improved. If this number cannot be sent then the sick should be substituted for healthy from Singapore so that the numbers of men for duty may be maintained to as great an extent as possible . . .'[44] The unhealthy situation continued to be a problem and on 1 May 1864 the Officer Commanding the Troops in the Straits wrote to the Governor saying that he proposed '. . . to despatch a detachment of LI men [*sic* – 34th Regiment] with the proper complement of officers to Labuan and request that tonnage may be taken up for that purpose . . .' In the same letter he advised the Governor that he had received instructions from Army Headquarters at Madras to proceed, personally, to Labuan to enquire into the cause of the sickness amongst the troops stationed on that island and also stated that he proposed to take Surgeon Major W. H. S. Burn of the 34th Regiment MLI, with him.[45] The Governor agreed to this measure in order to try to resolve this recurring problem of sickness of the troops stationed in the Labuan garrison. However, the situation of constant sickness among the troops continued to be a problem and one which was never satisfactorily resolved. The Sepoy contingents were withdrawn from Labuan Island in 1872 when the responsibility of providing an infantry battalion for the Straits Settlements was passed to The War Office and British Army battalions replaced the Indian army units.

The British battalions continued to provide a small detachment on the island and it is assumed, although no written evidence has been found, that the barracks were modified to meet the standard required for housing European troops. The whole defence was, at this time, merely an infantry detachment and there is no record of any fortification having been built on the island.

Labuan was incorporated into the Straits Settlements from 1906, although this union had been discussed as early as 1875.[46] On 1 December 1912 the status of Labuan Island and Dependencies was confirmed as being part of the Straits Settlements, ranking with Penang and Malacca.

The small presence on the island continued until the British infantry unit was withdrawn from the Straits Settlements at the outbreak of the First World War, which then left this Settlement without any means of defence. During the whole period, from the takeover by the British battalion until 1914, the entry in Army Lists shows, under 'East Indies', Labuan as having a Governor and Commander-in-Chief although for many years the appointment was shown as being unfilled.

With the outbreak of war and with the withdrawal of the detachment of infantry the few Europeans living on the island felt the need to have some form of defence and so formed a small volunteer unit, the Labuan Defence Corps, which came under the care of the Singapore Volunteer Corps. It was reputed to be the smallest volunteer unit in the Far

Labuan: street scenes, *c*1930. Photographs by courtesy of L. Ransley.

East. The arms issued to this small defence unit were twelve old single-loading Long Enfield rifles.[47] The unit continued to use the title Labuan Defence Corps from its formation in 1915 until 1922 when, after the reorganisation of the Local Forces it was renamed the 'Labuan Volunteer Detachment, Singapore Volunteer Corps'.[48] The unit was regularised by the new Volunteer Ordnance of 1923 which was published in the *Straits Settlements Government Gazette* Notification No 1997 of 21 December 1923. The new volunteer establishment included 'The Labuan Volunteer Detachment'. Winsley, in his history of the Singapore Volunteer Corps, recorded that in 1924 the Labuan Volunteer Detachment was reduced to four Colour Service Volunteers on its establishment.[49] The detachment continued to survive with just enough volunteers to justify its existence and in 1927 the Commandant of the Singapore Volunteer Corps visited the detachment which revived interest among the Europeans living on the island.[50] The unit was armed with one Lewis gun and twelve short rifles. Administrative costs involved in travel between Singapore and Labuan brought only infrequent visits by the NCO instructors and even less frequent visits by the Commandant of the Corps. Winsley recorded that in 1926 and 1927 the detachment suffered from a fall off of interest in training and a lack of recruits caused by the apathy among the local European residents. During these two years drills were only carried out at very infrequent intervals and in August 1928, because of the complete lack of interest, and also in order to reduce costs, the unit was disbanded. It was said that it was uneconomical and too small to warrant continuance. The twelve rifles were left for the purpose of forming a Labuan Rifle Club.[51] The unit was never reformed and Labuan existed without any military presence on the island during the peaceful years from 1928 until the outbreak of the Second World War.

NOTES TO CHAPTER NINE

1. Maxwell Hall, *Labuan Story, Memoirs of a Small Island*, (published by Chung Nam Printing Co, North Borneo, 1958).
2. Ibid, p6.
3. John Crawfurd, *A Descriptive Dictionary of the Indian Islands and Adjacent Countries* (reprinted, Kuala Lumpur, Malaya, 1971), p203.
4. Nicholas Tarling, *Britain, The Brookes and Brunei*, (published Oxford University Press, Kuala Lumpur, 1971), p55.
5. Ibid, p56.
6. Ibid, p57.
7. *Singapore Free Press*, dated 27 March 1845.
8. Tarling, op cit.
9. Nicholas Tarling, *British Policy in the Malay Peninsula and Archipelago 1824–1871*, (published Oxford University Press, Kuala Lumpur, 1969).
10. Hall, op cit, p115.
11. Ibid, p119.
12. Charles Burton Buckley, *An Anecdotal History of Old Times in Singapore 1819–1867*, (printed by University of Malaya Press, 1965), p454.
13. Hall, op cit, p129.

14. The inscription on the memorial stone, which still stands at Victoria, Labuan Island, reads:

This Island
Was taken Possession of
December 24th 1846
In the Name of Her Majesty
VICTORIA
Queen of Great Britain and Ireland
Under the Direction of
His Excellency Rear Admiral
Sir Thomas Cochrane CB
Commander in Chief
by
Captain G. R. MUNDY
Commanding
H.M.S. IRIS

15. Hall, op cit, p132.
16. The original Christian cemetery, which was known as the Naval Cemetery was situated on the shoreline and because of the expansion of the oil company pipe depot it was agreed that the memorial stones would be removed. A site was selected in the Botanical Gardens some three miles inland and the memorial stones were moved and relocated in their original pattern in the new setting.
17. The memorial inscription reads:

Sacred to the Memory of
Colour Sergeant Adam C Sutherland of the Royal Marines
late of Her Majesty's Brig Columbine
who fell mortally wounded in the noble
discharge of his duty in an attack on Balanini
pirates in the cutter of this vessel
on 30th May 1847.

18. *Singapore Free Press and Mercantile Advertiser,* vol13, No51, Thursday morning 21 December 1848, (National Library, Singapore).
19. Hall, op cit, p142.
20. *Singapore Free Press,* vol13, No18, 11 May 1848, (National Library, Singapore).
21. Ibid, vol13, No30, 27 July 1848.
22. Ibid, vol14, No9, 1 March 1849.
23. Hall, op cit, p149.
24. Straits Settlements Governors Military Diary 1852-1857. Entry No4 for 18 July 1853, (National Library, Singapore).
25. Ibid. Entry No3 for 6 August 1853.
26. Ibid, Entry No3 for 30 June 1853.
27. Ibid, Entry No8 for 20 September 1853.
28. Ibid, Entry No2 for 29 November 1853.
29. Ibid, Entry No3 for 29 November 1853.
30. Ibid, Entry No7 for 4 April 1854.
31. *Fort St George Gazette,* No2906 dated 18 May 1855, (National Library, Singapore).
32. Straits Settlements Governors Military Diary. Entry No5 for 25 May 1858, (National Library, Singapore).
33. Ibid, Entry No6 for 25 May 1858.
34. Straits Settlements Governors Diary - Foreign and Military, 1857-1858. Entry No3 for 7 September 1858, (National Library, Singapore).
35. Straits Settlements Governors Foreign and Military Diary. Entry No10 for 25 February 1859, (National Library, Singapore).
36. Ibid, Entry No12 for 25 February 1859.
37. Ibid, Entry No16 for 26 March 1859.
38. Ibid, Entry No11 for 21 May 1859.
39. Ibid, Entry No12 for 23 May 1859.
40. Ibid, Entry No34 for 30 April 1860.
41. Ibid, Entry No8 for 10 May 1860.

42. Ibid, Entry No8 for 10 July 1861.
43. Ibid, Entry No8 for 23 January 1862.
44. Ibid, Entry No6 for 6 June 1863.
45. Ibid, Entry No6 for 9 May 1864.
46. Straits Settlements Governors Register. Entry dated 17 November 1875, (Public Record Office, Kew. CO426/3).
47. Captain T. W. Winsley, SVC. *A History of the Singapore Volunteer Corps,* (printed by Government Printing Office, Singapore, 1938), p74.
48. Ibid, p86.
49. Ibid, p96.
50. Ibid, p99.
51. Ibid, p103.

CHAPTER TEN

The Perak and Sungei Ujong Campaigns of 1875–1876

Under the Pangkor Engagement of 1874 James Wheeler Woodford Birch was appointed the first British Resident to Perak. Only twelve months after he had taken up the appointment Birch was assassinated. This was on 2 November 1875 and his death led to the outbreak of the fighting which was later to be known as the Perak War. The British force was to bring about the pacification of Perak, the arrest, trial and execution of those who were found to be directly responsible for his death and also a number of Perak chiefs were exiled.

British policy changed towards the Malay States and this led to a Resident being appointed for Perak. In 1867 the Straits Settlements became a Crown Colony and the Colonial Office had generally tried to follow the Indian Government's policy of non-interference in the affairs of the Malay States. However there were occasions when it was deemed necessary to take action when the interests of the Straits Settlements appeared to be threatened.

Tin had been produced in the Malay States for centuries and during the nineteenth century the demand for tin grew and this demand was particularly keen in Britain, the result of this was the opening up of new mines in Selangor where the old fields at Lukut and Ulu Langat were superseded by new and richer mines at Kanching and Ampang. In Perak the old workings in the Ulu Perak and Kinta valley regions gave way to new developments at Larut on the north-western coast.[1]

With the development of the industry so the struggle for control began and this included Malay faction against Malay and also between the Malays and the Chinese who were working the tin mines. In Larut the fighting in the 1870s spilled over from the mainland onto the island of Penang where many of the Chinese leaders lived. This violence was an obvious threat to the security of the Straits Settlements and many of the merchants who had invested in the tin industry were adversely affected by the continued conflict between the rival factions. The Colonial Office resisted appeals for intervention until 1873 when the appointment of a new Governor, Sir Andrew Clarke, gave the office an opportunity to investigate a way to bring order to what was in effect the frontier of one of the settlements.[2]

Clarke was instructed to investigate and report on the problems in the Malay States and to submit his recommendations for the easing of the situation. The Colonial Office wanted peace, order and the protection of trade and commerce in the area. The new Governor chose to act and on 20 January 1874 he persuaded a number of Perak chiefs to sign a treaty which became known as the Pangkor Engagement[3] in which he created a new Sultan and appointed a British Resident to reside at his court. At this time Clarke negotiated a settlement with the leader of the Chinese at Larut and ended the fighting in that area.

It was perhaps not surprising that Clarke acted instead of submitting endless reports. He had been born in 1824 and trained at the Royal Military Academy, Woolwich. He was commissioned in the Royal Engineers in 1844 and served in Ireland, Tasmania, New Zealand and Australia. In 1853 he was appointed Surveyor-General for Victoria and two years later became Minister of Public Lands. On his return to England he served from 1857 until 1864 as Commander Royal Engineers in Eastern and then Midland Districts. In 1864 he was appointed Director of Public Works for the Admiralty and following this post he moved to the East Indies and was appointed Governor of the Straits Settlements on 4 November 1873. By the time he took up his appointment he had been promoted to the rank of Colonel and was KCMG and CB.[4]

There was a delay of nearly nine months between the signing of the Pangkor Engagement and the appointment of a Resident to Perak. To cover this lack of British authority Sir Andrew Clarke appointed an Assistant Resident in Perak; for this appointment he chose a rather colourful character who was known as Captain Speedy. Tristram Charles Sawyer Speedy was born in India in 1836 and enlisted into the Indian Army in 1854. He served as a British Agent in Abyssinia and then returned to India where for two years, 1869–1871, he was District Superintendent of Police at Oudh. In early 1872 he moved to the Straits Settlements and took up the post of Deputy Commissioner of Police. In July the following year he gave up his police appointment and returned to India to recruit soldiers for the Mantri (Chief) of Perak.[5]

Captain Speedy however confined his actions to Larut and did not venture into Perak proper. The Governor in the meantime delayed appointing a permanent Resident until he had received formal agreement to his actions at Pangkor. He informed the Colonial Office of his actions in a telegram dated 23 January 1874 and he received a reply on 27 January requesting him not to appoint any Residents until more information had been received by the Colonial Office regarding the Pangkor Engagement. Clarke sent two despatches, one on 26 January and another on 24 February. On 29 May he was given permission to appoint Residents.

James Wheeler Woodford Birch had established himself as a leading contender for the post of Resident at Perak. He had written to Clark claiming that '. . . My whole life has been spent in opening up new country and in improving and enriching a country and in teaching the native chief good Government . . .'[6] and his former service all confirmed this statement. He had been appointed as Colonial Secretary to the Straits Settlements in May 1870 and this was an indication of his ability as an administrator. He did however appear to have a reputation of pressing through changes whenever he was appointed to a new job. He had only been in Singapore for six months when his wife died in England which then gave him the added burden of having to take care of four children, whose ages were fourteen, thirteen, ten and five years of age.[7] In some ways his becoming a widower helped him to be selected as Resident of Perak. The difficult conditions that prevailed in the Malay States made it an unpopular appointment among the married officers who were senior enough to be considered. The majority preferred the comfort, security and social life of the Colony of Singapore and it was therefore obvious that Birch, as a widower and an energetic administrator should be chosen for the post. He had, since being in office in Singapore had the opportunity to deal with various Malay rulers and gained a fair knowledge of the Malay States.

The situation in Perak was complicated in that Clark settled the dispute of Perak's succession by appointing Abdu'llah as Sultan during the Pangkor Engagement. Sultan Ismail was at that time retired on a pension and given the title 'Sultan Muda'. In reality there were now two Sultans, Ismail who had been elected by the chiefs of Perak in 1871 and

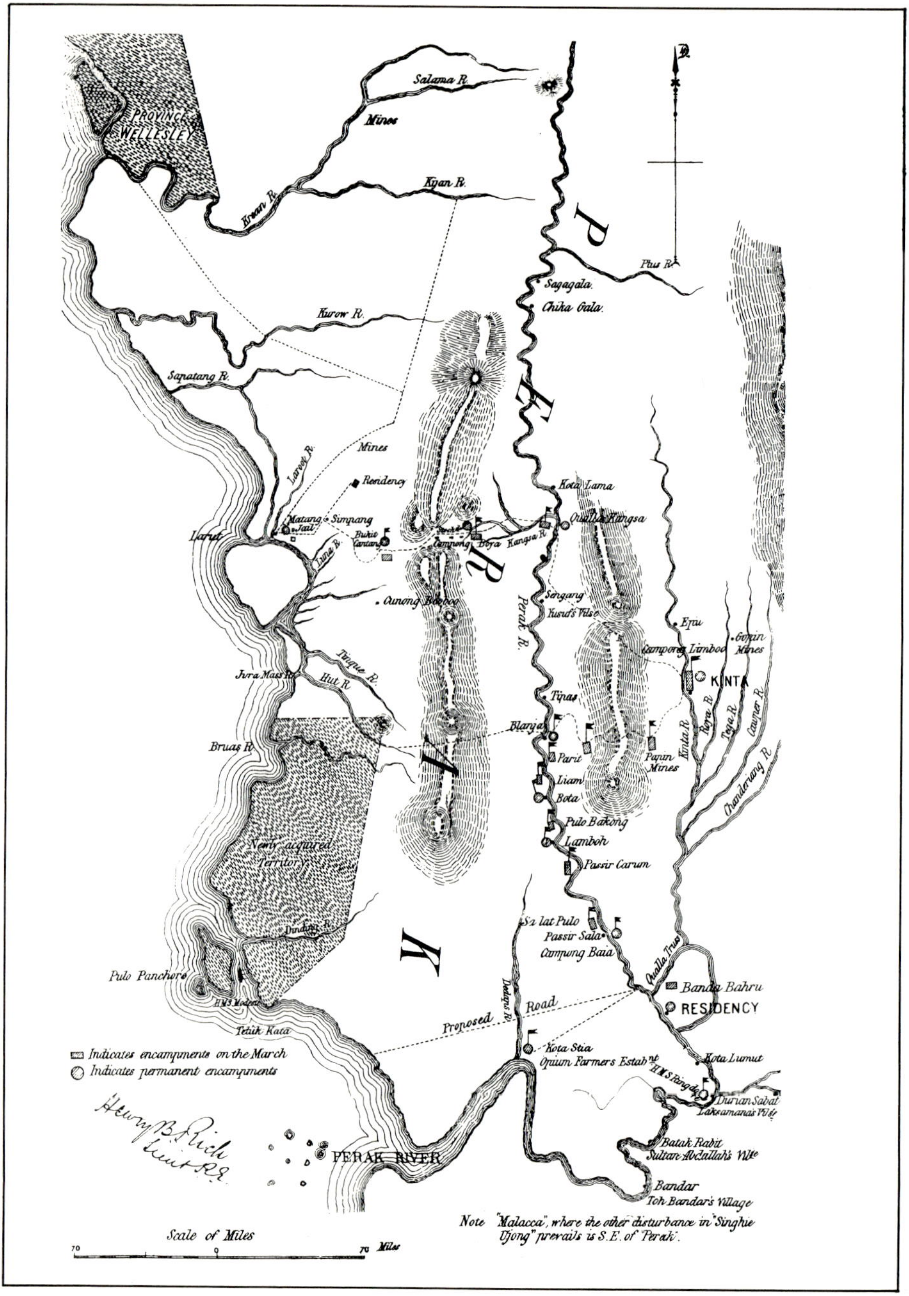

Map of the Perak Campaign, by courtesy of the Royal Engineers Museum.

Abdu'llah who had been appointed by the Pangkor Engagement. On Sultan Ismail's succession in 1871 the Straits Government approved his appointment, but when in 1872 Raja Abdu'llah also assumed the title, the Governor, who was at that time Major General Sir Harry St George Ord, CB, took the view that he would accept neither as Sultan. By the beginning of 1872 Sultan Ismail's authority ran throughout most of Perak with Sultan Abdu'llah having only two main supporters, the Laksamana (admiral of the fleet) and Dato Shahbandar. Ismail's ascendency was, however, threatened by two developments, the outbreak of the war at Larut between two Chinese factions and the campaign by Abdu'llah for recognition as Sultan. With the new Governor taking control and with the Pangkor Engagement, order was being restored and the mines reopened in Larut. Governor Clark had, by the end of February 1874, through his Commissioners made a good start on solving the mining problems and he then commenced trying to bring together the dissident chiefs. Birch took part in an expedition to Perak to interview Ismail and to attempt to obtain the royal regalia for the newly-created Sultan Abdu'llah. Although initially Ismail agreed to surrender the regalia, at a second meeting he refused to hand it over.[8]

On 3 October the Governor received the Secretary of State's approval for his actions in Perak and he provisionally appointed Birch Resident on the following day. The appointment of Birch as Resident and Captain Speedy as Assistant Resident was promulgated on 12 December 1874.[9] Three main tasks faced the Resident; the settlement of the boundaries between the Straits Settlements and Perak as defined by the Pangkor Engagement, the establishment of Sultan Abdu'llah's authority throughout the state and the setting up of a government for Abdu'llah, and the setting up of a police force with a legal and judicial system. On the future administration of Perak Clarke instructed Birch to '. . . use your best exertion to put down, by force, if necessary, all unlawful exactions of whatever nature . . .' and to also ensure that '. . . whatever revenue is collected shall be for the State alone . . .'[10] Birch made this order clear to Sultan Abdu'llah when he had his first meeting with him as Resident on 5 November which was the day after his arrival at Batak Rabit. He also discussed the boundary question and the setting up of a police force for the state of Perak. The long and complicated negotiations between Birch and the Perak rajas continued over the next eleven months and have been fully and adequately recorded in his Journal which has been edited by P. L. Burns and published by the Oxford University Press in their Oxford in Asia Historical Reprint series.

Opposition to Birch, to the Pangkor Engagement and to the concessions agreed by Sultan Abdu'llah began to bring about a mood of resistance among the Malay chiefs, who were still very much divided. A meeting held between the Sultan's supporters and some representatives of a number of the 'Ulu' chiefs agreed, on about 20 July 1875, that the Resident must be removed, quite how this was to be achieved was not agreed but it was considered at this meeting that it may have to be at the expense of an assassination. After a humiliating meeting between the chiefs and Birch the plot became more vital and arms and ammunition were imported into Perak. Sultan Abdu'llah is reputed to have given Maharaja Lela a written authority for the killing of Birch and on about 12 October it also received the support of Raja Ismail. Shortly before the end of October Sultan Abdu'llah sent Maharaja Lela the weapons whilst Ismail provided tin to make bullets and money to purchase gunpowder. On the eve of Hari Raya Puasa[11] Adbu'llah, who was at Pasir Panjang, sent a kris to the Maharaja Lela. With the obvious unrest within the state, Birch was determined to introduce direct rule and he did not appear to consider it necessary to alter his plan. The plan was that government would be conducted by the Resident in the Sultan's name with the assistance of an advisory council of chiefs while the day-to-day administration would be carried out through the datos and penghulus. Birch hoped that he would convince the chiefs to accept his plan, in particular the Laksamana and Dato Shahbandar. Just prior to leaving Bandar Bahru for the Bernam River he wrote to the new Governor, Colonel Sir William F. D. Jervois, KCMG, CB as follows:

> '. . . Nothing can exceed the general good feeling; and this Yusuf and Dris do all they can to foster.
>
> Everything is quiet.
>
> Yusuf is most confident that he can bring in Ismail.
>
> Your Excellency's visit has done an immense amount of good, and the people are only waiting for your Proclamation . . .'[12]

Birch then spent four days at Bernam to arrange for the future administration of the district where the Bernam River marked the boundary between Perak and Selangor, and at the same time check on the rumours that the Raja of the latter state was planning an attack on Perak. He was reassured on this point and returned to Lower Perak where he discovered rumours to the effect that the Laksamana and Dato Shahbandar, together with Maharaja Lela, were planning 'passive resistance'. Raja Yusuf advised Birch that it might be necessary for Birch to make an example by warning one or two kampongs of the consequences of resistance. Birch received the Governor's Proclamation of 15 October 1875 which announced that '. . . in compliance with the request of the Sultan and Chiefs, Britain had determined to administer the Government of Perak in the name of the Sultan by Commissioners and Assistant Commissioners of Her Majesty the Queen who would govern under his Excellency's instructions . . .'[13] Birch received at the same time new instructions as acting Commissioner[14] and five letters from the Governor to the rajas inviting them to join the new Malay Council. Birch travelled to Kota Stia and on the evening of 1 November returned upstream to Pasir Salak where he moored for the night. Whilst he was at Kota Stia he was warned by Nakhoda Trang not to trust the datos as there were rumours that there was to be trouble but Birch is reputed to have replied: '. . . I will take good care; what is the use of being frightened at every report; if one Mr Birch is killed, ten Mr Birchs will take his place . . .'[15]

Birch arrived at Pasir Salak shortly before midnight on Hari Raya Puasa, which on that year was on 1 November, and had with him his own boat, *Naga*, two other boats which were manned by twelve crewmen and fourteen sepoys. He was accompanied by Lieutenant T. F. Abbott of HMS *Thistle* and his own interpreter Mat Arshad. Birch and his party spent the night moored opposite the kampong of Maharaja Lela, and on the following morning moved across the river, with his own boat being moored to the floating bath house of a Chinese goldsmith's shop. Lieutenant Abbott crossed over the river with two other men to shoot whilst the remainder breakfasted on the shore near to the boats. Birch was suffering from a badly sprained ankle and so sent his interpreter, Mat Arshad, to see Maharaja Lela but he was refused admission. On Birch's instructions Mat Arshad posted three proclamations on the goldsmith's shop. Dato Sagor arrived with fifty armed men and Birch told the Dato the reasons for the posting of the proclamation and what he intended to do when he commenced his administration of Perak. He then went to the bath house to bathe. The proclamations were immediately torn down and Birch ordered Mat Arshad to post others. As he attempted to do so the Malays under Dato Sagor's command attacked the interpreter and rushed the bath house stabbing at Birch through the palm-leaf walls. Birch's sentry at the bath house leapt into the river without attempting to defend Birch.

Mat Arshad fell into the river and was picked up later by those who escaped. A boatman and a sepoy were killed and six others were wounded. Mat Arshad, although rescued from the river, died on the boat. Lieutenant Abbott and his two companions who had been across the river shooting were able to make good their escape to Bandar Bahru where they raised the alarm. Birch had been killed immediately and his body was left to float down river. His personal possessions were taken from his boat *Naga* and the boat then sent to Ismail to show that Birch had in fact been assassinated. The body of Resident Birch was recovered and brought to Bandar Bahru on 6 November where he was buried with full honours on the spot that he had planned to build a house.[16] Ismail, not wishing to be seen as having taken part in the assassination, sent the *Naga* back down-stream again and Abdu'llah also denied any implication in the affair.[17]

News of the death of Birch reached Penang on the evening of 3 November by the commander of the *Pluto*. The message was passed first to the Hon H. Plunket, the Superintendent of Police and then to the Lieutenant Governor, Lieutenant Colonel the Hon Archibald E. H. Anson, RA, who then called Captain William Innes, RE who was at that time employed as the Colonial Engineer. At the Central police office at Penang the Lieutenant Governor appointed Captain Innes Assistant British Commissioner and issued orders to the effect that he was to '. . . enquire into the circumstances of the murder, [and] to call upon the chiefs to hand over all those implicated . . .'[18] It was agreed that the ship *Pluto* would depart from Penang at 6.00 am on 4 November with two officers and sixty men of the 10th Regiment and thirty policemen led by the Hon H. Plunket, an apothecary and a Mr Kynnersley as the interpreter. Captain Innes was however not in command. There was some slight confusion as the Governor of the Straits Settlements received a telegram from Birch saying that everything was all right at the same time as he received the telegram that had been sent by Lieutenant Colonel Anson advising him that Resident Birch had been assassinated. The matter was not cleared up until 4 November when Anson was able to explain the situation to the Governor in Singapore.

The position was also complicated by the absence of Frank Swettenham who, as Assistant Resident of Selangor, had been helping Resident Birch. Nothing had been heard from Swettenham since 28 October although it was expected that he would go to Kuala Kangsar where he had arranged accommodation for a number of sepoys, but rumours had reached Penang that he too had been killed. Anson ordered Captain Speedy, the Assistant Resident, to despatch every available armed man to Kuala Kangsar which Speedy organised immediately.

In the meantime Captain Innes, had departed from Penang with a very divided force as Plunket controlled the police, Captain Thomas George Booth had command of the soldiers of the 10th Regiment and Lieutenant Abbott was in charge of the four sailors who were with the party. The only control that Innes might have had was of the small party of Sikhs from the Resident's Guard. The party reached Bandar Bahru and at the residence joined up with Lieutenant Abbott and his sailors and also found that Frank Swettenham was alive and at the Residency.

It was learned that two stockades had been built at Passir Salak, one by Maharaja Lela and the other by Dato Sagor, and it was decided that the British party would attack on 7 November by sending a party up river on each bank and that the boats would advance along the river with a 9-pounder field gun, a 12-pounder howitzer and rockets being mounted and fired from the boats when required. It was decided that the party should rise at 4.30 am and move up river for two miles by boat, land the military and then proceed the remaining

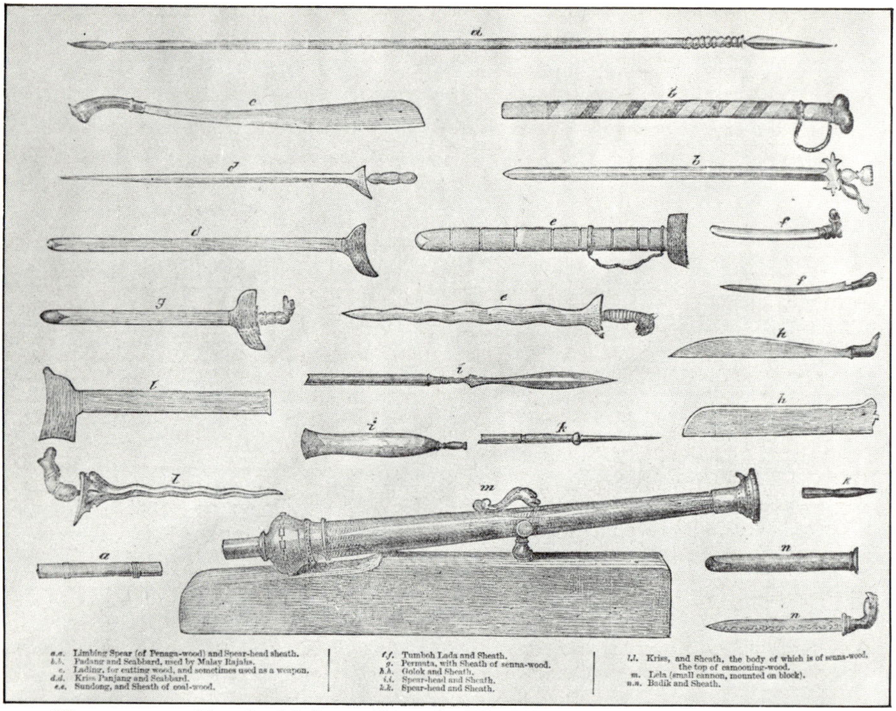

Weapons used by the Malays of Perak. ***The Illustrated London News*, 1875.**

The British Force ascending the Perak River. ***The Illustrated London News*, 1875.**

View of the Encampment, Bandar Bahru, from the back of the Residency.
***The Illustrated London News*, 1875.**

Camp at Pasir Salak, Perak River. *The Illustrated London News*, 1875.

three miles with the boats keeping level with the land force and giving covering fire when necessary. There had been a disagreement between Swettenham and Innes over the use of the guns. It was 7.30 am before the party set off and in the event the howitzer was left behind, as was the 9-pounder. The force landed at Passir Panjang at 8.00 am and formed up in the following order '. . . Malay Scouts, Booth's platoon of the 10th; Innes, Abbott, and the sailors; the Sikhs; the Penang police under Plunket; and the other platoon of the 10th under Elliott . . .'[19] The force set off not expecting to meet any resistance until they reached Passir Salak. They were advancing in single file when they were fired upon from a stockade which had been built some two miles from the kampong. The scouts emerging from cover were faced with:

> '. . . a wide and deep ditch backed by a strong stockade of trees and logs built on to a bank of earth. The timbered top of the stockade was full of loopholes, and as we appeared from the corn stalks we were greeted by a volley from pivot guns and muskets . . .'[20]

and as fire opened up, Nacodah Orlong, a Sumatran Malay who had been recruited by Swettenham, was killed immediately and the Scouts scattered. Captain Innes moved to the head of the party to examine the problem. Innes and Swettenham agreed that a flanking movement would need to be made and Swettenham with twelve men of the 10th Regiment and two Malays, set off but was almost immediately fired upon by the Sikhs who were mostly untrained and completely unused to this type of action. The Penang Police under Plunket had, as soon as firing had started, deserted and to add to the confusion the naval party on the boats fired the rockets, which according to Abbott, '. . . were of an obsolete pattern and used in wooden troughs, with paper primers stuck in one of the holes in the base, and ignited by a common match . . .', but as most of the projectiles fell well beyond the stockade there was no obvious damage to the enemy.[21]

After a period of shooting, Captain Innes, who was by this time on the extreme right, gave an order to rush the stockade. The result was described by Swettenham as follows:

> '. . . That was done, with the result that Innes was killed, the two officers of the 10th Regiment were seriously wounded, and there were other casualties. We got to the edge of the ditch, but it was impossible to cross it and climb the fence above the bank in full view of an unseen enemy a few yards distant and completely protected. We had to retire, and if the defenders of the stockade had known how to use their weapons no one should have got away. I did not know that Innes had been killed, for I was in the centre and delayed by helping to carry a severely wounded sergeant of the 10th, so when we reached the path, and joined Abbott and Plunket, who were waiting for us, the other casualties had already been removed. We had no surgeon and no stretchers, and the journey back was far from pleasant. We expected our enemies to follow us; but later we learned that they had retired when we did, as they did not appreciate being rushed . . .'[22]

Later there were recriminations as to the leadership and it was generally considered that had Captain Innes been given over-all command the whole event may well have ended differently. As it was, out of a force of 147 officers and men, there had been seventeen casualties including four killed. This all occurred in a brief two-hour skirmish and with the three military officers out of action, Innes had been killed with his sword in his hand, Captain Booth wounded in the foot and Lieutenant Elliott wounded in the side and arm, there was no one in authority to lead a determined attack, added to which the Sikhs had been virtually leaderless and the police party ineffective.

The setback to the British plans had far-reaching effects and the Governor at Singapore, Sir William Jervois, resolved to act immediately and embarked on a chartered steamship

Staat Rochussen, together with Major Dunlop, Captain Peyton (his ADC), Captain W. Whitla, of the 10th Regiment and twenty artillerymen, whilst on HM Gunboat *Fly* eighty soldiers of the 10th Regiment sailed for the troubled area. Jervois had a problem as there was unrest in Selangor and whilst he had hoped to use Mr J. G. Davidson as the Resident at Perak he was unable to do so, as to leave Selangor without a British presence would be to invite further trouble.

Jervois left Selangor and moved north to the Perak River where he met with the *Thistle* and the *Pluto* on 8 November when he learned of the defeat on the previous day of the British force at Passir Salak. The Governor, realising the seriousness of the situation, telegraphed to the Acting Colonial Secretary at Singapore and requested him to ask for 300 men of the 80th Regiment to be sent from Hong Kong as reinforcements and that the naval ship HMS *Modeste* be sent from Labuan.[23]

It was considered essential to hold Bandar Bahru and to keep open communications with Durian Sabatang which was the highest point on the Perak River that could be reached by gun boats. Jervois proceeded upriver to a point between Batarabit and Durian Sabatang where he met the Malay chiefs, the meeting was attended by Sultan Abdu'llah, the Laxamana and Dato Shahbandar and at this meeting the Sultan agreed to give assistance to the Governor. On the conclusion of the meeting, Jervois returned to Singapore having, prior to his departure, appointed Major Dunlop[24] Commissioner and Frank Swettenham as Deputy Commissioner at the Perak Residency with instructions to proceed up river. Major the Hon J. F. A. McNair, Madras Artillery was appointed Commissioner with W. E. Maxwell as his Deputy and was instructed to move to '... Qualla Kangsa ...'[25] and secure that area.

The request for help had been passed to the Colonial Office and it was agreed that help would be provided from Hong Kong and it was suggested, that if necessary, additional reinforcements could be provided from India. The Governor had not as yet arrived back at Singapore and so Irvine took the problem to the military Commandant. After consultation Irvine telegraphed the Secretary of State as follows:

> '... Commandant recommends thousand men from Calcutta with artillery and mountain guns. Have telegraphed accordingly. General Colborne offers three Companies. Have telegraphed to accept. Ringdove and Modeste on way from Hong Kong, Egeria will follow ...'[26]

The effect of the telegraphs sent by Irvine was to cause apprehension not only in London but also in India and Hong Kong. Jervois confirmed the requirement as soon as he had returned to Singapore. A message of 12 November 1875 was received from Lord Carnarvon as follows:

> '... Secretary of State [for India] tells me you now ask for two regiments of European infantry, or 1,500 bayonets, with artillery, 50 miles telegraphic apparatus, and a million cartridges ...'[27]

which went on to ask what policy Jervois had in mind and to give a report on the actual situation and to say whether he thought the situation was likely to worsen. Lord Carnarvon concluded '... Please send quick and full answer ...' Jervois despatched an answer which only partially satisfied Lord Carnarvon but nevertheless he reluctantly agreed to provide further reinforcements.

The Command at Singapore was requested to despatch a further eighty men from the 10th Regiment to Perak. It was at this stage of the planning that the telegraph cable had broken and that caused further confusion. Jervois had no idea as to whether his request for troops from India had been agreed and the error that had occurred in an earlier message which had resulted in the request for ammunition being relayed as a million cartridges was also uncorrected. In fact troops were on the way from India and a body of irregular soldiers had been raised in Singapore to improve the immediate military situation. The local Singapore newspaper, the *Straits Observer*, commented on the raising of the local force by Jervois as follows '... 120 picked men, chiefly Arabs and Seedie boys, who are expected to proceed across country, with Mr Davidson, to the "seat of war". If this little force only carries out one half of what is promised we may expect grand results. At present we say nothing ...' The auxiliary force, which was commanded by Mr de Fontaine, a former midshipman of the French Navy, did not depart from Singapore until 19 November which was the same day that Captain Morison took a flotilla of boats up to Perak. On 16 November the P&O steamship *Kashgar* arrived at Singapore from Hong Kong carrying 300 men of the 80th Regiment and Major General the Hon F. Colborne.

The Governor discussed the situation fully with General Colborne on the 16th and 17th and then requested the Senior Naval Officer on station, Commander Singleton, to embark troops on the following day for Perak. The vessels HMS *Ringdove* and HMS *Egeria* were to be used.

At this stage the actual objective of the campaign was defined. This was to bring to justice those who had committed the murder of Resident Birch and, as it was considered that this was probably Maharaja Lela, the force was to occupy his kampong and to take into custody the Maharaja. The force was then to display a show of strength by advancing '... up the Perak River to Passir Salak, an advance through Larut to Kuala Kangsar and, possibly, an advance from Bruas to Blanja ...'[28] It was also considered necessary to construct a road from Kota Stia to Bandar Bahru.

The force, under command of General Colborne, embarked on 18 November and proceeded towards Perak, although it was not known at this time where the reinforcements from India were to arrive as the telegraph was still out of action. The force from India commenced embarkation on 20 November with the final portion of the force, with the Brigadier General, leaving on 26 November. The force consisted of:

Commanding	Brigadier-General J. Ross
A.Q.M.G.	Major Mark Heathcote
Brigade Major	Major H. L. Hawkins
A.D.C.	Lieutenant J. J. Preston
D. A. Commissary-General	Captain A. R. Badcock
Royal Engineer	Major J. T. Twigge

Headquarters and 600 men of the 3rd Regiment of Foot (the Buffs).

Headquarters and 400 men of the 1st Ghoorkhas.

3/5th Royal Artillery, with four 7-pounder mountain guns, two $5\frac{1}{2}$-in mortars, 500 rounds per piece, 200 rockets.

A Signal detachment of 11 men and 100 miles of wire.

One Company Madras Sappers and Miners

In addition Medical Officers, orderlies and stretcher bearers together with camp equipment, tents, etc were embarked together with '... Shore provisions for ten days ...'[29]

On the eve of Major-General Colborne's departure from Singapore a message was received from Major Dunlop, which was dated 16 November and read:

'... Yesterday morning we made a combined land and water attack on the enemy. After a long day's fighting, the enemy making an obstinate and prolonged resistance we took and destroyed four stockades, including the Maharaja Lela's house and campong [*sic*], the Dato Sagor's house and campong, in fact all Passir Sala and Campong Gaja. We captured five guns, recovered Birch's two boats, brass gun and papers and much of his property, in the Maharaja Lela's house . . .'[30]

Dunlop ended his message with a sentence which stated that '. . . the campaign is by no means finished yet . . .' and it was this statement that was Jervois's lifeline as what Dunlop had achieved could well have been expected to be the end of the campaign. Dunlop had, on 14 November, agreed with Commander Stirling and Frank Swettenham that the attack should be mounted on 15 November. The force was to land at Passir Panjang and it was agreed that most of the Sikhs and all the policemen were to remain at camp. Major Dunlop's report records:

'. . . Everything being in readiness, we roused up the force at 3 a.m. on the 15th., and began getting the troops into the boats at 5 a.m. It was 6.30 a.m. before the whole party embarked, the Naval Brigade under Captain Stirling, being in charge of the gun-boats. The larger portion of the force had already started with Mr Swettenham to show them the place of disembarkation, and I followed in the steam gig 'Perak' with Captain Stirling . . .'[31]

On this occasion the force was directed by Dunlop and Stirling on the boats, and the land force was under the command of Captain Whitla, of the 10th Regiment. The total force consisted of 255 officers and men and this was made up as follows:

Landing Party	Commanded by Captain Whitla.
Advance Guard	Hon. H. Plunket and Mr Swettenham with 15 marines, 5 Sikhs, Raja Mahmood and 2 of his men, and two locals belonging to Mr Swettenham.
Main Guard	Three officers and 131 men of the 1/10th Regiment, One officer and 20 gunners of the 9th Battery, 2nd Brigade Royal Artillery, with one brass 12-pounder mountain howitzer.

The Naval detachment consisted of Commander Stirling and HQ in the steam-gig. Six boats led by Commander Bruce mounting one 7-pounder MLR gun, two 12-pounder howitzers, one small Coehorn mortar, and two rocket tubes. There were ten officers and sixty seamen.[32]

With the action by Major Dunlop and his party having been successful Sir William Jervois decided to leave for Perak and investigate the situation personally. He departed on 19 November on the *Pluto* with his ADC, the Attorney-General and Lieutenant H. E. McCallum, RE, who was acting as Private Secretary. The party arrived in the Perak River on 22 November and on meeting General Colborne at Durian Sabatang, asked him to suspend operations for the time being. On the same day he issued proclamations calling on the inhabitants of Perak to support the Government in its attempt to restore law and order. Notices were also posted on which rewards were offered for the apprehension of the murderers of Resident Birch.

In his book *British Intervention in Malaya* Parkinson gives an account of the way in which Sir William Jervois persuaded Major Dunlop to alter his report of the action on the 15th by adding to the start of the report the term 'In accordance with instructions from His Excellency, dated the 9th inst . . .'[33] thus implying that the action had been at the direct instigation of the Governor, whereas in actual fact Major Dunlop had acted as he considered best, he being the man on the spot. Sir William again met with Sultan Abdu'llah and other chiefs at which time all the Malay chiefs professed friendship but nevertheless

Jervois left Perak and went to Penang and on reaching there began to make preparation for the invasion of Perak. The force was to invade through Larut and Captain Speedy was engaged in making supply camps at nine-mile intervals between Larut and Kuala Kangsar. Major Dunlop was placed in charge of the road-making between Kota Stia and Bandar Bahru as it was intended that all the preparatory work should have been completed before the arrival of the troops from India. On 27 November the SS *Arabia* arrived in Penang Harbour with 200 men of the 3rd Regiment (The Buffs) on board. On the same day the naval ship HMS *Modeste* arrived from Labuan and on 29 November the first detachment of the Buffs left Penang for Larut. The advance up country had begun and a number of letters passed between Sir William Jervois and Ismail, the latter expressing his concern over the '. . . great many white soldiers [who] have come into the District of Perak . . .' Jervois wrote to Ismail calling on him to meet the British troops in Lower Perak showing a white flag and making contact with Major Dunlop. He was also told that if any of his people built stockades with a view to showing resistance to the British force he, Ismail, was to have them destroyed. He was also to provide as much accommodation as he could for the British troops. On 14 December Dunlop wrote to the Governor telling him that he had not had contact with Ex-Sultan Ismail and that he understood that Ismail had in fact gone to Kinta.

In the meantime the British force was moving in a pincer movement with General Colborne's column moving northwards with fifty boats, 240 soldiers and seventy seamen. The northern column under command of Brigadier-General Ross was, by 8 December, already approaching Kuala Kangsar. On 14 December, General Colborne's force was in Blanja which they had found deserted and on the same day a small force was sent six miles along the track towards Kinta. The path was blocked by trees and the party was fired upon by the Malays. Dunlop wrote '. . . from all the information we can obtain the Maharaja Lela is at present at Kinta, and from the opposition offered to our advance it is clear that ex-Sultan Ismail is determined to shelter him . . .'[34] General Colborne decided to press on and take Kinta and expecting to meet up with the column led by Brigadier Ross he moved forward and reached Kinta on 17 December, finding it deserted. The Maharaja Lela had been there but had fled northwards as had ex-Sultan Ismail, and with them, on elephants, had gone Panglima Kinta, Rajah Ngah (the Tunku Panglima Besar), three other chiefs and a number of faithful followers. They fled towards the southern provinces of Siam and Jervois was then forced to ask the Siamese to deny them entry. This move did, however, regularise the whole position for Sir William Jervois as it now permitted him to regard ex-Sultan Ismail as the leader of a band of assassins, which he did and it was therefore his task to hunt them and take them into custody. At this stage Sir William Jervois was censured for his action and for bringing such a large force into the Malay States. He was informed by London that he could not retain such a force in the States for any lengthy period. It was perhaps to his advantage that at the critical time there was a further outbreak of trouble at Sungei Ujong. In his reply to the Colonial Office he was now able to add '. . . Since writing the foregoing despatch I have received a telegram from Singapore of to-day's date, stating that the Resident at Sungei Ujong apprehends a general uprising of Malays in the States in his neighbourhood . . .'[35]

In early December as a result of a series of telegrams received which gave the indication that there was a general unrest in the Malacca and Sungei Ujong areas, a force of '. . . Three hundred and fifty Goorkhas and thirty Artillery gone to Malacca and Sungei Ujong, I think accounts thence much exaggerated. Have sent Colonel Anson[36] there instructed to enquire and report and to prevent hasty movement . . .' The Acting Assistant Resident at Sungei

Ujong was Commander Patrick James Murray, RN (Retired) who had been appointed by Sir Andrew Clarke and approved by Lord Carnarvon and by the Admiralty. He was praised on his appointment for his energy, tact and discrimination and it was through his service on the Malay coast that he became known to the Governor and had been appointed.

It was through no fault of his that the trouble of 1875 developed. He was in a difficult position as he was accredited to the Dato Klana, one of two rival Chiefs in a State which was not completely independent. The Klana owed his position in Sungei Ujong to British support who gave him recognition as the ruler. Hostilities broke out over the control of the district of Terrachi, which was on the eastward side of the range of hills separating Sungei Ujong from the Sri Menanti States. Terrachi was claimed by Dato Klana and also by Tunku Antah, with his claim being more effective as the disputed territory was on his side of the Bukit Putus pass. The Klana had appointed his own Punghulu (chief) to Terrachi in 1875 but later discovered that the man he had chosen was in fact working for Tunku Antah and had become his 'chief'. Murray visited Terrachi on 26 November 1875 with an escort of twenty men of the 10th Regiment under command of Lieutenant Henry Charles Hinxman; thirty police; Doctor Hoysted and Mr Daly an Australian surveyor. The party crossed the Bukit Putus pass and encamped at Terrachi about sixteen miles from Seremban where the Resident and Dato Klana normally resided. Not knowing that the Punghulu had changed sides, Murray and Hinxman stopped at his house for the night and on the morning of 27 November concluding that all was quiet Murray decided to return to his Residency. He left with Mr Daly and six policemen so that he would have some protection when he began his survey. Daly began his survey but as he proceeded towards Kuala Pilah he was threatened by armed Malays, which he estimated numbered about 200 men. As a result of a message Captain Murray returned to effect his escape.

After an exchange of shots a regular engagement ensued and Lieutenant Hinxman tried to drive the Malays from their position but as his own flank positions were threatened, and as his men were running short of ammunition, he had to withdraw and the whole party returned to Rassa arriving there at 6.00 pm but of course this left the Bukit Putus pass in the hands of the enemy. At this stage of the conflict there were no casualties.

The strength of the force at Sungei Ujong numbered forty-five men of the 10th Regiment and sixty-eight police but of these only fifty had rifles. Murray warned Tunku Antan to keep out of Terrachi in future and having acquainted the Lieutenant-Governor at Malacca of the situation he was reinforced by a party of a further twenty-three men from the 10th Regiment led by Lieutenant Charles Talbot Peyton. On the night of 2 December Tunku Antah moved his force to Paroe, which was only six miles from Seremban. Lieutenant Hinxman attempted to reconnoitre Paroe on 5 December but was driven back by the Malays who advanced further and established themselves at Ampangan. Hinxman reported to the Lieutenant Governor at Malacca that his detachment had to be divided between the barracks and the Residency, and that he was short of ammunition and food. He advised that the Malay force was only three miles away and that the road to Lukut, his way out, was unsafe. With the use of the brass cannon of Dato Klana the enemy force was dislodged and withdrew as far as Paroe. The force recruited locally in Singapore and under command of Mr de Fontaine landed at Lukut and reached Seremban on 6 December and with this additional force Lieutenant Hinxman resolved to attempt to dislodge the enemy from Paroe. Having detached a small group to hold the Residency and barracks at Seremban, Lieutenant Hinxman took to the field with the following force:

	Officers	*Other ranks*
10th Regiment	2	46
Arabs	2	85
Police	2	46
	6	177[37]

Thus giving a total force of 183. The party paraded at daylight on 7 December and marched to within a mile of Paroe. Lieutenant Hinxman detailed Mr Robinson to take twenty of the Arabs to the right and a Arab sergeant to take a further party of twenty to the left. Hinxman then advanced with the remainder of the force and halted under cover of a bank some 170 yards from the enemy-held stockade. When he thought he had his flanking parties in position he opened fire and at once began to suffer casualties from the return fire. Unfortunately his left flanking party had become lost and as the right party led by Mr Robinson was attacking he had little choice but to proceed. He planned to move to the left and attack from that direction but as he did so his party came under enfilade fire from the stockade and a sizeable part of his force failed to follow. He managed to secure the left-hand outwork with a small force consisting of Lieutenant Peyton, ten soldiers of the 10th Regiment, Mr de Fontaine and six Arab soldiers. The outwork had to be finally taken by bayonet. Mr Robinson had secured the right-hand outwork but was unable to hold it but fortunately at this critical time a 9-pounder brass gun arrived under the charge of a Mr Skinner. Skinner, Murray and a few police managed to bring the gun into action against the main stockade. After only firing six rounds the gun capsized but it had had a good effect against the main stockade. Lieutenant Peyton was able to return and collect as many of the stragglers as he could find and brought them forward to join Lieutenant Hinxman. With this party Hinxman charged the main stockade, from which the Malays fled; he then burned the stockades and the village and then returned to Rassa that afternoon while the Malay force withdrew to the Bukit Putus pass.[38]

The skirmish had been fairly heavy but the action had been won due to the courageous leadership by Lieutenant Hinxman and Lieutenant Peyton who were given credit for their actions. The casualties given in various accounts differ, with the press reporting the Malay losses as being '. . . 80 killed and 200 wounded . . .'[39] whereas Parkinson quotes '. . . three Rajahs and over 100 men killed . . .',[40] the Lieutenant Governor of Malacca put the figure at '35 Malays killed [and] great numbers wounded . . .' and Murray recorded that some sixty to eighty had been seen dead by the Chinese. The British casualties were recorded as being

	Killed	*Wounded*	
10th Regiment	2	12	
Arabs	5	12	(including both officers)
Police	4	5	
	11[41]	29	

Lieutenant Peyton received praise in the local press for his gallant 'dash' with his small party of men from the 10th Regiment at the stockade at Paroe, whereas Murray received criticism from the press.[42]

Lieutenant Colonel the Hon A. E. H. Anson was prepared to move into the field from Penang and although anxious to join the field force was unable to do so until 6 December and at that time he found that Sir William Jervois had redirected 358 officers and men of the 1st Ghoorkhas, (*sic*) under command of Lieutenant Colonel E. B. Clay, with Messrs Kynnersley and Neubronner acting as interpreters. Half a battery of artillery consisting of

The Sungei Ujong Cemetery, 1977. This site is now being allowed to 'revert to nature'. The cannon have been removed to the Malacca State Museum. Photograph by Mrs J. Harfield.

The 10th Foot Memorial at the Sungei Ujong Cemetery. Photograph by Mrs J. Harfield, 1977.

two guns, Coehorn mortar and fifty rockets, were embarked on SS *Abyssinia* and a further reserve of 200 men of the Buffs were to follow south. Jervois however hoped that the latter would not be required and could be diverted back to their original destination, Perak. The SS *Malda* arrived off of Malacca and Lieutenant Colonel Anson and his force were met by Captain Erasmus H. Vaughton who was deputising for the Lieutenant Governor, The Hon C. B. Plunket who had gone to Sungei Ujong to join Murray. Vaughton informed the reinforcements that Sungei Ujong might be attacked at any time, that the mines at Chedras were in danger, that wells were being poisoned and that Malacca could be attacked at any time. This danger was not generally agreed and Anson soon resolved that Plunket had acted incorrectly in moving to 'the front' with a platoon of infantry, which was in any case outside his area of jurisdiction. He had left Captain Vaughton, a soldier with no experience in political matters, to deal with the political situation in Malacca and with wild rumours being circulated Vaughton must have been relieved at the arrival of Colonel Anson. Colonel Anson despatched a message requesting Plunket to return to Malacca to do his duty there and himself proceeded to Lukut and commenced unloading his force. The Buffs he decided to return to Penang and arranged for their immediate return.

The unloading was a slow process due to the lack of coolies and it was therefore necessary to despatch the troops to Rassa in groups, which he did, the first party moving on 9 December and the remainder on 10 and 11 December. In actual fact Colonel Anson was there as the Governor's representative and not as the military commander but in effect he simply took command and organised things. He reached Rassa on 12 December travelling with the rearguard which was provided by Commander Stirling, RN and thirty seamen. On arrival he divided the force into two divisions and resolved to attack the Malay stockades which had been built in the Bukit Putus pass. Colonel Anson's force, which consisted of 540 men, was organised as follows:

1st Division
Lieutenant Colonel Hill
Captain Murray
Commander Stirling, RN

Goorkhas	3 Officers	120 men
Seamen	2 Officers	30 men
Artillery	1 Officer	10 men

Total 9 officers and 160 men, with one 24-pounder gun and 83 rockets.

2nd Division
Lieutenant Colonel E. B. Clay
Lieutenant North, Royal Engineers

10th Regiment	2 Officers	41 men
Goorkhas	2 Officers	232 men
Arabs	1 Officer	70 men
Artillery	1 Officer	22 men

Total 8 officers and 365 men, with a 9-pounder gun and 36 rockets.[43]

Lieutenant Colonel Hill and his division moved off on 19 December and marched to Dato Klana's house. Halting there for one night, he moved on the next day towards Pantai, and from there he crossed the Langkap pass and entered the Terrachi valley on 21 December. It had been a tiring march through a difficult terrain of dense jungles and swamps and his route had caused him to cross the Muar river on seventeen occasions. On reaching his destination he found that the enemy had already departed. The 2nd Division, under Lieutenant Colonel Clay made the frontal approach to the Bukit Putus pass, leaving on 20 December with a vanguard of fifty men accompanied by Lieutenant W. North RE. The vanguard encountered felled trees which halted the advance and whilst North and a party stopped to clear the obstruction, the remainder of the advance party moved off to the left and soon encountered a strong stockade which was guarded by sharpened bamboo, and to back-up this post there were two other stockades spaced out at 80-yard intervals. Captain George Nicholas Channer, of the Bengal Staff Corps, led a party of twenty-five Goorkhas and a report on his action by his Commanding Officer reads as follows:

> '... He proceeded cautiously with a party of 25 Goorkhas as if he were on a sporting expedition. He advanced, himself, to within a few yards of the stockade – the enemy had kept no look-out, they were cooking at the time. Captain Channer seized the opportunity, and followed immediately by two men named in the margin, was the first to jump into the stockade. He found it occupied by 25 or 30 men. The first man, Captain Channer shot dead with his revolver, and the two Goorkhas each shot his man. They were soon followed by the rest of Captain Channer's party. Having expelled the Malays, Captain Channer opened a smart fire on the other two stockades, and doubtless some were wounded in that and the other stockades; but the Malays always carry off their dead (if they have time) and wounded but they must have suffered, as proved by the track of blood they left behind ... [of the Gurkhas, one was killed, one seriously wounded and two slightly wounded] ... Of Captain Channer's gallantry, coolness, and intrepidity I cannot sufficiently express my approval ...'[44]

The two Goorkhas mentioned in the marginal note were 'Sepoy Bulbeer Ghurtie, and Sepoy Geltman Thappa', both of B Company, 1st Goorkhas. Captain Channer was subsequently awarded the Victoria Cross for his gallant action on 20 December 1875. It was promulgated in *The London Gazette* of 12 April 1876.[45 46] This action concluded the actual fighting and the 2nd Division moved on to Bandole on 21 December, burning stockades *en route*.

Colonel Anson moved his headquarters from Rassa to Terrachi and ordered his force to move on to Sri Menanti, which they did and once having established police posts in the territories now under military control the British force withdrew to Rassa and Malacca. Anson returned to Penang leaving behind Murray who had the problem of deciding which chiefs had been friendly and which had been hostile to the British administration. The Governor, Sir William Jervois, was unable to make any decision as to which chiefs should be placed in charge as he too was uncertain as to their allegiance. He visited Malacca on 18 January 1876 and met with the Dato Klana and the Dato of Rembau.

On 29 January, Captain Murray visited Terrachi again with Captain Channer, Mr Skinner, ten police and a force of thirty Goorkha soldiers. He arranged for the collection of arms and ammunition from the Malays and in fact collected thirty-six iron guns and lelas and 198 muskets in Sri Menanti and Muar, whilst in Jumpole he collected a further eight guns and 150 muskets. The situation remained peaceful during February and March but there was a further outbreak of violence in April when a number of police stations in the State of Sri Menanti were attacked and taken and then destroyed. The situation was quickly dealt with and order restored once again to the territory. The ship *Ringdove* was involved in the suppression of this latest outbreak of violence and moved up river taking a number of Goorkha troops to deal with the situation.[47]

The Perak problem had still to be concluded and on 13 January a reward was offered for the capture of Maharaja Lela, this being $6,000, and $3,000 each for Dato Sagor and Pandak Indut. It was hoped that with the virtual defeat of ex-Sultan Ismail a number of the tribes would turn against him and his followers and assist the British in apprehending him and his dissident leaders. The strength of the British force in Perak and in Malacca in early 1876 was:

> '... 3rd Regiment - 950; 10th Regiment - 700; 80th Regiment - 300; 1st Gurkhas - 600; One battery, R.A. - 120; One company Bengal Sappers - 80. Total approx 2,750 ...'[48]

Early in January 1876 it was learned that Ismail and about 200 of his followers were at Jarnai in Perak and that the Maharaja Lela was at Bada. The information was supplied by Inche Karim of Salama who was reputed to have 300 men ready to assist the British in the capture of Ismail, however when a British force under command of John E. Hewick, the Assistant Superintendent of Police, Province Wellesley arrived the number was found to be only twenty-eight. Although the trap was ready, Ismail received a warning and escaped but had to leave behind seventeen of his elephants. The small British party killed four or five of ex-Sultan Ismail's men, captured three, and then burned the houses which Ismail and his followers had been using.

After this incident Ismail appealed to the Sultan of Kedah for protection who, in turn, asked Anson for advice. Anson who was by this time back at his post as Lieutenant Governor of Penang replied '... I advised him to recommend the Sultan to go into the State of Kedah and then to allow himself to be handed over to me ...'[49] however the Sultan of Kedah was not too happy with this suggestion but eventually agreed to the plan.

Whilst these various negotiations and attempts to arrest the ex-Sultan were going ahead Brigadier-General Ross's northern column was operating out of Kuala Kangsar. On 4 January a detachment of this force suffered casualties at Kota Lama. The kampong had been occupied and the Malays disarmed without any sign of resistance and General Ross had landed with his staff and a small escort. Suddenly, out of the jungle, rushed some fifty or sixty Malays and attacked the party, but the seamen in the landing party held firm and repulsed the attack. Unfortunately the Brigade Major, Major H. L. Hawkins was killed with a spear and Major the Hon J. F. A. McNair, Madras Artillery, who was with the party and was Colonial Engineer of the Straits Settlements at the time, recorded that Major Hawkins:

> '... received a frightful spear wound, the blade passing through his chest. A sailor named Sloper ran to his help, and shot two Malays who were running up to continue the attack, when Major Hawkins is reported to have exclaimed: "Save yourself, you can do me no good now." The officers who had gone on towards the river now returned, and tried to move him, but they were compelled in turn to fall back towards the river. Surgeon Townsend being the first to be assailed by three Malays with spears. One he shot with his revolver, but the man struck him down in falling, and his two companions dashed in to spear him, when they were bayonetted by a couple of seamen ...'[50]

The body of Major Hawkins was recovered and taken to Kuala Kangsar where he was buried in the small cemetery at Bukit Chandran.[51]

A small expedition was mounted on 20 January which reached Kota Lama and having driven the Malays from the kampong the whole kampong was then destroyed. The Kota Lama Malays were pursued for a while and then the British force withdrew having suitably avenged the death of Major Hawkins and the others who had been killed on 4 January.

At the end of February a further plan was evolved which it was hoped would bring about

Grave of
Major H. L. Hawkins,
Bengal Staff Corps,
at the Bukit Chandran
Cemetery, 1976.

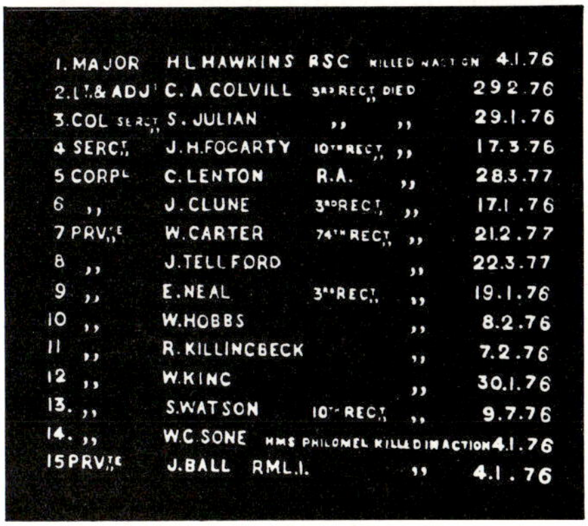

Above and right:
The Perak Memorial at Bukit
Chandran Cemetery, 1976.

the apprehension of ex-Sultan Ismail. The Governor wrote:

'... Before this expedition was organised, however, Mr Hewick, who had been sent to watch the movements of the ex-Sultan, reported that Ismail was going to cross into Quedah. I therefore ordered preparations to be suspended ...'

Hewick, however, did not wait for orders and on 4 February arranged a plan with the Sultan of Kedah and using two Kedah chiefs, Tunku Ibrahim and Haji Aboo, set about locating and contacting Ismail. Hewick later wrote:

'... On the 9th [March] Hadjee Aboo came to me saying that Ismail was on the frontier waiting. I directed him to return, and get him into Kedah territory, as far as Cheeah, about 10 miles from Koopang, so as to be away from any chance of molestation on the part of the Patanis, who had sent parties out. This was done, and on the 9th instant I proceeded to Cheeah to see Ismail. On arriving I found him encamped in the jungle with his men, and with a number of women and children, in all about 174. I had an interview, and he promised to proceed quietly to Kedah, and then go on to see the Lieut-Governor at Penang ...'[52]

Hewick reported that Ismail was in fact destitute and that many of his people were sick and that he had already lost a number of his people in the jungle. He was accorded little respect by his own followers and had to build his own house and care for his own elephants. Ismail surrendered on 13 March at Kuala Ketee and moved to Kuala Muda on 17 March. Hewick did not pursue the Maharaja Lela, who was near at the time, as he did not wish to lose Ismail. At the time of surrender the Sultan of Kedah had 1,000 of his men dispersed in the jungle, out of sight, behind Ismail so as to cut off his retreat should he change his mind, this precaution proved to be unnecessary. Ismail was taken to Penang on the steam-launch *Mata Mata* on 20 March and Colonel Anson wrote:

'... On the 23rd [March], Sultan Ismail was brought to my office, in Penang, and handed over to me the Regalia (except of course the elephants that had formed part of it) [the elephants and his women and children remained in Kedah under the protection of the Sultan] I found him a very gentlemanly and pleasant old man, and I felt much sympathy with him. His possession of the Regalia showed that he had been the actually recognised Sultan ...'[53]

The regalia was later exhibited at the Colonial Exhibition in London and the display included a sword, said to have belonged to Alexander the Great, a chain of office, armlets, a gold-sheathed kris, and the seal of Sultan Mohammed Shah, as well as other items of historic significance to the Perak Malays. Colonel Anson sent Ismail and the regalia to Singapore and on arrival on board the ship HMS *Ringdove* one of the officers on the ship commented that Ismail was '... a fine intelligent looking old man, and in every respect a great contrast to Abdu'llah ...'[54] This comment supports a description used by Colonel Anson who referred to Abdu'llah as '... the good-for-nothing Abdullah ...' On arrival at Singapore, Ismail handed over the regalia to the Governor and then was taken by HMS *Ringdove* to Johore Bahru where he was placed under the custody of the Maharaja of Johore.

During early March, Dato Sagor was apprehended and three of those actually accused of the murder of Resident Birch were also caught and brought to trial, these were named as Seputum, Che Gondah and Ngah Ahmat. The trial took place at Bandar Bahru on 3 and 4 March 1876 and all three pleaded not guilty to the charge. All three were found guilty and were sentenced to death by hanging, subject to the Sultan's confirmation of the sentence. Jervois, however, advised the Sultan, through the Deputy Commissioner, to commute the sentences given to Che Gondah and Ngah Ahmat which he did. They were given penal servitude for life and only Seputum was hanged; his sentence was carried out at Bandar Bahru on 20 May 1876 in the presence of the Sultan and the Perak chiefs.[55]

This now left only the Maharaja Lela free with a few of his followers. Eventually he agreed to surrender to the Maharaja of Johore and consequently the Maharaja of Johore's steamship *Pantai* went to Penang and then onto Larut. The Maharaja Lela embarked on the ship on 13 July accompanied by Pandak Indut and Ngah Jabbor who were also wanted by the authorities. The party reached Johore on 20 July. They eventually surrendered to the Governor who wrote:

> '. . . I guaranteed them a fair and impartial trial, and that, if found innocent of the crime with which they are charged, they will be set at liberty. They in consequence have surrendered themselves today and they are now lodged in the Civil Prison in Singapore . . .'[56]

Sultan Abdu'llah was alarmed at this turn of events as he had written to the Maharaja approving of the murder of Birch and if this evidence came to light during the trial then the Governor would have little option but to arrest the Sultan of Perak, Abdu'llah, for complicity in the murder of Resident Birch. The whole trial was long and complicated and has been given excellent coverage in the book *British Intervention in Malaya 1867–1877* by C. Northcote Parkinson. The final outcome was that Sultan Abdu'llah was found to be implicated and that he was deported, along with the Laxamana, the Shahbandar and the Mantri to the Seychelles. They were sent to the islands in 1877 but later Sir Andrew Clarke interceded on behalf of Sultan Abdu'llah and gained permission for him to return to Singapore, which he did in 1894. He later was allowed to visit England and learned the English language. The Mantri was also permitted to return to Singapore where he eventually died.

The trial of Maharaja Lela, Pandak Indut, Ngah Jabbor, Dato Sagor and three others was held at the Mantri's house at Martang and lasted from 14 December until 22 December. The court was guarded by a company of the 10th Regiment. All were found guilty and sentenced to death. The Maharaja Lela, Dato Sagor and Pandak Indut were all hanged near Larut jail on 20 January 1877, their sentences having been confirmed. The remaining four were imprisoned for life, although one, Che Tuah, was declared to be an imbecile.[57]

The British administration came under criticism and as a result Mr Davidson was refused promotion and resigned, and Captain Speedy was in fact declared redundant. The Perak Police were formed and the War Office agreed to release Lieutenant Paul Swinburne of the 80th Regiment on 29 January 1877 to act as the Superintendent.

The British forces who had taken part in the Perak campaign were awarded the Indian General Service Medal (1854–1895) with the bar 'Perak', those who had already qualified for the medal were entitled to the bar. The qualification period was given as from 2 November 1875, or in the case of the troops who had embarked from India, 27 November 1875, to 20 March 1876. Gordon in his book *British Battles and Medals* records: '. . . The first sentence of General Order No 111, dated 1st September 1879, reads: "The Queen having been pleased to command that the grant of the India Medal of 1854, with the clasp inscribed 'Perak 1875–76' . . ."' The date was omitted from the bar when it was produced although Gordon states that he was unable to find a reason for the omission. The naming of the medal is in large sloping capitals.[58]

An examination of the Perak Medal Roll at the Public Record Office at Kew has revealed that the original rolls were subsequently amended and those members of the various regiments who remained at Singapore were deleted from the roll and an annotation refers to the recipients having to return the medals.[59] The medals and bars were issued to those who took part in the campaign and who were still serving in the Straits Settlements during 1880. The Governors Despatches for 1880 show entries dated 15 October as issuing the medal to Major the Hon J. F. A. McNair, Mr A. M. de Fontaine, Mr Frank Swettenham, etc, and on 27 October to Mr W. E. Maxwell.[60] The issue of medals can be traced throughout the next two years. Commander P. J. Murray RN (Retired), the Resident at Sungei Ujong was sent his on 14 April 1881,[61] Mr W. A. Pickering's was sent on 29 August 1881,[62] and one recipient did not receive his until the latter half of 1884, this was Sergeant G. Singh, who was shown on the medal roll as Corporal Gugh Singh of the Perak Police.[63]

After the close of the Perak Campaign of 1875–1876 two memorials were erected, one at the Bukit Chandran Cemetery and the other in Malacca town. The Bukit Chandran memorial is an imposing memorial standing on ground overlooking the river. The inscriptions on the memorial read on one side:

In
Memory
of
The Officers and Men
who fell in the Perak War
1876–1877

and on the other, showing individual details:

1.	Major H. L. Hawkins	B.S.C.	Killed in Action	4.1.76
2.	Lt & Adjt C. A. Colvill	3rd Regt	Died	29.2.76
3.	Col Sergt S. Julian	”	”	29.1.76
4.	Sergt J. H. Fogarty	10th Regt	”	17.3.76
5.	Corpl C. Lenton	R.A.	”	28.3.77
6.	” J. Clune	3rd Regt	”	17.1.76
7.	Prvte W. Carter	74th Regt	”	21.2.77
8.	” J. Tellford		”	22.3.77
9.	” E. Neal	3rd Regt	”	19.1.76
10.	” W. Hobbs		”	8.2.76
11.	” R. Killingbeck		”	7.2.76
12.	” W. King		”	30.1.76
13.	” S. Watson	10th Regt	”	9.7.76
14.	” W. C. Sone	HMS Philomel	Killed in Action	4.1.76
15.	” J. Ball	R.M.L.I.	Killed in Action	4.1.76[64]

Within the same cemetery are the individual graves of the fifteen officers and men listed on the memorial. The 74th Highlanders joined the Straits Settlements garrison in 1877 and remained there until 1878. A note in the Governor's Despatches dated 5 March 1877 showed that '. . . Reinforcements of 74th Regt . . .' had been requested so as to bring it '. . . to its full estab of 923 all ranks . . .' and were due at the next trooping season.[65]

The Malacca Memorial obelisk is situated in the town of Malacca at the car park on Jalan Fort, slightly to the right of the Malacca Museum. The Malacca Memorial is to the soldiers who had been killed in the Sungei Ujong campaign of 1875–1876. The memorial inscription is interesting as it contains details of men of a British Regiment of Foot, a

Goorka (*sic*) regiment and the Arab Contingent. The spelling of the name Sungei Ujong has been engraved as 'Sungie Ujong'. The inscriptions are on four sides of the obelisk and read as follows:

1st Batt H.M.S. 10th Foot.

Sergeant F. Owen
Private J. Ball
J. Newman
H. Smith

1st Goorka Light Infantry
Naik Bucktring Rav
Sepoy Duljeet Poon

Arab Contingent
Corporal Ali San
Private Mahomet Awal
Mahomet Nizam
Ferozef Mahrout
Ali Musjie
Mariano Volady

Erected by their
Comrades in memory
of those who lost
their lives on active
service against the
Malays in
Sungie Ujong
1875–6[66]

The following record shows those British Army officers and men who were listed on the 'Perak' Medal Roll. The list has been transcribed in the order in which names appear on the medal roll and the abbreviations used are those shown on the roll. The only change that has been made is the move of the position of the initials of the recipient from before to after the surname.

ANNEXURE A

NOMINAL ROLL OF ARMY PERSONNEL WHO WERE AWARDED THE INDIAN GENERAL SERVICE MEDAL (1854–1895) WITH BAR 'PERAK'

The medal and bar were awarded to members of a force sent to Perak as a result of the murder of the British Resident Mr J. W. W. Birch. The qualifying period for this medal and bar is given as 2 November 1875 until 20 March 1876. The start of the qualification period relates to the troops despatched from either Hong Kong or the Straits Settlements. For personnel who embarked from India the commencement date was 27 November 1875.

Number	*Rank*	*Name and Initials*	*Unit*
340	Private	Ash G.	1/10th Foot
1434	"	Aston J.	"
1860	Lce Corp	Anderson M.	"
266	Private	Andrews V.	"
1697	Private	Askey J.	"
1087	"	Allen P.	"
1081	"	Adams G.	"
1764	"	Adams J.	"
1978	"	Argust J.	"
78	"	Arthur J. J. R.	"
1491	"	Aylin E.	"
1152	"	Arkell J.	"
1809	Lce Serjt	Atuhurst G.	"
83	Pte I	Allen J.	"
1734	"	Arbour S.	"
2025	Private	Adlam U.	"
1206	"	Ainsworth H.	"
1531	"	Aitkin R.	"
291	"	Alderson W.	"
231	"	Anderson F.	"
1916	"	Annalls A.	"
1521	"	Appleton J.	"
93	"	Arthurs W.	"
1452	"	Atkinson J.	"
2033	"	Atkins C.	"
1907	"	Austin E. S.	"
—	Hosp Appce	Allison A. A.	Bengal Sub Medl Dept
—	—	Ashton M. A.	Teleg Dep, India
2462	Private	Ablewhite W.	A. H. Corps
—	Lieut	Armstrong T. F. C.	1/3rd Foot
1382	Private	Allaway H.	"
1906	"	Aylan G.	"
2085	"	Andrews J.	"
1444	"	Atkins H.	"
33	"	Anderson G.	"
207	Drumr	Addis T.	"
1669	Serg	Allen W. J.	RA
—	Major Genl	Anson (Administr of the Govt)	RA
1916	Bomb	Adams E.	3/5 RA
—	Major	Amiel C. F.	80th Foot
4202	Col Sgt	Allen J.	"
159	"	Auscombe J.	"
3559	Sgt	Archer J.	"
1968	Private	Attrude G.	"
1741	"	Abraham J.	"
1505	"	Adair H.	"
546	"	Adey J.	"
1155	"	Booth E.	"
1845	"	Bowden A.	"
1440	"	Bradley F.	"
1877	"	Braze T.	"
3733	"	Brennan E.	"
273	"	Brereton J.	"
1381	"	Brisbane W.	"

Number	*Rank*	*Name and Initials*	*Unit*
1664	Private	Broadhurst A.	80th Foot
1853	"	Brooks J.	"
1688	"	Brown J.	"
4227	"	Brown J.	"
1553	"	Brown W.	"
1766	"	Brunton R.	"
1785	"	Buchanan J. R.	"
1728	"	Bunday H.	"
2054	"	Burgess W.	"
169	"	Burgwin J.	"
2055	"	Burke D.	"
1856	"	Burnett J.	"
1810	"	Burt A.	"
1609	"	Burkenshaw H.	"
1984	"	Burton J.	"
2071	"	Burnett W.	"
278	"	Butler F.	"
1693	"	Buxton P.	"
998	"	Bailey F.	"
—	Capt	Brackenbury M. C.	RE
—	Bk Sergt	Bradley N.	OS Dept
2508	Gunr	Bacon J.	3/5 RA
2025	"	Boothroyd A.	"
2022	"	Brown C.	"
1639	"	Bulmer W.	"
944	"	Burgin J.	"
2036	"	Butcher C.	"
2024	"	Byrnes J.	"
2531	Private	Batley J. J.	A. H. C.
1770	Pte	Banks H.	1/10th Foot
440	"	Bridle G.	"
1470	"	Belcher S.	"
1806	"	Brown I.	"
1849	Gunner	Buckingham J.	9/2 Bde RA
2707	"	Brooks W.	"
2487	"	Bryan M.	"
2610	"	Buckley T.	"
Capt and Paymaster		Bailey I. I.	1/3rd Foot
2001	Serg Ins of Mus'ty	Boon G.	"
3852	Colr Serg	Brooman J.	"
1646	Sergeant	Brown W.	"
30	"	Brennan J.	"
1881	Private	Barkin O.	"
399	Corpl	Burke T.	"
1695	"	Butler J.	"
3818	Drumr	Byrne P.	"
1795	"	Birmingham C.	"
2034	Private	Baker G.	"
1852	"	Baker James	"
1062	"	Baker Joseph	"
1399	"	Bain J.	"
1092	"	Bancroff J.	"
1426	"	Barnes G.	"
2007	"	Berwick G.	"
1185	"	Blucher J.	"

Number	Rank	Name and Initials	Unit
1362	Private	Baker W.	1/3rd Foot
34	”	Baker W.	”
1986	”	Banks J.	”
907	”	Beeston F.	1/10th Foot
1449	Lce Corp	Berry G.	”
1029	Private	Burne A.	”
1843	”	Buskin J.	”
2093	”	Banchop J.	”
2158	”	Barry J.	”
370	”	Belworthy S.	”
375	”	Baylis J.	”
897	Corpl	Barber J.	”
213	Private	Brand V.	”
1601	Corporal	Byrne J.	”
1751	Private	Briscoe B.	”
1657	”	Bardon T.	”
145	”	Bryson J.	”
1812	”	Bushell G.	”
1542	”	Brennan T.	”
1572	”	Burke W.	”
393	”	Ball G.	”
1792	”	Bartlett J.	”
1666	Corpl	Bird H.	”
125	Private	Brown C.	”
1822	”	Bowen G.	”
190	”	Burns J.	”
423	”	Braddick W.	”
1302	”	Best W.	”
292	”	Buckingham J.	”
1045	”	Batten G.	1/3rd Foot
1961	”	Beckett W.	”
1625	”	Benson W.	”
2108	”	Bowers J.	”
2154	”	Braham J.	”
37	”	Brooksbank J.	”
1570	”	Bround C.	”
1624	”	Ball L.	”
85	”	Brooks G.	”
1504	”	Brownbeard G.	”
264	”	Bryan P.	”
217	”	Buchanan A.	”
1991	”	Bull J.	”
1373	”	Butler J.	”
1275	”	Byrne L.	”
2041	”	Barnes E.	”
35	”	Bennett W.	”
1626	”	Blackburn J.	”
1413	”	Boswell J.	”
191	”	Brown H.	”
248	”	Buckley J.	”
2061	”	Bryan J.	”
2116	”	Beaumont W.	”
2099	”	Bentley T.	”
208	”	Boyland J.	”
1850	”	Benson F.	1/10th Foot
1956	”	Barry S.	”
165	”	Beaumont R.	”
1962	”	Brams C.	”
2099	”	Boyle H.	”
99	”	Baker R.	”
107	”	Biles H.	”
—	”	Brain C.	”
298	”	Boote W.	”
71	”	Barnett J. J.	”
2170	”	Boutoft H.	”
—	Lieut	Booth T. G.	”

Number	Rank	Name and Initials	Unit
1496	Private	Balmer J.	1/10 Foot
295	”	Barry G.	”
611	”	Bishop J.	”
84	”	Barrett B.	”
2	”	Bennett H.	”
1980	”	Bevan W.	”
1887	”	Batts A.	”
1539	”	Braithwaite J.	”
1789	”	Brown J.	”
32	”	Bailey H.	”
1784	”	Beales G.	”
1970	”	Brabrook T.	1/3rd Foot
1734	”	Bracken T.	”
1336	”	Brown T.	”
1686	”	Burridge B.	”
2084	”	Burridge M.	”
1240	Corpl	Benson W.	”
1376	Pte	Bird J.	”
1091	”	Bowes L.	”
1210	”	Brady J.	”
1396	Private	Brenchley H.	”
2026	”	Bridgland C.	”
2044	”	Brock E.	”
—	Major	Badcock J. R.	SAC (General 1st Cl Capt)
—	Capt	Bradshaw J. L.	80th Foot
—	Surgeon	Bomford G.	Bengal Medical Service
—	Assist Apothy	Bell T.	Bengal Sub Medl Dept
1680	Serg	Battys N. F.	Army Hosp Corps
2478	Gunr	Bagnall G.	RA
2083	Drmr	Barnwell W.	80th Foot
1346	Private	Ball D.	”
202	”	Banks A.	”
1928	”	Barr T.	”
1448	”	Barber A.	”
1382	”	Barklie M.	”
1514	”	Barnes J.	”
2053	”	Barrett D.	”
1442	”	Baxter J.	”
2018	”	Beale W.	”
1709	”	Beatty J.	”
1126	”	Bedford W.	”
105	”	Bentley J.	”
275	”	Bentley J.	
1556	”	Betts J.	”
29	”	Birch W.	”
109	”	Bird F.	”
16	”	Bond W.	”
1897	”	Booker A.	”
—	Surgeon Maj	Collis W.	AM Dept
—	Surgeon	Campbell W. J.	”
1440	Sergt	Charlton E.	9/2nd Bde RA
2612	Gunr	Carroll J.	”
2619	Trumpr	Cole G. R.	”
2611	Gunr	Camp J.	”
2561	”	Cato J.	”
2480	”	Conway H. J.	”
1853	”	Coring W.	”
1854	Gunner	Crockard W.	”
—	Captn	Crawford C.	RE
—	Lt Col	Cox T. A. P.	1/3rd Foot
—	Sub Lieut	Chippendall G. H.	”
—	Lieut & Adj	Colvill G. A.	”
—	Quartermaster	Colley J. W.	”

Number	Rank	Name and Initials	Unit
991	Quartermaster Sergeant	Coombes J.	1/3rd Foot
1729	Col Sergt	Carter J.	,,
1910	,,	Cook J.	,,
1840	,,	Cumber H. C.	,,
1023	Sergt	Cook A.	,,
1791	,,	Callaghan D.	,,
28	Private	Colley G.	,,
1432	Corpl	Crawley A.	,,
1268	Private	Clarke J.	,,
1836	Corpl	Clarke W.	,,
801	Private	Cuthbert J.	1/10th Foot
947	Sergeant	Clarke C.	,,
2134	Private	Cody M.	,,
280	,,	Chandler A.	,,
938	,,	Caverley M.	,,
1248	,,	Collard E.	,,
1522	,,	Cummings P.	,,
1842	,,	Cattell W.	,,
344	Corp	Cutler J.	,,
1746	Private	Crawley J.	,,
1535	Corpl	Cousins W.	,,
1745	Private	Clarke F.	,,
1267	,,	Carey J.	,,
88	,,	Christmas G.	,,
240	,,	Cotes J.	,,
1253	,,	Callaghan C.	,,
1147	,,	Curram J.	,,
1869	,,	Cloake W.	,,
1759	,,	Conway M.	,,
1965	,,	Clawson A.	,,
1602	,,	Connelly E.	,,
657	,,	Coleman P.	,,
967	Lce Corp	Cox W.	,,
617	Private	Curtis W.	,,
249	,,	Clarke J.	,,
520	Corpl	Clune J.	1/3rd Foot
207	Private	Candlin M.	,,
2142	,,	Clarke J.	,,
1342	,,	Clarke N.	,,
1850	,,	Clarke T.	,,
2038	,,	Corney W.	,,
244	,,	Cracknell G.	,,
1742	,,	Crockett W.	,,
1423	,,	Crump J.	,,
1849	,,	Crouch J.	,,
2000	,,	Crawley J.	,,
1805	,,	Cullinan P.	,,
2069	,,	Carver J.	,,
1698	,,	Cass H.	,,
1197	,,	Casey E.	,,
1251	,,	Clyde A.	,,
1688	,,	Collins H.	,,
1449	,,	Conroy D.	,,
73	,,	Conroy R.	,,
1924	,,	Cottle R.	,,
1664	,,	Cropper W.	,,
1898	,,	Campbell J.	,,
1476	,,	Chilton J.	,,
1441	,,	Clarke W.	,,
1931	,,	Cook J.	,,
2043	Corpl	Clarke J.	1/10th Foot
2135	Private	Cafferty J.	,,
1965	,,	Clawson A.	,,
1580	,,	Cleary D.	,,
17	,,	Clarke W.	,,

Number	Rank	Name and Initials	Unit
2157	Private	Cahill M.	1/10 Foot
1750	,,	Caselton C.	,,
1481	,,	Chuk J.	,,
1735	,,	Clarke J.	,,
2035	,,	Collins P.	,,
1888	,,	Cook F.	,,
1966	,,	Clayton I.	,,
450	Corpl	Cassidy B.	3/5 RA
1501	Gunr	Cable W.	,,
1803	,,	Casey L.	,,
299	Private	Cullinan J.	1/3rd Foot
302	,,	Coughland J.	,,
41	,,	Callon M.	,,
1436	,,	Cullow T.	,,
628	,,	Casey J.	,,
1390	,,	Connelly J.	,,
241	,,	Cook E.	,,
1243	,,	Cook J.	,,
2090	,,	Cooper T.	,,
197	,,	Cranham J.	,,
196	,,	Cole C.	,,
1438	,,	Cullinan H.	,,
1112	,,	Creigh J.	,,
1453	,,	Coldwell M.	,,
42	,,	Carter J.	,,
1744	,,	Claydon N.	,,
1862	,,	Coleman J.	,,
1743	,,	Coleman J.	,,
2161	,,	Cornwall C.	,,
2078	,,	Cowley W.	,,
1473	,,	Cullen J.	,,
704	,,	Curnock W.	,,
1867	Corpl	Cheeseman C.	,,
2135	,,	Clare W.	,,
1512	Gunr	Clark W.	3/5 RA
1807	,,	Cochrane T.	,,
2042	,,	Conboy P.	,,
2054	,,	Coller J.	,,
1811	,,	Crassey T.	,,
—	Captain	Creagh C. A. F.	80 Foot
—	Lieut	Cole C. C.	,,
—	,,	Cameron S. W.	,,
1270	Serg	Cross J.	,,
1989	Corpl	Colyer W.	,,
1435	,,	Cook D.	,,
1340	Prte	Carey F.	,,
112	Private	Carpenter J.	,,
1578	,,	Carroll J.	,,
1389	,,	Carter H.	,,
1052	,,	Case T.	,,
1656	,,	Chadwick H.	,,
187	,,	Chare C. E.	,,
1992	,,	Charles T.	,,
1914	,,	Chase R.	,,
1377	,,	Chesterton J.	,,
1797	,,	Christy J.	,,
1884	,,	Clarke A.	,,
325	,,	Clarke S.	,,
3694	,,	Clarke T.	,,
1497	,,	Clarke T.	,,
1827	,,	Clay I.	,,
1658	,,	Clayton H.	,,
1733	,,	Clemson J.	,,
1379	,,	Clift R.	,,
1478	,,	Collier W.	,,
954	,,	Collins H.	,,

Number	Rank	Name and Initials	Unit
1324	Private	Collinson R.	80 Foot
1241	”	Comaskey J.	”
243	”	Conlon J.	”
545	”	Constable T.	”
1913	”	Cook G.	”
82	”	Cook G.	”
1866	”	Cooper H.	”
1874	”	Coy W.	”
123	”	Coyne J.	”
1480	”	Crane L.	”
1716	”	Crankshaw W.	”
194	”	Crudgington R.	”
1523	”	Cullen P.	”
232	”	Cullen M.	”
1650	”	Culley F.	”
1786	”	Cunningham F. C.	”
502	”	Cowdrell R.	”
631	”	Curtin J.	”
—	Colonel	Clarke Sir A., KCMG,CB,CIE	RE
—	Major-General	Colborne Sir F., KCB	Staff
—	Asst Comy	Campbell C. G. L.	Control Dept
—	Sergt	Craig J.	Control Dept
—	Mr	Cartier W. E.	Indian Teleg Dept
—	Mr	Connelly	”
—	Mr	Cross J. H.	”
—	Surg	Churchill G. F.	AM Dep
—	Surgeon	Connolly B.	”
—	Asst Apothy	Carroll C.	Bengal Sub Med Depart
2565	Gunner	Ducker H.	9/2 Bde RA
2701	”	Duggan T.	”
32	Sergt	Davis G.	1/3 Foot
968	Corpl	Downs J.	”
1170	Drmr	Dewan D.	”
270	Private	Deadman F.	”
1067	”	Deadman W.	”
1296	”	Devaney P.	”
1406	”	Dillon J.	”
1581	”	Duggan H.	”
1575	”	Dooley J.	”
1794	”	Donlan T.	”
235	”	Donnell J.	”
1681	”	Dumphy R.	”
1297	”	Dudley A.	”
1491	”	Daley J.	”
1177	”	Durrand F.	”
1594	”	Dawson G.	”
1494	”	Dickson J.	”
74	”	Doyle T.	”
3706	”	Day W.	”
84	”	Delaney J.	”
1007	”	Dooland J.	”
1587	”	Durrand J.	”
65	”	Davis A.	”
1104	”	Dunkley W.	1/10 Foot
19	”	Dobbie W.	”
1356	”	Dawd D.	”
1773	”	Duncan G.	”
1891	”	Dickarty C.	”
65	”	Davis L.	”
2004	”	Duggan P.	”
90	”	Dixon J.	”
1683	”	Dumberlane P.	”
1493	”	Doyle J.	”

Number	Rank	Name and Initials	Unit
2126	Private	Devis H.	1/10 Foot
1875	”	Doolan J.	”
1954	”	Dowling P.	”
2122	”	Donovan J.	”
1345	Gunr	Dalton M.	3/5 RA
1186	”	Daly T.	”
2059	”	Dowens W.	”
1155	”	Driscoll T.	”
1641	”	Duke T.	”
2481	”	Deacon J.	RA
2699	”	Dewey J.	”
—	Major	Dunlop	Inspr Genl of Police
3790	Private	Davis J.	1/3 Foot
1518	”	Derlin J.	”
1424	”	Dillon M.	”
1940	”	Douglass W.	”
45	”	Doyle P.	”
198	”	Darcy N.	”
1618	”	Davis J.	”
2102	”	Denham R.	”
81	”	Dillon J.	”
2043	”	Dowe J.	”
1325	”	Dowds C.	”
75	”	Duggan M.	”
201	”	Dunne J.	”
1048	Sergt	Day H.	80 Foot
488	”	Dickey W.	”
2101	Corpl	Davis C.	”
1409	”	Dicken E.	”
1076	”	Dollin R.	”
209	Private	Daniels W.	”
3653	”	Darlin T.	”
1312	”	Davis W.	”
1930	”	Day W.	”
1089	”	Dell J.	”
1361	”	Dermott J.	”
1571	”	Dermott M.	”
1429	”	Dillon B.	”
327	”	Dixon H.	”
1971	”	Doe J.	”
1732	”	Donnelly J.	”
101	”	Donovan T.	”
269	”	Doogan W.	”
1330	”	Doyle A.	”
243	”	Drinkwater E.	”
2005	”	Dudley A.	”
1343	”	Duff J.	”
90	”	Duggan M.	”
2079	”	Dumbleton B.	”
507	”	Dunn C.	”
260	”	Dutton J. H.	”
—	Serg Maj	Davie G. S.	AM Dept
—		Daly D. D.	Supt of Works
698	Pte	Dakin S.	1/10 Foot
48	”	Drew C.	”
1891	”	Darling I.	”
1144	”	Danacher M.	”
1737	”	Darkins I.	”
1998	”	Dowd M.	”
1862	Gunner	Ewins G.	9/2 RA
1410	Private	Edmunds C.	1/3 Foot
998	”	Eastman T.	”
1329	”	Ecclestone W.	”
1544	”	Edwards T.	”
1412	”	Egan R.	”
1577	”	Elliott E.	”

Number	Rank	Name and Initials	Unit
1145	Private	Elliott H.	1/3 Foot
1572	”	Englefield H.	”
66	”	Evans P.	”
500	”	Ebrill W.	”
1865	”	Enfield C.	”
1233	”	Eyres R.	”
1722	”	Elson W.	1/10 Foot
220	”	Ely J. T.	”
718	”	Eveleigh C.	”
201	”	Evans J.	”
244	”	Elwick W.	”
1893	”	Emmerson A.	”
—	Lieut	Elliott A. W.	”
1828	Corpl	Eason J.	”
1963	Private	Edgar R.	”
66	”	Englefield J.	”
1936	”	Ewart R.	”
2178	”	Everett J.	”
895	Corpl	Else W. H.	80 Foot
1581	Private	Eccles T.	”
1624	”	Edwards R.	”
1543	”	Ellis G.	”
84	”	Ellison R. H.	”
214	”	Evans G.	”
1662	”	Evans J.	”
—	Surg Major	Evatt Geo J. H., MD	AMD
—	Apothecary	Elliott J. A.	Bengal Sub Medl Departt
—	Lieut	Elwes E. V.	3/5 RA
2519	Trumpr	Foley W. J.	9/2 RA
2567	Gunr	Flamery W.	”
1865	”	Forrester J.	”
—	Sub Lieut	Ferguson H. I.	1/3 Foot
160	Sergt	Furnish J.	”
1315	”	Fairbrother J.	”
1951	”	Filmer A.	”
2032	Corpl	Fisher J.	”
278	”	Field O.	”
2017	Private	Foster G.	”
272	Drummer	Francis P.	”
1967	”	Fuller J.	”
24	”	Finn T.	”
1119	Private	Fearnley J.	”
202	”	Flynn A.	”
1718	”	Fitzgerald T.	”
574	”	Foster H.	”
1479	”	Farr N.	”
2059	”	Fenton T.	”
1481	”	Fletcher J.	”
2016	”	Frazer W.	”
1576	”	Foreman J.	”
1474	”	Fox M.	”
1675	”	Fowler W.	”
1215	”	Fenton G.	”
985	Corp	Freeman J.	1/10 Foot
110	Col Serg	Fraud G.	”
973	Privte	Fox J.	”
1557	”	Fay J.	”
1076	Sergt	Fogerty J.	”
98	Private	Fitches A.	”
210	”	Forrest W.	”
1583	”	Feehilly B.	”
2002	”	Freeman J.	”
1434	Sergeant	Flynn J.	”
926	Pte	Feeley W.	”
1797	”	Feeley F.	”

Number	Rank	Name and Initials	Unit
2108	Sergeant	Flegg R.	1/10 Foot
1894	”	Flint I.	”
—	Mr	de Fontaine A. M.	Inspr of Police
882	Bat Sgt Maj	Fowkes J.	3/5 RA
1645	Gunr	Flynn J.	”
1243	”	Foley J.	”
1465	”	Francis W.	”
1821	Private	Fuller E.	1/3 Foot
1958	”	Foster J.	”
4013	”	Franks E.	”
592	”	Fitzgerald T.	”
1333	”	Fairey J.	”
739	”	Farrell P.	”
243	”	Fisher A.	”
1707	”	Fisher C.	”
1641	”	Fogarty R.	”
1362	Sergt	Faulks J.	80 Foot
469	”	Fredericks H.	”
1947	Corpl	Fisher T.	”
1918	”	Fricker J.	”
1830	Private	Farmer W.	”
2019	”	Farrell P.	”
1643	”	Faulkner W.	”
677	”	Feighney R.	”
1832	”	Felton R.	”
2052	”	Ferris J.	”
1860	”	Finnigan P.	”
1228	”	Fisher J.	”
230	”	Fisher J.	”
1286	”	Fitzmorris T.	”
604	”	Fitzpatrick P.	”
235	”	Fletcher J.	”
1695	”	Flewitt J.	”
3828	”	Foden J.	”
1820	”	Foster R.	”
1621	”	Ford G.	”
340	”	Ford T.	”
1449	”	Foulton W.	”
1465	”	Fox W.	”
1748	”	Foy M.	”
2080	”	Francis J.	”
1925	”	Furnaux J.	”
2607	Corpl	Goode J.	9/2 RA
2706	Gunr	Gibbons J.	”
2613	”	Godfrey J.	”
2568	”	Griffiths W.	”
—	Captain	Graham E. G.	1/3 Foot
1735	Sergt	Gleeson J.	”
1820	Corpl	Green J.	”
1699	Sergt	Giles T.	”
1843	Corpl	Gatrell F.	”
1952	”	Green R.	”
1169	Private	Gurton R.	”
2015	”	Gent J.	”
1548	”	Goodwin J.	”
1440	”	Gravey T.	”
2067	”	Green G.	”
1019	”	Grant J.	”
1841	”	Green T.	”
2076	”	Green G.	”
1965	”	Gander L.	”
349	”	Gardiner H.	”
2049	”	Godwin G.	”
2030	”	Goldsmith G.	”
2003	”	Goldsmith H.	”
3668	”	Grainger J.	”

Number	Rank	Name and Initials	Unit
1503	Private	Gregory W.	1/3 Foot
245	"	Green H.	1/10 Foot
456	"	Gardner J.	"
1856	"	Gracie J.	"
1154	"	Grierson G.	"
236	"	Grayson J.	"
1725		Griffiths J.	"
1523	Sergt	Goodwin M.	"
344	Private	Goff W.	"
55	"	Gillard E.	"
1943	"	Gardner J.	"
357	Arm Sgt	Green J.	"
1994	Private	Gilligan J.	"
2173	"	Gain G.	"
1611	Corpl	Graham H.	"
70	Lce Corp	Green N.	"
20	Private	Green W.	"
58	"	Greenacre C.	"
1896	"	George C.	"
2064	"	Goldie J.	"
403	"	Greenwood J.	"
1708	Corpl	Goodson I.	"
2171	Pte	Giles A.	"
27	"	Gilmore F.	"
1668	Private	Gailey J.	1/3 Foot
1899	"	Gleeson J.	"
1513	"	Grant J.	"
1953	"	Gleeson T.	"
2153	"	Giles R.	"
1411	"	Griffiths W.	"
1704	"	Green C.	"
193	"	Gill J.	"
1380	"	Gilpin R.	"
1149	"	Grant J.	"
2015	"	Gardiner C.	80 Foot
173	"	Gibson J.	"
1390	"	Goddard W.	"
1901	"	Godden T.	"
668	"	Godwin J.	"
1545	"	Goodwin F.	"
3642	"	Graham E.	"
1765	"	Graham J.	"
1279	"	Grant J.	"
1915	"	Green H.	"
1697	"	Green J.	"
1800	"	Gretton W.	"
1550	"	Griffiths H.	"
836	"	Gubbins N.	"
1999	"	Guttridge A.	"
—	Sergt	Guest T.	Comt Dep India
—	Corpl	*Gujn Singh	Perak Police
—	Surg	Gormley J. A.	AMD
—	Surg Maj	Gaye A. C.	"
—	Captn	Gwynn H. L.	3/5 RA
346	Sgt	Glasby W.	"
1123	Gunr	Garrod E.	"
2400	"	Gollock T.	"
2398	"	Gould J.	"
2402	"	Gough T.	"
2025	Sergt	Horwood E.	9/2 RA
2703	Gunr	Harrison J. H.	"
1839	"	Hall G.	"
1872	"	Hardy R.	"
2484	"	Hogan J.	"
—	Lieut	Hare J.	RE
—	Lieut	Howarth W. C.	1/3 Ft

Number	Rank	Name and Initials	Unit
—	Lieut	Howey E. W.	1/3 Ft
—	Sub Lieut	Hind I. W.	"
—	"	Holme B. F.	"
1256	Sergt	Herlihy J.	"
1830	"	Hodges A.	"
1873	Corpl	Holland G.	"
1221	"	Howes J.	"
1629	"	Hallows W.	"
1654	Private	Haley H.	"
1914	"	Hugill	"
1615	Drumr	Headley G.	"
1842	Private	Hamilton H.	"
1639	"	Harrison R.	"
1684	"	Herbert H.	"
1301	"	Herrick J.	"
47	"	Hewitson J.	"
1229	"	Hopgood W.	"
1710	"	Howells E.	"
—	Lieut	Huntley H. C.	1/10 Foot
1765	Sergt	Houlahan T.	"
1444	Private	Harrison H.	"
196	"	Harris J.	"
1380	"	Healey P.	"
861	"	Haggett W.	"
558	"	Hamilton W.	"
1961	"	Harvey E.	"
2025	"	Heaton J.	"
1848	"	Hall J.	"
66	Sergt	Hall W. J.	"
1951	Prvte	Henderson G.	"
398	"	Heath W.	"
1983	"	Hook A.	"
1232	"	Hogan T.	"
2152	Corpl	Howes R.	"
2080	Prvte	Hand J.	"
1761	"	Hart G.	"
1882	"	Hart E.	"
1604	Sergt	Heathcote J. R.	"
1473	Prvte	Hayward T.	"
2127	"	Hughes P.	"
239	"	Hughes R.	"
1953	"	Hall J.	"
1993	"	Hand M.	"
701	Private	Hubbert P.	1/3 Foot
1529	"	Hughes J.	"
427	"	Humphries G.	"
1302	"	Hunt A.	"
232	"	Hunt W.	"
1844	"	Hemmings W.	"
1388	"	Hurley D.	"
46	"	Harrington J.	"
568	"	Hilton J.	"
83	"	Hagan T.	"
1690	"	Harrington A.	"
1978	"	Hickey J.	"
1533	"	Hobbs W.	"
1917	"	Hobins J.	"
1456	"	Hughes F.	"
1939	"	Humphrey G.	"
1672	"	Harris J.	"
230	"	Hartfield W.	"
2087	"	Hawthorne G.	"
1228	"	Henley P.	"
1808	"	Henry J.	"
1265	"	Hersey T.	"
48	"	Hill R.	"

Number	Rank	Name and Initials	Unit
1173	Private	Hartley T.	1/3 Foot
300	"	Hennessey J.	"
1595	"	Headley W.	1/10 Foot
1763	"	Hockley F.	"
2182	"	Housam C.	"
867	"	Harwood W.	"
143	"	Harrison G.	"
2169	"	Hill A.	"
269	"	Hoyle J.	"
1902	"	Hodgson R.	"
153	"	Holmes W.	"
1671	"	Harman T.	"
2164	Corpl	Healey T.	"
—	Lieut	Hinxman H. C.	"
354	Corpl	Hullet M.	"
2182	Private	Hibbert J.	"
1900	"	Haynes W.	"
2160	"	Holbrook A.	"
1700	"	Hammond G.	"
2050	"	Hill J.	"
2066	"	Higgins T.	"
1749	"	Holliday F.	"
1808	"	Hogan M.	"
265	"	Hunt I.	"
31	"	Hickey N.	"
630	"	Hill I.	"
1095	"	Howard H.	1/3 Foot
1696	"	Hicklin T.	"
1536	"	Hughes H.	"
1721	"	Hutchinson H.	"
1754	"	Haghey J.	"
1162	"	Hancock H.	"
1003	"	Hall J.	"
1408	"	Handley J.	"
245	"	Harris W.	"
718	"	Harty J.	"
1638	"	Hefran W.	"
199	"	Hennessey E.	"
2050	"	Hill J.	"
200	"	Herring D.	"
2159	"	Hodgson S.	"
12	"	Hole B.	"
1160	"	Hoolan W.	"
2144	"	Hughes J.	"
2077	"	Humphreys J.	"
206	"	Hynes P.	"
1436	Sgt	Hollyhock T.	3/5 RA
2090	Bomb	Hacon W.	"
—	Lieut	Henriques E. N.	"
1628	Gunner	Hammersley T.	"
2098	Gunr	Henaghan J.	"
2082	"	Hillier G.	"
2096	Corpl	Hobcroft S.	"
2092	Bomb	Hodges W.	"
—	Sub Lieut	Host A. W.	80 Foot
1176	Private	Hacker W.	"
1502	"	Hackman S.	"
227	"	Haines G.	"
1777	"	Hall J.	"
926	"	Hall J.	"
1476	"	Hall W.	"
—	Captn	Huskisson S. G.	"
—	Lieut	Howard W.	"
1547	Private	Hammersley J.	"
1859	"	Hammerton T.	"
1573	"	Hammonds W.	"

Number	Rank	Name and Initials	Unit
1891	Private	Hammonds W.	80 Foot
3	"	Hancott S.	"
572	"	Harbottle G.	"
121	"	Harper J.	"
1824	"	Hart G.	"
2008	"	Hart J.	"
135	"	Hartrick W.	"
1816	"	Harvey J.	"
288	"	Hammer G.	"
134	"	Hawkes C.	"
147	"	Hawkins W.	"
1951	"	Hayden G.	"
1823	"	Hayes C.	"
1966	"	Hayes D.	"
235	"	Haywood J.	"
1975	"	Hazelton W.	"
3399	"	Healey D.	"
974	"	Hearn P.	80 Foot
1576	"	Hearly B.	"
529	"	Henzil M.	"
1906	"	Herman G.	"
709	"	Hodges T.	"
1557	"	Hollard H.	"
1032	"	Holman C.	"
97	"	Holmes A.	"
829	"	Hopkins G.	"
1994	"	Hopkins J. S.	"
2094	"	Hopkins W. F.	"
2083	"	Hopson J.	"
1498	"	Horrocks J.	"
1720	"	Howard C.	"
543	"	Howard H.	"
1715	"	Howard J.	"
2067	"	Howes J.	"
129	"	Hoult E.	"
1499	"	Hughes J.	"
4008	"	Hulks G.	"
925	"	Hunt J.	"
1450	"	Hunter R.	"
—		Hewick John C.	Asst Inspr of Police
—	Surg	Hoysted J.	AMD
—	Hospl Appce	Hayden G. W.	Bengal Sub Med Departt
—	Major	Heathcote M. H.	Bengal Staff Corps
—	Mr	Haywood E.	Spt of Police
3184	Drummer	Hearkins T.	1/10 Foot
2101	Pte	Hendry N.	"
2125	"	Holmes W.	"
1901	"	Hitchman I.	"
466	"	Hanlon I.	"
1464	"	Hudson W.	"
2046	"	Harris I.	"
1047	"	Hudson W.	"
647	"	Hudson W.	"
2702	Gunr	Ingrams A.	9/2 RA
1771	Pte	Isaacson W.	1/10 Foot
2022	Pte	Irwin L.	"
1838	Private	Irvine W.	80 Foot
—	Captn	Innes Wm.	RE
—	Major Genl Sir	Jervois W. F. D., CC, KCMG	RE
2485	Gunr	Jackson J.	9/2 RA
2570	"	Johnson F.	"
1822	Orderly Room Clerk	Julian S.	1/3 Foot
1295	Color Sergt	Johnson A.	"

Number	Rank	Name and Initials	Unit
1694	Private	Johns J.	1/3 Foot
2079	Sergeant	Judd B.	”
1381	Private	Jumps A.	”
1937	”	Johnson R.	”
1037	”	Johnson T.	”
203	”	Johnson W.	”
1126	”	Jackson H.	”
229	”	Jeffries J.	”
240	”	Jeffries W.	”
2068	”	Jones G.	”
242	”	Johnson E.	”
1607	Sergt	Johnson H.	”
1549	Private	Johnson H.	1/10 Foot
166	”	Johnson J.	”
1813	”	James H.	”
144	”	Jones B.	”
1904	”	Judge S.	”
1243	”	Jenkins J.	”
252	”	Johnson W.	”
1566	Corpl	Jenkins G.	”
1695	Private	Jones G.	”
1728	Corpl	Jones W.	”
291	Private	Jones C.	”
2051	”	Jones J.	”
2390	Corp	Jones M.	3/5 RA
1083	Sergt	Jefferson W.	80 Foot
997	Private	Jones J.	”
442	”	Jervis J.	”
1919	”	Jacobs T.	”
272	”	Jackson T.	”
285	”	Jenkins H. C.	”
1710	”	Jetton T.	”
1679	”	Johnson J.	”
10	”	Johnson R.	”
2058	”	Johnson W.	”
279	”	Johnson W.	”
1691	”	Jones J.	”
1391	”	Jones W.	”
1814	”	Jones W.	”
1964	”	Jones W. J.	”
1251	”	Joyce M.	”
—	Dep Comy	Johnstone J.	Control Dept
959	Corpl	Kelly T.	9/2 RA
2608	Trumpr	Keeling J.	”
2620	Gunner	Kelly T.	”
2614	”	Kennard J.	”
—	Captain	Kerr W. F.	1/3 Foot
—	”	Kelly W. E. R.	”
—	Lieut	Kekewitch A. G.	”
1826	Serg	Kent G.	”
2148	Private	Kilfoyle P.	”
1200	”	Kinner G.	”
470	”	Keefe J.	”
1788	”	Kenny J.	”
1977	”	Kenton E.	”
1553	”	Kingsland J.	”
1147	”	Kirby H.	”
1144	”	King G.	”
1973	”	King W.	”
1242	”	Killingbeck R.	”
1739	”	Kettle J.	”
1316	”	Kenny J.	”
17	”	King T.	”
2152	”	Ketson W.	”
869	Private	Kemlen M.	1/10 Foot
1138	”	Kenna M.	”

Number	Rank	Name and Initials	Unit
180	Private	King T.	1/10 Foot
2028	”	Kelly M.	”
2132	”	Kelly T.	”
197	”	Kirk T.	”
2015	”	Kirwin F.	”
1552	”	Kemp E.	”
2107	”	Kennedy M.	”
328	”	Kavanagh P.	”
2072	”	Kenny J.	”
1277	Gunr	Kersley G.	3/5 RA
2111	”	Kingsley J.	”
1711	Corpl	Kelly T.	80 Foot
1357	Private	Kettle J.	”
1344	”	Kelsill H.	”
182	”	Knowles H.	”
—	Comt Offr	Kerr W. G.	Control Dept
—		Kinnersley C. S.	Magistrate Singapore
2584	Pte	Kirby W.	AH Corps
—	Surgeon	L'Estrange A. H.	AM Dept
2704	Gunr	Loughlin P.	9/2 RA
—	Lieut	Leslie A. T.	1/3 Foot
1693	Sergt	Locke J.	”
1989	Private	Lipscombe G.	”
2037	”	Llewellyn J.	”
1955	Corporal	Lewis H.	”
1980	Private	Lovell F.	”
1552	”	Leary G.	”
643	”	Linahan M.	”
1855	”	Liddle H.	”
1999	”	Lane M.	”
1670	”	Leaper H.	”
1597	”	Leary D.	”
86	”	Leary J.	”
1377	”	Locke W.	”
1505	”	Loughlan W.	”
1709	”	Lyshall W.	”
2028	”	Lambourne G.	”
79	”	Lacey P. M.	”
1222	”	Lovejoy W.	”
1634	”	Lund I.	
1322	”	Lawler T.	”
1621	”	Law E.	”
196	”	Loughlin T.	”
—	Lieut	Louth F. R. H.	1/10 Foot
76	Private	Lambden J.	”
284	”	Long W.	”
998	Drummer	Leo W.	”
28	Private	Leete A.	”
979	”	Linford J.	”
2009	”	Lawless W.	”
498	Sergt	Leggett F.	”
237	Private	Long W.	”
1740	Sergt	Lavey W.	”
81	Private	Lappin P.	”
1906	”	Larkins J.	”
1908	”	Liley A.	”
1182	”	Lowdes T.	”
2083	”	Lovegrove J. T.	”
2189	”	Lacey R.	”
2009	”	Lindsay J.	”
1910	”	Loose R.	”
1250	”	Lines W.	”
1909	”	Ling W.	”
1907	”	Lee J.	”
602	”	Lawless W.	”
—	Lieut	Lang J. W.	”

Number	Rank	Name and Initials	Unit
322	Pte	Longman J.	1/10 Foot
1545	Private	Ladd W.	1/3 Foot
2075	Drummer	Lawrence H.	"
1611	"	Lester H.	21 Bde
1376	Private	Lacey J.	80 Foot
2065	"	Lawrence R.	"
1902	"	Lawton H.	"
1795	"	Leigh R.	"
1461	"	Lennon W.	"
1506	"	Lewis P.	"
1612	"	Lloyd E.	"
2115	Pte	Lowe G.	1/10 Foot
2615	Gunr	Lane W. F.	RA
1990	Private	Lodge J.	80 Foot
374	"	Lockett J.	"
1991	"	Locker W.	"
1903	"	Loney T.	"
1657	"	Lord W.	"
185	"	Lowbridge W.	"
271	"	Lowe S.	"
1321	"	Lynch T.	"
1771	"	Lyons T.	"
1647	Dr	Leather J.	"
1110	Dr	Lewis J.	"
—	Asst Apothy	Lowman W.	Bengal Sub Medl Departt
—	"	Luke A.	"
—	Depty Comy	Leech A. S. B.	Control Dept
—	Bt Major	Lloyd C. W.	16th Foot
2562	Private	Lawrence W.	A.H. Corps
—	Lt	Monckton P. E.	9/2 RA
1969	Gunr	McCullough J.	"
—	Lieut	McCallam H. E.	RE
—	Major	Morley F.	1/3 Foot
—	Lieut	Martin M. N.	"
1225	Sergt	Mooney J.	"
1509	Corpl	Meek W.	"
1259	Sergt	Mitchell A.	"
425	Private	Mansfield G.	"
1335	"	May G.	"
20	"	McDonough J.	"
2096	"	Maher J.	"
1662	"	Masterson T.	"
2062	"	Meadows J.	"
227	"	Meehan F.	"
52	"	Murphy D.	"
1074	"	McGrath J.	"
1404	"	McGuire T.	"
1810	"	McLean W.	"
4070	"	McSherry W.	"
1834	"	Madden P.	"
1703	"	Massey H.	"
1428	"	Mason W.	"
2011	"	Meehan M.	"
1635	"	Midgley C.	"
—	Lieut	Meyrick H.	1/10 Foot
575	Private	Murray F.	"
912	Drummer	Mick J.	"
1126	Corpl	Mulhall J.	"
1912	Private	Monger G.	"
2138	"	Mahony J.	"
500	"	Mitchell C.	"
1858	"	McMillan D.	"
1952	"	McDonald F.	"
1270	"	Murphy J.	"
531	"	McGreil J.	"
1777	Corpl	Mehigan J.	1/10 Foot
1947	Private	Moore J.	"
302	"	McSweeney R.	"
2141	"	Murphy J.	"
229	"	Miller J.	"
918	"	Marshall R.	"
1532	"	Miller T.	"
2053	"	McDonald J.	"
1753	"	Mann A.	"
323	"	McCorry R. W.	"
1795	"	McKay T.	"
1658	"	McGuirk T.	"
1075	Drumr	Mick G.	"
1836	Private	Moore D.	1/3 Foot
1913	"	Martin J.	"
1455	"	McGuire J.	"
2053	"	McGuirk M.	"
4066	"	McKeen R.	"
1434	"	McLean J.	"
3438	"	Morrow J.	"
1515	"	McIntyre J.	"
450	"	McGrady W.	"
1006	"	McMahon M.	"
1897	"	McMenemy C.	"
4211	"	McQuade P.	"
1079	"	Martin T.	"
1005	"	Meade J.	"
555	"	Mills J.	"
1472	"	Moran H.	"
1454	"	Moran E.	"
1673	"	Moseley J.	"
1351	"	Mullins P.	"
1291	"	McLoughlin T.	"
3653	"	Moore R.	"
188	"	McCarter J.	"
1195	"	McCormick J.	"
523	"	McGarry J.	"
2029	"	McKerr M.	"
684	"	Murray E.	1/10 Foot
809	"	Moss J.	"
2119	"	Moffatt R.	"
1925	"	McNalty T.	"
1946	"	Murray W.	"
1992	Lce Corp	McCarroll W.	"
1699	Private	Mees E.	"
1844	"	Morrissey M. M.	"
2076	"	Murray T.	"
1845	"	Murtagh B.	"
1975	"	McGolly T.	"
1703	"	McIndol J.	"
148	"	Mason H.	"
1949	"	Meehan J.	"
1793	"	Merefield A.	"
2021	"	Mills J.	"
1578	"	McLaughlin J.	"
13	"	McEwen J.	"
1950	"	Meehan P.	"
1972	"	Morgan J.	"
1045	"	Murphy P.	"
197	"	McNamara M.	1/3 Foot
1781	"	Mahoney T.	"
1294	"	Massey G.	"
1417	"	McGregor J.	"
1143	"	Mitchell J.	"
1263	"	Moore J.	"
303	"	Murphy P.	"

Num-ber	*Rank*	*Name and Initials*	*Unit*
1893	Private	Mulcahy M.	1/3 Foot
641	”	Maddigan J.	”
1964	”	Middleton J.	”
231	”	Meads W.	”
1888	”	McFadden J.	”
1560	”	McParland T.	”
116	”	McNamara T.	”
650	”	Mahoney J.	”
51	”	Martin T.	”
1217	”	Milson E.	”
189	”	Monaghan H.	”
190	”	Morgan W.	”
1712	”	Muldoon H.	”
1839	Corpl	Murray T.	”
1374	Drumr	Mullineux J.	”
2013	Pte	Muspratt H.	”
—	Mr	Maxwell W. E.	Asst Resident Perak
—	Commr	Murray P. J.	RN
—	Surgeon	McNamara I.	AMD
1891	Bomb	Mayne G.	RA
—	Lieut	Monck-Mason G. G.	3/5 RA
—	Major	McNair	RA (Colonialr) Engineer)
918	Sergt	Machin F.	80 Foot
1761	Corpl	McDonald N.	”
1525	”	McCready R.	”
1618	”	McMullins H.	”
1534	”	Murch S.	”
300	”	McCaffery H.	”
2007	”	McCall M.	”
2017	”	McCann P.	”
1419	”	McCart N.	”
62	”	McCleary D.	”
1314	Private	McCormick M.	”
1349	”	McCullough J.	”
2020	”	McCullough M.	”
1303	”	McDonald J.	”
3969	”	McElwene M.	”
1858	”	McGilland T.	”
1799	”	McGookin J.	”
1855	”	McGuickam J.	”
1778	”	McGuire P.	”
1719	”	McGuirk P.	”
1546	”	McKee J.	”
1775	”	McKenny W.	”
309	”	McKenzey J.	”
4090	”	McLoughlin H.	”
124	”	McMullin J.	”
1247	”	McMullin J.	”
1778	”	McSherry B.	”
1981	”	Macklin J.	”
2092	”	Maden G.	”
1487	”	Major R.	”
1459	”	Mallinson W.	”
1568	”	Malloy J.	”
181	”	Manson A.	”
1927	”	Mantill H.	”
1742	”	Marmer J.	”
2000	”	Martin C.	”
1712	”	Massey T.	”
1933	”	Matthews A.	”
1846	”	May J.	”
1580	”	Meachin A.	”
1492	”	Meers E.	”
432	”	Melsop J.	”
1683	”	Mercer G.	”

Num-ber	*Rank*	*Name and Initials*	*Unit*
1532	Private	Mieken J.	80 Foot
83	”	Middelton E.	”
2063	”	Middow A.	”
49	”	Millar J.	”
1481	”	Minshull A.	”
1976	”	Mitchell G.	”
986	”	Mitchell H.	”
1669	”	Mitchell M.	”
1395	”	Mitchell T.	”
268	”	Moffatt J.	”
277	”	Mollart J.	”
2106	”	Monaghan P.	”
1372	”	Moore R.	”
2048	Private	Moore R.	80 Foot
1745	”	Moran J.	”
1746	Corpl	Morris G.	”
180	Pte	Morris J.	”
1642	”	Morris T.	”
161	”	Morris W.	”
1211	”	Morris W.	”
131	”	Moss J.	”
315	”	Mountford W.	”
1678	”	Mulgrew J.	”
579	”	Mulholland J.	”
127	”	Mullens T.	”
1606	”	Murphy P.	”
1730	”	Murphy P.	”
1501	”	Murphy W.	”
1109	Gunr	Mallen R.	3/5 RA
1658	”	May J.	”
2399	”	McDonald D.	”
1656	”	McIntyre J.	”
970	”	Morton R.	”
1000	”	Munce R.	”
1403	Serjt	McPherson J.	23 Bde RA
—	Asst Comy	McLeod W. T.	Control Dept
—	Mr	Munro R. A.	Teleg Dep Indian
—	Mr	Mitchell S.	Engineer
2456	Lce Serg	Mould G.	Army Hospl Corps
—	Bt Lt Col	Nicolls O. H. A.	RA
—	Lieut	North W.	RE
1134	Private	Newett J.	1/3 Foot
2012	”	North T.	”
1659	”	Nield W. E.	”
1146	”	Nickless J.	”
1470	”	Newkirk H.	”
239	”	Norman W.	”
1929	”	Neal E.	”
1661	”	Niell J.	”
1769	”	Noyle E.	”
14	”	Nelson R.	1/10 Foot
384	”	Nurse C.	”
1303	”	Nertney P.	”
2010	”	Newman G.	”
63	”	Nightingale W.	”
494	”	Norton I.	”
1957	”	Newman I.	”
2140	Gunr	Noon J.	3/5 RA
201	Sergt	Norton W.	80 Foot
157	Private	Navan J.	”
1868	”	Newton H.	”
1753	”	Nicholls J.	”
1926	”	Night H.	”
263	”	Nixon J.	”
1773	”	Nolan J.	”
1339	”	Nolan P.	”

Number	Rank	Name and Initials	Unit
1681	Private	Nolan T.	80 Foot
1819	"	Norman F.	"
—	Surg Maj	O'Halloran W.	AM Dep
—	Lieut	Ohmanney A. E.	1/3 Ft
—	Sub Lieut	Offard H. H.	"
94	Armr Sergt	Orr G.	"
865	Private	O'Farrell M.	"
68	"	O'Keefe C.	"
1046	"	O'Maley J.	"
—	"	O'Donnell J.	"
1458	"	O'Brien W.	"
581	"	O'Grady T.	"
123	"	O'Brien M.	"
1089	"	Oglesby F.	"
403	"	O'Donnell P.	"
1860	"	Ohlson A.	"
1312	"	Olive H.	"
749	Sergt	Owen F.	1/10 Foot
491	Private	O Gorman J.	"
87	"	Oberstall A.	"
2081	"	Osbond G.	"
1072	"	O'Connor D.	"
2104	"	O'Neil P.	"
2168	"	O'Connel J.	"
1736	"	Organ W.	"
2017	"	O'Brien P.	"
1731	"	O'Donnell J.	"
1989	"	Osborne C.	"
1149	"	O'Brien P.	"
2146	Gunr	O'Hara J.	Perak
2396	"	O'Hare F.	"
119	Private	Oakley T.	80 Foot
1751	"	O'Hara T.	"
1456	"	O'Neill J.	"
166	"	Owen E.	"
2087	"	Owen H.	"
1840	"	Owen J.	"
165	"	Owen W.	"
119	"	Oakley T.	"
—	Asst Comy	Offley G. W.	Control Dept
—	Mr	O'Brien J.	Teleg Dept Indian
—	Surg Maj	Orton T. J.	AMD
—	Surg Major	Oldham C. F.	Bengal Med
—	Capt	Paton G. (Staff ADC)	24 Ft
2489	Gunr	Phillips J.	RA
1632	Sergt Major	Pepper J.	1/3 Foot
258	Private	Parfitt W.	"
1950	"	Parkinson R.	"
1981	"	Peake H.	"
498	"	Pickett S.	"
236	"	Privett G.	"
233	"	Pain G.	"
1358	"	Pullen F.	"
1287	"	Parsons T.	"
1947	"	Patterson G.	"
1874	"	Perry W.	"
80	"	Phelan M.	"
1288	"	Prince G.	"
1309	"	Purcell J.	"
1542	"	Page T.	"
2072	"	Pass T.	"
53	"	Payne W.	"
2027	"	Pearson S.	"
3575	"	Parsons E.	"
1988	"	Pearce J.	"
1705	"	Pearson W.	"
1687	Private	Paine E.	1/3 Foot
2065	"	Pedder J.	"
594	"	Poore G.	1/10 Foot
3	Drumr	Plowright J.	"
1153	Private	Pale J.	"
1615	"	Powell R.	"
155	"	Price J.	"
2056	"	Payne C.	"
42	"	Philmore W.	"
1928	"	Peacock A.	"
—	Lieut	Peyton C. T.	"
—	"	Paton G. B.	"
1979	Private	Price D.	"
2014	"	Patterson J.	"
1476	"	Power W.	"
2172	"	Pyzer E.	"
—	Lieut	Palmer H. W.	"
39	Pte	Pash G.	"
2150	Gunr	Patterson W.	3/5 RA
340	Private	Powell J.	1/3 Foot
1466	"	Power M.	"
77	"	Power J.	"
—	Lieut	Prior I. E. H.	80 Foot
1401	Private	Pritchard S.	"
1446	"	Parker S.	"
2028	"	Parker W.	"
1256	"	Parks H.	"
1885	"	Parrott G.	"
172	"	Parry H.	"
1726	"	Pendebury J.	"
116	"	Perkins J.	"
2027	"	Phillips J.	"
2046	"	Phillips A.	"
186	"	Preece G.	"
2031	"	Prince J.	"
2081	"	Pritchard C.	"
1163	"	Pummell A.	"
—	Hon	Plunket H.	Supt of Police Penang
—	Lieut	Preston I. I.	4th Bn Rifle Bde
—	Asst Apoth	Prichard T. B. F.	Madras Sub Med Dept
—	Mr	Pickering W. A.	Protector of Chinese
—	Hon	Phillipps George	—
2131	Private	Quinn J.	1/10 Foot
2047	"	Quick J.	80 Foot
1496	"	Quick J.	"
1752	"	Quinn R.	"
2616	Gunr	Ricketts W.	9/2 RA
2705	"	Rixham J. B.	"
2492	"	Routledge G.	"
2577	"	Rowe W.	"
—	Lieut	Rich H. B.	RE
—	Captain	Reeves C. E.	1/3 Foot
1369	Sergt	Rawlings J.	"
1631	Corpl	Rowley J.	"
1276	Sergt	Ryan T.	"
1829	Corpl	Reid J.	"
1656	Private	Rogers J.	"
1230	"	Rourke T.	"
1324	"	Ray J.	"
1421	"	Reddin J.	"
1320	"	Roberts W.	"
57	"	Rogers W.	"
1231	"	Read J.	"
2086	"	Rogers E.	"
1116	"	Ross S.	"

Number	*Rank*	*Name and Initials*	*Unit*
1637	Private	Reynolds W.	1/3 Foot
949	”	Roberts T.	”
1234	”	Rankin J.	”
1563	”	Reilly E.	”
2071	”	Rogers J.	”
1944	”	Roberts F.	”
525	”	Russell W.	1/10 Foot
1180	”	Ryan M.	”
856	Sergt	Rooney D.	”
1442	Private	Rooney T.	”
1115	”	Ryan J.	”
57	”	Richardson J.	”
1654	”	Richards R.	”
1218	”	Reeves G.	”
296	”	Rattle J.	”
1948	”	Ray G.	”
86	”	Robinson J.	”
560	Corpl	Rooks W.	”
94	Private	Russell W.	”
33	”	Rhodes W.	”
1872	”	Robinson J.	”
1847	”	Rock M.	”
213	”	Rowe W.	”
1818	”	Revis G.	”
1053	”	Rock T.	”
1865	”	Robertson R.	”
1527	”	Robins E.	”
577	”	Rodgers J.	”
1135	”	Ray J.	”
1981	”	Reavey H.	”
1926	”	Robertson H.	”
1346	”	Robinson T.	1/3 Foot
1954	”	Rose T.	”
—	Brig Gen	Ross J.	Staff
2098	Serg	Richardson W.	80 Foot
1974	Private	Ralphs F.	”
201	”	Ranesey G.	”
189	”	Rand E.	”
1953	”	Raymond C.	”
158	”	Renfrey A.	”
59	”	Rainbird J.	1/10 Foot
1986	”	Rees J.	”
1870	”	Robinson H.	”
1108	Sergt	Rutledge W.	”
1100	Corpl	Reddy P.	”
2203	Gunr	Riely A.	R.A.
—	Lieut	Rigg H. A.	”
2491	Gunr	Ross G. F.	”
1348	”	Russell B.	”
1506	Corp	Robinson R.	”
2401	Gunr	Russell E.	”
1894	Private	Rhoades T.	80 Foot
1586	”	Rich E.	”
2012	”	Richardson G.	”
1108	”	Rickman G.	”
1963	”	Ring E.	”
1960	”	Roberts H.	”
1311	”	Robinson J.	”
259	”	Robinson J.	”
1483	”	Robinson R.	”
1296	”	Rodgers J.	”
1337	”	Rubie G.	”
1085	”	Rubie R.	”
2010	”	Russell J. T.	”
1254	”	Ryan D.	”
1912	”	Ryan H.	”

Number	*Rank*	*Name and Initials*	*Unit*
1423	Private	Ryan J.	80 Foot
3985	”	Ryan M.	”
1385	”	Ryan T.	”
2004	”	Ryell C.	”
—	Surgeon Major	Randell H. L.	AMD
—	Asst Controller	Robinson W. G. W.	Control Dept
1914	Dr	Robertson A.	1/10 Foot
1915	Dr	Russell R.	”
1698	Pte	Rouse Henry	”
—	Capt	Shippard W. O. C.	9/2 RA
1913	BS Maj	Shaw J.	”
1914	Corpl	Spencer T. H.	”
1918	Sergt then Corp	Summerhayes H.	”
1912	Gunner	Smith J.	”
2493	”	Smith C.	”
1831	”	Snell J.	”
2494	”	Stubbersfield J.	”
1375	Sergt	Sweet D.	1/3 Foot
1467	”	Sexton M.	”
1724	Private	Sheppard H.	”
1725	”	Smith F.	”
1700	Corpl	Smith W.	”
373	Sergt	Snow J.	”
1645	Private	Shea A.	”
1190	Corpl	Sullivan J.	”
1975	Private	Stonebridge J.	”
1636	”	Sheppard W.	”
1722	”	Saunders W.	”
741	”	Scott J.	”
1421	”	Shanks T.	”
1883	”	Sheward E.	”
569	”	Slinger J.	”
1872	”	Smith E.	”
1394	”	Smith W.	”
913	”	Sharples T.	1/10 Foot
1	”	Speakman W.	”
1185	”	Spellman A.	”
405	”	Somers J.	”
1663	”	Shepherd J.	”
1165	”	Shortle M.	”
357	”	Stoodley E.	”
30	”	Staunton T.	”
2058	”	Smith W.	”
79	”	Stradwick R.	”
349	”	Sartin J.	”
871	Lce Corp	Sedhouse E.	”
407	Private	Stokes D.	”
1095	”	Stewart H.	”
1867	Sergt	Starks E.	”
2964	Private	Sleighter G.	”
2059	”	Smith R.	”
270	”	Smith J.	”
1744	”	Stacey J.	”
1877	”	Stanbridge J.	”
1463	”	Sharrard S.	”
1921	Corp	Styles J.	”
2166	Private	Scanlon J.	”
1500	”	Stafford S.	”
181	”	Simons J.	”
367	”	Sumph R.	”
238	”	Smith W.	1/3 Foot
1531	”	Spear G.	”
1892	”	Spring P.	”
69	”	Stone S.	”
1737	”	Salt F.	”

Number	Rank	Name and Initials	Unit
1153	Private	Sherridan T.	1/3 Foot
1530	"	Skinner F.	"
1909	"	Smith J.	"
1104	"	Smith J.	"
2020	"	Smith T.	"
1024	"	Stancombe G.	"
1201	"	Sullivan J.	"
1409	"	Shaughnessy A.	"
1523	"	Seeker J.	"
2098	"	Spencer E.	"
1608	"	Stevens J.	"
2145	"	Stonebridge G.	"
59	"	Stratford G.	"
1871	"	Stone G.	"
1801	"	Saunders J.	"
1918	"	Scott R.	"
134	"	Sheehan E.	"
63	"	Short J.	"
4088	"	Stevenson J.	"
168	"	Smith W.	"
2157	"	Sullivan T.	1/10 Foot
1467	"	Smith W.	"
1929	"	Sergen J.	"
50	"	Smith G.	"
278	"	Sheldon T.	"
2079	"	Seymour J. E. (alias Warren)	"
666	"	Shine C.	"
1012	"	Smith P.	"
273	"	Somers R.	"
1651	"	Steeple G.	"
1710	"	Smith P.	"
1554	"	Stenson C.	"
1955	"	Stenson J.	"
2143	"	Smart J.	"
1598	"	Smart C.	"
2140	"	Stenson M.	"
1916	"	Say T.	"
286	"	Shed J.	"
1838	"	Smith H.	"
1706	"	Stanley J.	"
1935	"	Stevenson W.	"
1853	"	Swain J.	"
993	"	Scarrott H.	"
1558	"	Smyth T.	"
—	Capt	Scott T.	"
771	Corpl	Smith F.	"
1672	Pte	Shaw W.	"
1562	Private	Seaton J.	1/3 Foot
1564	"	Smith E.	"
1864	"	Smith F.	"
1065	"	Smith W.	"
1537	"	Stone J.	"
2094	"	Scanlon J.	"
1508	"	Sloane J.	"
1992	"	Smith T.	"
1588	"	Spires J.	"
3795	"	Sabin H.	"
2095	Drumr	Sullivan S.	"
1992	Corpl	Sumner W.	3/5 RA
2304	Trump	Shannon V.	"
1078	Gunr	Smith W. D.	"
1084	"	Stansfield G. P.	"
1094	"	Spencer W.	"
—	—	Syal Singh (Late Store Lascar)	3/1 CPD RA

Number	Rank	Name and Initials	Unit
—	Capt	St Clair T. T. Bde Maj China	49 Foot
—	Lt	Swinburne P.	80 Foot
3124	Col Sgt	Shore E.	"
814	"	Squires C.	"
1503	Corp	Stevenson J.	"
375	Private	Salt G.	"
1889	"	Salter C.	"
439	"	Sands W.	"
1533	"	Sutherley W.	"
962	"	Saville E.	"
2074	"	Scott E.	"
1360	"	Scott J.	"
1842	"	Scullion J.	"
1253	"	Seddons T.	"
262	"	Senior J.	"
183	"	Sharp W.	"
970	"	Shaw F.	"
323	"	Shea C.	"
2035	"	Shellock J.	"
1659	"	Shipley J.	"
1770	"	Silock J.	"
234	"	Simcock J.	"
2022	"	Simms G.	"
1815	"	Skelley J.	"
610	"	Slack E.	"
376	"	Smith C.	"
1101	"	Smith G.	"
570	"	Smith H.	"
1063	"	Smith J.	"
1350	"	Smith J.	"
1828	"	Smith J.	"
1948	"	Smith J.	"
144	"	Smith J.	"
258	"	Smith J.	"
1749	"	Smith L.	"
61	"	Smith M.	"
1240	"	Smith M.	"
1494	"	Smith W.	"
2045	"	Smith W.	"
1805	"	Speers J.	"
253	"	Spellman W.	"
1526	"	Stacey A.	"
2016	"	Stanley A.	"
1274	"	Stansbury H.	"
214	"	Stansfield J.	"
2062	"	Stephens E.	"
1107	"	Stevenson G.	"
2009	"	Stobbs J.	"
1353	"	Stock J.	"
138	"	Stokes F.	"
1987	"	Stovas W.	"
225	"	Stover W.	"
2060	"	Summersby H.	"
1988	"	Sutton J.	"
2021	Drumr	Smith J.	"
—	Commy	Stanes R.	Control Dept
—	Asst Commy	Skinner E. G.	"
—	Captn	Salterthwaite J. H.	RE
—	Mr	Swettenham F. A.	Dep Commissr
1394	2nd Corp	Smith J.	A.H. Corps
—	Corporal	Singh Gujn (Listed also under "G")	Perak Police
2574	Gunr	Skam W.	RA
—	Condr	Sharpe J. H.	Comt Dept
1921	Gunr	Taylor W.	9/2 RA

Number	Rank	Name and Initials	Unit
2495	Gunr	Thompson E.	9/2 RA
2617	”	Thomas W. J.	”
—	Major	Twigge J.	RE
1837	Sergeant	Tressler T.	1/3 Foot
2024	Corporal	Thomas J.	”
1691	Private	Townsend G.	”
2004	”	Tubbs B.	”
1354	”	Toast T.	”
239	”	Thompson J.	”
1313	”	Taylor F.	”
1281	”	Thatcher W.	”
1072	”	Thompson W.	”
2110	”	Thompson G.	”
464	”	Turner C.	”
1935	”	Thomas E.	”
1357	”	Thomas W.	”
204	”	Towers S.	”
1345	”	Townhill J.	”
1619	”	Turner J.	”
87	”	Taylor G.	”
1985	”	Theakstone J.	”
1982	”	Tillbrook W.	”
1974	”	Turner R.	”
1791	”	Tanswell E.	1/10 Foot
3217	”	Travers J.	”
1042	Corpl	Turvey B.	”
2116	Private	Thompson T.	”
1707	”	Turner T.	”
2102	”	Tait R.	”
1455	”	Thompson W.	”
590	”	Taylor T.	”
19	”	Toyne J.	”
289	”	Templeman R.	”
1922	Sergt	Tuffin T.	”
1470	Pte	Todd W.	”
293	”	Tookey J.	”
1743	”	Tomlin W.	”
1787	”	Tinklin S.	”
222	”	Theobald W.	”
171	”	Timms C.	”
215	”	Tillett R.	”
64	Corpl	Tomlin W.	”
776	Private	Taplin D.	”
204	”	Thompson S.	”
808	Col Serg	Tait C.	”
294	Serg	Tucker W.	”
1419	Drummer	Timmins E.	”
40	Pte	Taylor W.	”
1677	”	Townsend E.	”
82	”	Tutty B.	”
2008	”	Toole J.	”
1748	”	Taylor G.	”
324	”	Thompson W.	”
—	Capt	Tatham W. J.	RA
2007	Sgt	Tough C.	3/5 RA
1927	”	Thompson M.	”
849	Gunr	Templeton R.	”
—	Captain	Tucker C.	80 Foot
—	”	Tyler C. I. R.	”
2104	Serg	Tatum C.	”
2081	Corpl	Thompson H.	”
2096	Private	Thompson G. E.	”
120	”	Tallice G.	”
335	”	Tatton S.	”
1331	”	Taylor R.	”
1895	”	Taylor T.	”

Number	Rank	Name and Initials	Unit
1023	Private	Terry R.	80 Foot
1844	”	Thomas J. E.	”
1794	Pte	Thompson A.	”
2002	”	Thompson W.	”
162	”	Thorley T.	”
1529	”	Tomlinson W.	”
1162	”	Topham G.	”
1322	”	Tyldesley T.	”
—	Surg	Townsend E.	AMD.
2583	Private	Tray R.	AH. Corps
—	Mr	Towers G. L.	Comt Dept
1527	Colr Sergt	Underhill I.	1/3 Foot
1789	Corpl	Urmson T.	80 Foot
188	Private	Upperdine J.	”
—	Sub Lt	Ventris E. P.	1/3 Foot
1469	Private	Vanderlynde P.	”
2395	Gunr	Veale J.	3/5 RA
211	Private	Venables S.	80 Foot
1080	”	Vickery S.	”
1873	”	Voss J.	”
1284	”	Vallender B. G	1/10 Foot
2496	Bomb	Wells W.	9/2 RA
1928	Gunr	Walder J. J.	”
2700	”	Wall H.	”
2576	”	Wheeler H.	”
1931	”	White J.	”
2347	”	Wilson A.	”
2131	Pay Master Serg	Warrington H. W.	1/3 Foot
1957	Corpl	Whitmore R.	”
1352	Serg	Williams G.	”
1682	Private	White W.	”
1652	”	Whitwith J.	”
1833	Corpl	Wells W.	”
1471	Pte	Wakeford W.	”
1546	”	Wallace J.	”
233	”	Weller C.	”
1604	”	Williams W.	”
1445	”	Williamson F.	”
207	”	Walsh J.	”
246	”	Withey R.	”
187	”	Williams J.	”
1665	”	Williams T.	”
1538	”	Winters E.	”
1571	”	Wood G.	”
1547	”	Woods W.	”
1068	”	Whiting E.	”
—	Capt	Whitla W.	1/10 Foot
81	Pte	Ward R.	”
1645	Corpl	Williamson J.	”
1596	Private	Watson S.	”
1687	”	Welsh J.	”
1530	”	Widdows E.	”
1548	”	Wynne E.	”
230	”	Wake J.	”
279	”	Wiles J.	”
299	”	Whitehouse A.	”
13	”	Wrench C.	”
1884	”	Watts S.	”
46	”	Wickham E.	”
1814	”	Walsh E. J.	”
1536	”	Wakeley J.	”
2096	”	Waugh J.	”
15	”	Woods J.	”
2061	”	Warder W.	”
2062	”	Way E.	”

Number	*Rank*	*Name and Initials*	*Unit*
61	Private	Woods J.	1/10 Foot
2026	"	Watkins J. W.	"
1693	"	Walsh W.	"
1688	"	Wilson F.	"
1202	"	Wallace R.	"
1815	"	Whale G.	"
1213	"	Walker H.	1/3 Foot
1851	"	Williamson E.	"
228	"	Whitfield B.	"
1938	"	Wood W.	"
1702	"	Walker J.	"
1125	"	Williams J. T.	"
88	"	Westone W.	"
1663	"	Wilkinson S.	"
15	"	Wratton J.	"
1541	"	Wratton T.	"
1859	"	Woods D.	"
2157	"	Windgate A.	"
1132	"	Warrington R.	"
1941	"	Webb J.	"
205	"	Wright C.	"
1430	"	Wood G.	"
1339	"	White J.	"
2018	"	Whiting R.	"
708	"	Williams E.	"
2151	"	Whatley R.	"
1555	"	Whitehead J.	"
2033	"	Whiting W.	"
62	"	Wallace T.	"
1697	"	Wilkinson C.	"
1468	"	Wilson J.	"
1727	"	Walker J.	1/10 Foot
1480	"	Walsh J.	"
1674	"	Willins G.	"
1924	"	Winfield C.	"
35	"	White W.	"
1780	"	Watts I.	"
77	"	Wells G.	"
156	Dr	Wheatley W.	"
2187	Bomb	Winkworth A.	3/5 RA
1993	Trump	Walker W. N.	"
2189	Gunr	Wood R.	"
2190	"	Wratten J.	"
1930	"	Wardle J.	RA
2497	"	White W.	"
1681	Pte	Wright A.	1/10 Foot
—	Sub Lt	Walmsley W. H.	80 Foot
1665	Sergt	Walker H.	"
1217	"	Ward H.	"
336	"	Waters J.	"
1911	Private	Wakeling G.	"
1569	Private	Waldron G.	80 Foot
2155	"	Woolley T.	1/3 Foot
1990	Drumr	Wright J.	"
1468	Pte	Wilson I.	"
—	Mr	Wilson A. (Clerk)	Ind Comt Dept
131	Pte	Walsh J.	80 Foot
219	"	Walsh J.	"
1818	"	Ward J.	"
1851	"	Ward I.	"
264	"	Warner C.	"
122	Private	Wassell W.	"
1630	"	Watts J.	"
1924	"	Waxham T.	"
206-	"	Way J.	"
1813	"	Weir M.	"
294	"	Westwood J.	"
228	"	Westwood T.	"
2061	"	White G.	"
949	"	White W.	"
60	"	Whitehouse J.	"
2091	"	Whittaker W.	"
2097	"	Whyman T.	"
137	"	Wiley T.	"
24	"	Wilkins H.	"
1759	"	Wilkinson J.	"
361	"	Williams T.	"
573	"	Williams T.	"
2064	"	Willison T.	"
1491	"	Willott S.	"
2013	"	Wilmott W.	"
1846	"	Wilson C.	"
1175	"	Wilson H.	"
192	"	Wilson S.	"
1375	"	Wiltshire S.	"
2076	"	Wiltshire S.	"
622	"	Wise J.	"
1539	"	Woodford J.	"
1486	"	Wooding I.	"
440	"	Woods J.	"
1871	"	Woodlams F.	"
2078	"	Worly J.	"
2057	"	Wright J.	"
566	"	Wrae J.	"
701	"	Watson C.	"
2573	"	Warsie E.	A.H. Corps
—	Lieut	Warton R. G.	1/10 Foot
2005	Pte	Ward W.	"
1993	Private	York J.	1/3 Foot
1888	"	Young T.	80 Foot
1260	Lce Corp	Young H.	1/10 Foot

Note: the above medal roll includes British Army personnel, Indian Army personnel included in service units, and civilians. It does not, however, include the medal roll of the '1st Goorka Light Infantry'.

NOTES TO CHAPTER TEN

1. P. L. Burns (ed), *The Journals of J. W. W. Birch*, (Oxford University Press, Oxford in Asia Historical Reprints, 1976), pp 2 and 3.
2. Ibid, pp 4 and 5.
3. W. G. Maxwell and W. S. Gibson, *Treaties and the Engagements affecting the Malay States and Borneo*, (London, 1924), pp 28–30.
4. *Harts Army List*, 1875.
5. Burns, op cit, p6.
6. R. H. Vetch, *Life of Lieutenant General the Hon Sir Andrew Clarke*, (London, 1905), p174.
7. Public Record Office, Kew. Straits Settlements, Original Correspondence. CO273/82.
8. Burns, op cit, p20.
9. Straits Settlements Government Gazette dated 12 December 1874.
10. Burns, op cit, p23.
11. Hari Raya Puasa – the ending of the fasting month.
12. Burns, op cit, p35.
13. Ibid, p36.
14. Straits Settlements Government Gazette dated 22 October 1875.
15. National Archives and Record Centre, Singapore. Perak Enquiry Papers, vol2.
16. Ernest Birch visited the grave of his father at Bandar Bahru in 1883 and at that time he also visited Passir Salak (Pasir Salak) where Resident Birch had been killed. He met many people who had known his father and during this visit was given his late father's watch and gun.
17. J. Kennedy, *A History of Malaya*, (MacMillan, 1962), p176.
18. C. Northcote Parkinson, *British Intervention in Malaya 1867–1877*, (University of Malaya Press, 1960), p241.
19. Ibid, p245.
20. F. A. Swettenham, *Footprints in Malaya*, (London, 1942), p62.
21. Parkinson, op cit, p245.
22. Swettenham, op cit: Parkinson comments that the attack must have been made by the '... rear platoon of the 10th, the only troops by then uncommitted ...'
23. Straits Settlements Papers No 327 dated 16 November 1875.
24. Major Samuel Dunlop, Royal Artillery.
25. Straits Settlements Papers No 327 dated 16 November 1875.
26. Parkinson, op cit, p252.
27. The demand for cartridges was excessive but this was due to an error by the telegraph operator.
28. Parkinson, op cit, p256.
29. Ibid, p257.
30. Ibid, p258.
31. Ibid, p260.
32. Ibid, pp 260 and 261.
33. Ibid, p264.
34. Extract of a report by Major S. Dunlop, RA, Chief of Police, Singapore.
35. Straits Settlements Papers No 335, dated 2 December 1876.
36. Lieutenant Colonel the Hon Archibald E. H. Anson, RA who was Lieutenant-Governor of Penang.
37. Parkinson, op cit, p278.
38. Ibid, p279.
39. *Singapore Daily Times* dated 7 January 1876.
40. Extract from Regimental Record of 10th Regiment of Foot.
41. As a result of the action Lieutenants Hinxman and Peyton, Sergeant Owen and Privates Adams and Haynes, all of the 10th Regiment, were specially commended for their distinguished conduct. Sergeant Owen was killed during the action and was buried at Sungei Ujong together with Privates J. Ball, J. Newman and H. Smith.

The small cemetery is situated about one mile to the south of the Garrison lines at the present day town of Seremban. At the edge of a small kampong, hidden between small maize plantations, is the secluded cemetery which contains the graves of the four soldiers of the 1st Battalion, 10th Regiment of

Foot who were killed during the action at Sungei Ujong. When I last visited the cemetery in 1977 the area was fenced and well kept and contained a memorial stone for all four soldiers. The inscription read:

HERE LIE BURIED
Sergt F. OWEN
Pte J. BALL
J. NEWMAN
H. SMITH
of the 1st Batt H.M.'s 10th Foot
who died on active service
in SUNGEI UJONG
in 1875 & 1876.

The memorial was flanked by three cannon, one a 1-pounder of the Hon East India Company service and two 12-pounders. As the cemetery is in a remote position it has subsequently been classed as an abandoned cemetery and will be allowed to revert to nature. The three cannon have been handed to the Malacca Museum Trust for preservation.

At one time there was an interesting notice displayed at the entrance of the small cemetery and was obviously aimed at the local inhabitants of the area, and read:

'BY ORDER OF THE DISTRICT OFFICER SEREMBAN
NO ONE SHOULD COME TO PRAY FOR NUMBERS
"LOTTERY" OR TRESPASS INTO THIS GRAVEYARD.
TRESPASSERS WILL BE PROSECUTED'

Alan Harfield, *Christian Cemeteries and Memorials in Malacca,* (British Association for Cemeteries in South Asia, 1984), pp72 to 74.

Letters from Tan Sri Dato' Mubin Sheppard, CMG, MBE, Hon Secretary, The Malaysian Branch of the Royal Asiatic Society to the author dealing with the condition of the Sungei Ujong cemetery.

42. *Straits Observer* dated 17 December 1875.
43. Parkinson, op cit, p285.
44. Straits Settlements Papers No 18 dated 14 January 1876.
45. *The London Gazette* dated 12 April 1876.
46. Brigadier The Rt Hon Sir John Smythe, Bt, VC, *The Story of the Victoria Cross,* (Frederick Muller, 1963), p104.
47. Parkinson, op cit, p289.
48. Ibid, p294.
49. Ibid, pp295 and 296.
50. Major the Hon J. F. A. McNair, *Perak and the Malays,* (London, 1878), p387.
51. Colonel B. A. H. Parritt, MBE, *Red with Two Blue Stripes,* (Charles A. Lusted, 1974). Illustration shows the author (Colonel B. A. H. Parritt) standing in front of the grave of Major H. L. Hawkins.

Notes from author's note-book dealing with the Bukit Chandran cemetery shows the inscription on the memorial stone as reading:

Sacred to the memory
of
Major H. L. Hawkins, B.S.C.
who fell at Kota Lama
4th Jany 1876

52. Straits Settlements Papers No129 dated 21 March 1876.
53. General Sir Archibald E. H. Anson, *About Others and Myself,* (London, 1920), p348.
54. Parkinson, op cit, p299.
55. Ibid, p300.
56. Straits Settlements Paper No 272 of 30 July 1876.
57. Parkinson, op cit, p306.
58. Major L. L. Gordon, *British Battles and Medals,* (Gale and Polden, 1962), pp 237 and 238.
59. Public Record Office, Kew. Medal Roll WO/100/45.
60. PRO Governors' Despatches CO/426/5 No 16091 dated 15 October 1880.
61. Ibid CO/426/7 No 6761 dated 14 April 1881.
62. Ibid CO/426/7 No 17494 dated 4 October 1881.
63. Ibid CO/426/8 No16347 dated 24 September 1881 and No16716 dated 30 September 1881.

64. Inscriptions taken from author's notebook listing all known military memorials and graves in West Malaysia.
65. PRO Governors' Despatches CO/426/5 No 2597 dated 5 March 1877.
66. As for note 64.

CHAPTER ELEVEN

Defences of the Straits Settlements 1859–1915

It was during the year 1859 that work commenced on Government Hill for the construction of a large fortification. The fort which had been part of the planned defence of the settlement was discussed as early as 1843 when Captain Samuel Best, Madras Engineers submitted his plan. The defence plan was shelved at that time and did not materialise until 1859 when a revised defence plan had been agreed. This included the construction of a new barracks for the Native Infantry regiment at Tanglin and the building of the largest fortification that Singapore was to know. The fort which was being built on the summit of what was then Government Hill necessitated the demolition of Raffles's House, which had by then become known as Government House, and the levelling off of the top of the hill. The fort was named 'Fort Canning' in honour of Viscount George Canning who was Governor-General and the first Viceroy of India from 1857 until 1862. The hill then became known as 'Fort Canning Hill'.

Two years prior to the commencement of the construction of Fort Canning a detachment of European gunners from the Madras Artillery were permanently stationed in Singapore and were then accommodated in the barracks located at Pearls Hill. The new fortification was intended to be the home and the headquarters of the artillery that were to be stationed in the Settlement.[1]

There were a number of changes in the officers holding key posts in the Straits Settlements during 1859 and these included Captain R. Macpherson, Madras Artillery, being appointed as Acting Resident Councillor at Malacca,[2] Lieutenant G. E. Borradaile, of the 14th Regiment Madras Native Infantry being appointed as Aide-de-Camp and Secretary to the Honourable the Governor of the Straits Settlements,[3] Captain Henry Man '... assumed charge of the Residency of Singapore on the 14th [July] ...',[4] and the Governor of the Straits Settlements became a military man from 8 August 1859 when Colonel Orfeur Cavenagh was appointed to that post.[5] In October Captain J. Burn, of the 40th Regiment Bengal Native Infantry was appointed Secretary and Aide-de-Camp to the new Governor,[6] and Lieutenant G. E. Borradaile was moved to the vacant post of '... Police Magistrate, Commissioner of the Court of Requests and Coroner in Province Wellesley ...'[7]

By May 1859 the construction of Fort Canning had progressed well and the Chief Engineer of the Straits Settlements, Captain G. C. Collyer, had seven 68-pounders mounted on the firing platform facing the sea. The work had been mainly carried out by a force of 400 Chinese coolies.[8] At the same time Fort Fullerton was being enlarged and was extended to nearly three times its former size with gun platforms for 56- and 68-pounders, a house for the officer in charge of the fort built in the centre and with new barracks for the soldiers built alongside the roadway within the fort. There were a number of trees planted within the fort to give shade around the parade ground. The total cost of the work carried out at Fort Fullerton was, according to the Singapore historian Charles Burton Buckley, '... said to be $840,000 ...'[9] When the construction of the gun platform at Fort Canning had been completed it was then noticed that the site was overlooked by Pearls Hill and, because of this, the Chief Engineer, with the aid of a Chinese coolie force, had the top of Pearls Hill taken off so that it no longer posed a possible danger to Fort Canning.[10]

During the same month the Officer Commanding the Troops at Singapore wrote to the Governor complaining that mistakes were frequently being made with the saluting of ships, etc visiting from foreign nations by the person responsible for the Saluting and Flag-staff station on Government Hill, and suggested that he '... appoint a European Signals Sergeant to [be in] charge of the Government Hill Flagstaff ...'[11] The Governor however replied that '... he has no means of attaching anyone to the Government Flagstaff as proposed nor can he apply for the Salary from the Government of India for one, but proposes the withdrawal of Conductor Taylor's allowance of Rs 30 a month ... and then assigning it to any intelligent soldier selected for the appointment. This person should be under the orders of the Executive Engineer of the Station ...'[12]

The *Singapore Free Press* published an article dealing with the building of Fort Canning in their paper of 12 May which read:

> '... Government Hill is undergoing a rapid metamorphosis from the peaceful seat of the Governor's residence into a strong and extensive fortification, to be called, we believe, Fort Canning. The top of the hill is being raised several feet so as to afford the required extent of level surface, and when the works are completed they will enclose a space of about seven acres, with barracks and other accommodations for a large number of troops. The batteries are to carry from 25 to 30 guns, most of them of very heavy metal, and the work has been pushed on so energetically by the Engineering Officer in charge that seven pieces (68-pounders) are already in position on the face of the hill towards the sea. The work is being executed almost entirely by free labour no fewer than 400 Chinese coolies being employed upon it and convicts are used only for certain very limited portions of the work which they are supposed to do better than the Chinese. Fort Fullerton is also in the course of being enlarged to nearly three times its former extent, and is to be armed entirely with heavy guns - 56- and 68-pounders.
>
> The estimated cost of these works already authorised is said to be about Drs [Dollars] 840,000 which will no doubt be increased to at least a million when the actual extras are brought in; and in addition to this, we have heard that similar but smaller works are contemplated on Pearl's Hill and Mount Sophia ...'[13]

The newspaper report then commented on the defence works, which it considered to be unnecessarily large and expensive:

> '... Even if we grant the possibility of a temporary emute from the Chinese, who form the bulk of our natives, and that it might be needful for the Europeans to betake themselves, for a time, to some defended point; such a place of refuge could have been abundantly supplied by one or two slight works on the same hill, with a dozen guns of ordinary calibre, all of which could have been thrown up at one twentieth part of the cost, and such a stronghold as that now coming into existence never could have originated in the mere idea of such a possibility. From the magnitude of these works, then, we infer that they have their origins not from any merely local cause, but in the general policy of H. Majesty's Government in this part of the world ...'[14]

The building of the fortifications continued throughout 1859 and into 1860 but despite the comments made in the *Singapore Free Press* there was never any intention at this time to

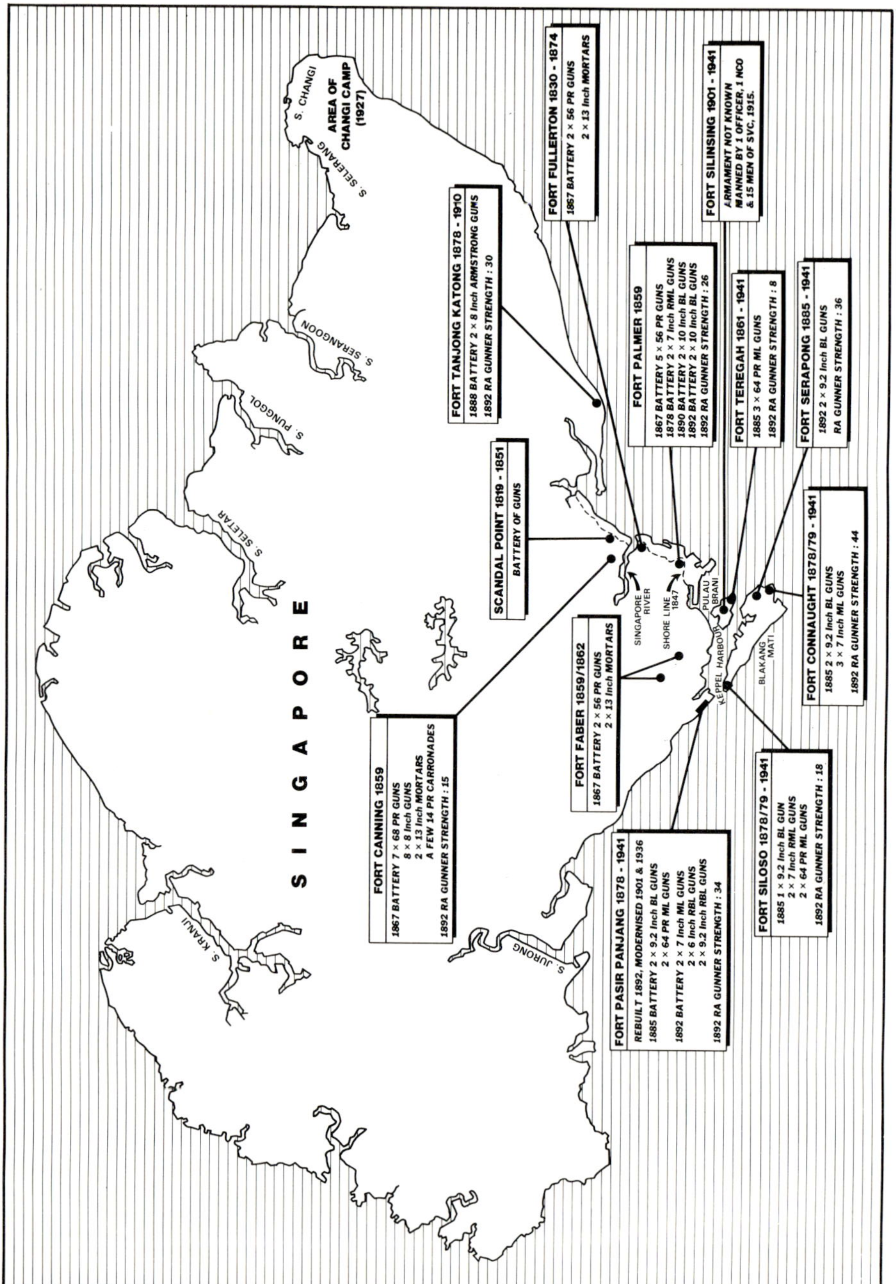

The nineteenth century defence positions on Singapore Island.

build additional fortifications at Pearls Hill and Mount Sophia. In April 1860 it was reported that the Company of European Artillery in the Straits had been brought up to its full establishment.[15] During September 1860, a report was submitted on the condition of the various barracks and installations throughout the three settlements and the report made a number of recommendations for the improvement of conditions, these being:

'*Penang*

considers as quite inefficient the Hospital accommodation for the European and Native Artillery, the building requires ventilation and more suitable provision is necessary for the Sepoy Dispensary.

The Sepoy barracks [were] considered in excellent order but recommends the ground around to be well drained and the vegetation growing thereon cleared.

Malacca

considers as defective in every respect the temporary barracks for the Native Infantry detachment and suggests the flooring being repaired and the Mess being provided with trestle beds. The Garrison Hospital is badly situated and badly ventilated and not large enough; it is only separated by a wall from the Artillery Barracks, there is a communicating door between the two back courts to enable the men from the Barracks to procure water from the only available well in the hospital compound and the washing of the Artillery court and cook rooms are carried past the Hospital along an open sewer and the privies are above and close to the well and close to the Roman Catholic Burial Ground. He strongly urges the necessity of better accommodation being provided for the garrison at Malacca and remarks that the Convicts there are far better off in this respect that the Sepoys.

Singapore

Highly approves of the Barracks and Hospital accommodation for the Sepoys but considers the new privies [were] badly planned and badly placed ... [also] suggests the grant of an additional allowance of nutritious diet [be given] to the European soldiers in the Straits ...'[16]

The problem of providing a new hospital at Penang was passed immediately to the Chief Engineer of the Straits Settlements for his urgent attention. The Chief Engineer had, during the early part of 1860, changed, as Captain G. C. Collyer had been granted fifteen months leave on a '... Medical certificate ...' with permission to proceed to Europe.[17] He had been relieved by Captain John F. A. McNair, as a temporary measure. McNair was Executive Engineer and Superintendent of Convicts at Singapore. Colonel Charles Edward Faber, who during his previous tour of duty in the Settlement had constructed the gateways to the cemetery on Fort Canning Hill, was appointed and arrived at the settlement on 20 April 1860 having been posted from India.[18] During the following month Lieutenant W. H. G. Palmer, of the 40th Regiment, Madras Native Infantry was appointed as an additional executive Engineer at Singapore.[19]

In addition to the building of the new hospital at Penang the Chief Engineer had to deal with the problems that had been highlighted in the report made in September 1860 which included the improvement and repair of the barracks at Malacca and it was also recommended that in the case of Malacca a more appropriate building be 'hired' for use as the garrison hospital.

Lieutenant G. E. Borradaile, who had moved to Province Wellesley in 1859 was moved twice in 1860, in February he was moved from Prince of Wales Island garrison to Malacca to take over the post of '... Magistrate of Police and Commissioner of the Court of

Requests at Malacca during the absence of Lieutenant Playfair on six months sick leave . . .',[20] and in June he was moved back to Singapore to '. . . officiate as Secretary and Aide-de-Camp to His Honorable the Governor, during the absence of Captain Burn on special duty . . .'[21] In August the '. . . services of Lieutenant G. E. Borradaile 14th Regt MNI are replaced at the disposal of the Madras Government . . .'[22]

In April 1861, it was notified that the 20th Regiment, Madras Native Infantry would leave Madras on 8 April by the steamer *Arracan* and on the transport *Jubal Cain* and that the regiment would relieve the 22nd Regiment, Madras Native Infantry serving at Penang. The 22nd Regiment, MNI would return to Madras on the same vessels.[23] Later in the same month a request was made by the Officer Commanding the Artillery to the Officer Commanding Troops, Singapore asking for additional ponies. His request commented on the:

> '. . . present inefficient strength of the establishment of ponies for the Mountain Train Battery at Singapore, the Battery consists of:
>
> 6 twelve pr Howitzers
> 2 three pr guns
> 4 wagons
> 2 stone carts
>
> For the transport of these carriages only 8 ponies are allowed, and barely sufficient for 2 pieces with three wagons and cart, also the present ponies are made use of in conveying rations from the Commissariat Stores . . .'[24]

and suggested that if the ponies were required to help move rations and stores from the Commissariat Store the whole establishment of ponies should be increased. The proposal was generally agreed and a notice appeared in the local press requesting that tenders be submitted for the provision of ponies to the military.

In June 1861, the Resident Councillor at Penang wrote to the Governor concerning the number of guard duties that the Native Infantry regiment were required to supply. The total number of men required to man three reliefs per day was excessive. He listed the Guards required on any one shift as follows:

'Treasury Guard	1 Havildar, 1 Naique and 12 sepoys.
Commissariat Guard	1 Naik [*sic*] and 6 sepoys
Court House	1 Naique and 3 sepoys
Convict Jail Guard	1 Havildar, 1 Naique, and 12 sepoys
HMs Jail	1 Naique and 6 sepoys
Province Wellesley	1 Havildar, 1 Naique, 13 Sepoys . . .'[25]

This was a total of sixty-one men required per shift, a total of 183 men on duty per day.

Not only was the Resident Councillor at Penang unhappy with the state of defence but the Officer Commanding the Troops at Singapore was also uneasy at the lack of a suitable force with which to defend the settlement. He wrote to the Governor on 28 September 1861 in which he brought to notice:

> '. . . the defenceless state of the Island of Singapore in case of a sudden attack from a well organised European Enemy, [and he did] not consider that merely coastal Batteries are sufficient for the defence of all the Island, as a well informed enemy would, of course, land his forces out of Battery range and consequently would have to be opposed in this field and [he] is therefore of [the] opinion that a large inland Field Force is required . . .'

and in respect of the number of troops garrisoned at Singapore he noted that:

'. . . after furnishing the necessary guards scarcely 200 efficient Native Infantry with the Singapore Volunteers are available for Field Service, the strength of the Artillery being only 40 Europeans and 70 Golundauz, these are not sufficient men to man the heavy guns at present in position, to say nothing of the improbability of employing Field Artillery with so small a force . . .'

He finally comments that he understands that the garrison is shortly to be augmented by a regiment of European Infantry.[26] The Governor replied on the same day to the effect that although it was agreed that a European Infantry regiment would be stationed in Singapore the overall policy was that the settlement was to be '. . . defended mainly at sea . . .'[27]

Buckley records, in his *Anecdotal History of Singapore* that:

'. . . In this year [1861] the fortification and barracks on Fort Canning were completed, and the European Artillerymen were removed from the buildings on Pearls Hill, which were from that time occupied by the Commissariat Department. The attap barracks at Tanglin were so far advanced as to be capable of affording ample accommodation for a European regiment . . .'[28]

Buckley wrote his book at the turn of the century and his reference to the time of occupation of Fort Canning differs slightly from that given in the official records as an entry in Governors' Diary dated 23 January 1862 reads: '. . . Fort Canning will soon be placed at the disposal of the Military authorities . . .'[29]

In February 1862, the Officer Commanding the Troops, Singapore requested that an office and a '. . .small office establishment of one clerk and one peon . . .' be provided for the Barrack Master, Straits Settlements.[30] An official reply to his request was, however, not given until 14 May when he was informed that: '. . . the recent decision of the Supreme Government relative to the abolition of the Barrack Mastership in the Straits makes any consideration of an office establishment unnecessary . . .'[31]

On 1 April 1862 the Officer Commanding the Troops in the Straits wrote to the Governor again regarding the critical situation that existed in the Penang Garrison. The Penang Garrison had provided a detachment of the 20th Regiment, Madras Native Infantry for Labuan and this then reduced the garrison to only 243 men, of whom thirty-five were in hospital. With 183 men required to do duty daily this only left twenty-five men free to carry out barrack fatigues, etc. It was requested that:

'. . . stowage may be at once provided for the relief of the 2 companies of the 20th MNI at Malacca and two companies of the 4th MNI at Singapore, so that the former may be brought on the strength of the Pinang [*sic*] garrison without delay . . .'[32]

The reply given was to the effect that:

'. . . in the event of the services of the "Hooghly" being available she will leave Pinang on the 5th instant for Singapore and on her return can take up a weak company [i.e. a company not up to its full established strength] of the 4th MNI to Malacca, bringing on the relieved company of the 20th MNI to Pinang, the other company must be relieved on her next trip from Singapore to Pinang . . .'[33]

This move of companies appears to have at least given temporary relief to the over-stretched Pinang Garrison. During the same month it was reported that troops at Pinang were not able to carry out 'Escalating drill' as Fort Cornwallis was considered to be unsuitable for the training of that type of drill and at the same time it was noted that the detachment at Malacca was too small to be able to carry out that drill. It would appear that official sanction was required for the discontinuance of any drills that were included in the Indian Army annual training programme.

By mid-1862 the Artillery had been moved from Pearls Hill into their new accommodation at Fort Canning and although the soldiers had moved, the ponies of the Mountain Train Battery were still stabled at Pearls Hill and it was suggested that the temporary stables at Pearls Hill be taken down and re-erected at a suitable site within Fort Canning.[34] The proposal was not agreed and it was suggested that until the authorised stabling was provided at Fort Canning, the Battery Sergeant remain at Pearls Hill and occupy the quarter '. . . recently occupied by Captain Blair . . .' or should that building be required by the Commissariat Department who had taken over Pearls Hill, then the Subaltern's quarter might be used by the Battery Sergeant.[35]

Although the troops at Fort Canning had settled into their new home by mid-1862 a report was submitted by the Senior Surgeon on Station commenting adversely on the conditions that existed at the Fort Canning Hospital. He complained that:

> '. . . the least noise from the Guard Room or Orderly Room or the shouts of a drunken man in the lock-up can be very injurious to a sick man and to remedy this, the Committee urges that the cells for refractory persons be at once erected and that the Guard, Orderly and Defaulters Rooms at present being converted into Quarters (be used for their correct purpose) . . .'[36]

The proposal was agreed and it was ordered that the section of hospital then being used as the Guard room, etc be returned to the hospital staff for use by them as originally intended.[37] With the completion of the accommodation at Fort Canning it was agreed that, as the establishment of a Barrack Master and staff had been disallowed, an appointment of a Barrack Sergeant be instituted at Singapore '. . . for the purpose of looking after the Buildings in Fort Canning . . .'[38] and, in addition, the Barrack Sergeant be given a Peon to assist him with his duties.

With the completion of the other fortifications on Singapore Island it was found necessary to form a 'Local Company' of Madras Artillery in order that there could be a general re-allocation of the artillerymen in the Straits Settlements. It was originally intended that the Local Company would be partially stationed in Pinang and partially in Malacca. It was suggested by the Officer Commanding the Troops in the Straits that '. . . as the detail for Malacca is so small and there are so many married men in it, it would not be advisable to send it to Malacca and recommends that it be detained at Pinang where there is plenty of accommodation for the women and children . . .'[39] The Governor agreed to the suggestion and as it was impossible to supply fresh rations for a small body of European artillerymen at Malacca he also agreed, on 16 April 1863, that a detachment of the Company of Golundauz stationed at Singapore should be stationed at Malacca. The whole of the 'Local Company' was therefore stationed at Pinang.[40]

With the question of the misuse of hospital accommodation having been settled at Fort Canning a further problem arose when the fort was inspected by the Commander Royal Artillery after which he submitted a report to the Officer Commanding Troops in the Straits saying that he was very unhappy that the '. . . General Magazine is only 30 yards from the Hospital . . .' and considered that should the hospital take fire '. . . the Magazine would go with it . . .'[41] There is no record of the Magazine, or the hospital having been moved as a result of this recommendation.

In September 1863 the establishment of the Mountain Train Battery was to be discontinued as the European Artillery had been increased in its establishment and it was considered that the former unit no longer fulfilled the purpose for which it was originally formed.[42]

Whilst it is apparent that the military authorities were dissatisfied with many of the aspects of the fortifications and the establishments in the Straits Settlements, and in particular Singapore, the civilian European population were even more dissatisfied. With the negotiations underway for the transfer of the management of the Settlements from the Government of India to the Colonial Office, they complained that the main defence force for Singapore should be naval and that:

> '. . . it was inexpedient, if not impracticable to rely upon the land works, as Fort Canning and Fort Fullerton would not protect the town from an attack by sea, and the probable result of their attempting to annoy a hostile fleet would be the destruction of the town . . .'[43]

The business factions in Singapore were quite convinced that there could be no military defence of the Settlement and that any land operation to defend the island would only result in the destruction of the city, their warehouses and of course the subsequent loss of their fortunes. Buckley records that the defences at that time consisted of the following works and commented on the unsuitability of the fortifications that had only been completed within the previous two or three years, as follows:

> '. . . Fort Canning had seven 68-pr guns, eight 8-inch shell guns, and two 13-inch mortars, with a few 14-pr carronades, at a height of about 200 feet, on the hill over a quarter of a mile from the beach. Fort Fullerton was on the beach, with nine 68-pr guns and one mortar . . . Fort Palmer was a small earthwork overlooking the eastern entrance to Keppel Harbour, and had five 56-pr guns. Fort Faber was also an earthwork, halfway up Mount Faber, overlooking Keppel Harbour, with two 56-pr guns and two mortars on the top of the hill. The total number of guns in the forts was thirty-six. The distance of the guns at Fort Canning from the beach put them at a great disadvantage against an enemy's ships at sea, and both that and Fort Fullerton would have drawn the fire right upon the centre of the town, while the two small works at Mount Palmer and Mount Faber were of very little use. The military expenditure was, in consequence, regarded by the mercantile community as very unsatisfactory, and it swallowed up in 1863 nearly one half of the revenue . . .'[44]

Whatever their grievances were, they had little effect on the defence policy at that time. The headquarters of the military in the Straits Settlements was at Singapore and the senior officer held the appointment of 'Officer Commanding Troops, Straits Settlements' and the officer holding that appointment was also given the acting rank of Brigadier.

The garrison at Singapore also had, as part of the headquarters staff, a Garrison Sergeant Major. The NCO holding that appointment in 1863 was Sergeant John Kearney, 102nd Regiment, but he was posted from Singapore to Cannanore in February 1864 again to act as Acting Garrison Sergeant Major.[45] His relief was Sergeant Major J. Shelley but he was in turn posted after a short tour of duty and in August 1865 Garrison Sergeant Major R. Matthews, of the 42nd Regiment, Madras Native Infantry became the appointment holder at Singapore.[46]

In February 1864, the findings of a General Court Martial were promulgated in the Madras Military Orders. The Court Martial had been held at Penang on 19 January 1864 when Subadar Abdul Guffor of the 20th Regiment Madras Native Infantry was charged for '. . . conduct unbecoming the character of an officer in having at Penang on or about the 14th October 1863 been drunk . . .' He was found Guilty and was sentenced to be severely reprimanded and '. . . further to be suspended from rank, pay and allowances for a period of three Calendar months . . .' The findings were, in fact, commuted as '. . . A reprimand is

Those asking for a reduction in military expenditure, which they considered to be excessive, commented as follows:

> '. . . Even if the present class of Native troops should be continued there seems no reason why one Regiment of sepoys might not be dispensed with. One Regiment, with an exceedingly limited number of Commissioned Officers with it, was thought sufficient before the Sepoy rebellion, and now that it has long been quelled, two Regiments, with above forty Commissioned Officers, cannot be indispensable . . .'

The garrison of the Straits Settlements was seen to be an internal defence force which could be called upon to quell any internal disturbances and it was not envisaged, by the civil authorities, as being a defence force against a foreign aggressor. One of the alternative suggestions that had been placed before the Governor was the provision of a military force consisting of six companies of the Ceylon Rifles. This being a 'local force', the cost in terms of pay and allowances, etc would be very much lower than that of a European regiment, or the two regiments of sepoys that were on station at the time of the Transfer. In addition it was suggested that the Ceylon Rifles should be supported by a local Police Force of about 800 all ranks and that this combined force would '. . . be a great saving . . .'[59] The suggestion was not accepted and the two regiments of sepoys continued to garrison the Straits Settlement for a further two years. The 7th Regiment, Madras Native Infantry relieved 35th Regiment, Madras Native Infantry on 17 March 1868 and in May 1869 the 8th Regiment, Madras Native Infantry departed from Singapore. On 12 May 1869 the 7th Regiment MNI moved from the Penang Garrison to Singapore and then became the garrison troops for the whole of the Straits Settlements.

Lieutenant Colonel R. Macpherson was appointed Colonial Secretary, Straits Settlements in November 1867 and held the appointment until his premature death at Singapore on 6 December 1869. He was buried with full military honours at the New Cemetery, on the Bukit Timah Road, Singapore,[60] and at a later date a memorial was erected in front of St Andrew's Cathedral, and in view of his connection with Malacca a brass memorial plaque was placed in Christ Church, Malacca, the inscription reading:

'In Memorium
RONALD MACPHERSON
Lieut Colonel RA (MADRAS)
Lieut Governor and
Colonial Secretary
Straits Settlements
Born June 4th 1817
Died Decr 6th 1869'[61]

In July 1868 the question of the cost of firing the morning, midday and evening gun at Fort Canning was considered excessive and the cost of firing the '. . . gun should not be thrown upon the Colony . . .'[62] and the discussion on this subject continued for almost a year when, eventually, a decision was promulgated to the effect that the firing of the 'morning and evening guns . . . should be continued . . .'[63] Another interesting note in the Governor's Diary for 1868 refers to the strength of the military force in the Straits Settlements. The 8th Regiment Madras Native Infantry had been warned to move from May 1869, thus reducing the garrison strength to one sepoy regiment, which was in accordance with the wishes of the civil authorities. In September 1868, the Garrison Commander reported to the Governor that the number of men '. . . at present allocated to the Straits is insufficient . . .'[64] In 1871 the 19th Regiment Madras Native Infantry arrived on 8 March and the 7th

Regiment Madras Native Infantry was relieved from duty in the Straits Settlements from April of that year. During the latter part of the year there was continued discussion in respect of the replacement of the sepoy regiment with a regiment of the British Army, and it was agreed that the Native troops '... will be at once replaced by Europ'n Infantry ...'[65] and in the following year, 1872, the 19th Regiment MNI returned to India, in fact to Trichinopoly. The first of the 'European Infantry' regiments replaced the sepoy regiment when the 80th (Staffordshire Volunteers) Regiment of Foot was posted to the Colony.

The transition from a Native Infantry garrison regiment to a European Infantry regiment was not without its problems and in particular the first British Army unit suffered due to lack of planning on the part of the ordnance organisation in the Straits Settlements. The 80th Regiment embarked on the HMS *Orontes* on 11 January 1872 and arrived at Singapore on Sunday 17 March. The regiment, under command of Major, Brevet Lieutenant Colonel H. C. Smith, had embarked 2 Field Officers, 6 Captains, 11 Subalterns, 5 Staff Officers, 40 Sergeants, 32 Corporals, 569 Privates together with 7 Officers' wives and 2 children and 67 soldiers' wives and 77 children. One Sergeant, 1 wife and 2 children had been disembarked at Malta on account of an outbreak of measles, and 1 Private at Suez, however the number had been increased with the birth of an officer's child during the voyage out to Singapore. On arrival the regiment disembarked and marched the three miles from the docks to Tanglin Barracks.[66] In March, D Company moved to Malacca with Captain Charles Tucker in charge and on 20 April A and B Companies moved to Penang with Major Charles F. Amiel in command. Five companies remained at Singapore, these being C, E, F, G and H Companies, and it was this section of the regiment that suffered due to the lack of adequate clothing being provided from the Ordnance Depot. The 'Digest of Service' of the Regiment records that:

> '... it took some time before the whole Regiment could be fitted with white clothing so that the men suffered a good deal from a disease called 'Pemfagus' which was brought on by their being obliged to wear their English uniform which was much too hot for the climate and raised immense blisters on their arms and legs which on bursting left hideous sores ...'[67]

The regiment only remained at Singapore for about eight months after which they were relieved by 1st Battalion, 10th Regiment. The 80th Regiment embarked four companies under command of Lieutenant Colonel H. C. Smith, on board HMS *Adventure* at Tanjong Pager wharf on 28 November 1872 and the vessel sailed at 3.00 pm. The remainder of the regiment moved to Hong Kong during December and the whole regiment were reunited at Hong Kong by 9 January 1873. The regiment, during its tour of duty in the Straits Settlements, lost nine soldiers in Penang and three in Singapore due to various illnesses, and one soldier, 4059 Private J. Walker jumped overboard during the sea voyage to Hong Kong and was never seen again.[68]

With the 80th Regiment having handed over to the 1st Battalion, 10th (North Lincoln) Regiment as the garrison battalion the actual infantry situation appears to have been settled, however there were, in 1873, further discussions in respect of the artillery defences of the Settlements and it was suggested that:

> '... Fort Canning, Singapore and Fort Cornwallis, Penang [were] useless against modern artillery in an organised attack by sea or land; on the other hand in civil disturbances the guns are too big to shoot the people in the streets and the country around; [it was considered] conveniently better to kill the Colonists with a few light field guns, however, as shooting H.M. Native subjects in the streets and defeating a

foreign enemy are intimately connected the subject should be considered together before re-modelling the existing works . . .'[69]

The disturbances in Malacca and Perak during the period 1875 to 1877 deferred the question of the modernisation of the defence works at Singapore and Penang, and during the height of the Perak Campaign there were in addition to the 1st Battalion, 10th Regiment, the 1st Battalion 3rd (East Kent, The Buffs) Regiment, and three Companies of the 80th Regiment. The full implication of the full strength of the Field Force that was mounted for the campaign is covered in Chapter Ten. During 1876 a further report was submitted to the Governor in respect of the unhealthy conditions prevailing at Malacca and Penang and the need for restoration work to be carried out on both the barracks and the hospitals at both stations. Once again the expenditure being incurred as a result of the campaigns at Perak and Sungei Ujong permitted only the essential repair work to be authorised.[70]

With the closure of the Perak campaign the troops in the Straits Settlements were again reduced and in 1878 the garrison battalion was being provided by the 74th (Highlanders) Regiment. In the following year that regiment was replaced by the 28th (North Gloucestershire) Regiment and in order to help bring the defence works in Singapore to a satisfactory standard after a certain amount of necessary neglect during 1875–77, the Colony received agreement from the War Office to spend £2,700 on immediate repairs to the defence works.[71]

The question of a revised plan for the land defences of Singapore Island was again discussed in 1879 and the matter remained unresolved although work did begin on the new defences planned to give cover for the harbour area. The locations of new batteries were given as Mount Siloso on Blakan Mati Island, Mount Serapong, also on Blakan Mati and Teregah Point on Pulau Brani. There were problems associated with the construction of these new sites and a report was submitted on 10 April 1879 by Lieutenant H. E. McCallum, RE, on the problems of the blasting operations at Mount Siloso. He gives the location of the new battery as follows:

'. . . Mount Siloso is the most westerly feature of the island of Blakan Mati, which overlooks the new harbour, and forms the southern side of the entrance thereto. It consists of a narrow ridge about 150 yards in length, falling at a gradient on one in nine in a N.W. direction, and on which the new battery is now being constructed. Beyond this ridge is a narrow neck of ground communicating with a tongue of very limited dimensions, bounded by the cliffs, and at a height of 115 feet above sea level . . .'

It was the ridge that caused the problem in that it gave a 'dead angle' for the one 7-inch and two 64-pounder guns that were to be mounted at the Fort Siloso battery. Lieutenant McCallum obtained from the Singapore magazine about 19,000 lb of condemned powder which he used in controlled blasting conditions to clear the paths of fire of the guns. Having completed his blasting the remainder of the construction and the clearing away of unwanted rubble was carried out by Chinese coolies working under contract to the Royal Engineers in charge of the new defence works.

The progress of the new defence works continued slowly and in October 1881 it was considered necessary that Captain H. Jekyll, Royal Engineers should survey sites for additional new works and that he should be given all necessary assistance for the completion of this work. The new sites included Pasir Panjang and Tanjong Katong.[72] By 1881 the garrison battalion had again changed and was, by then the 2nd Battalion, 3rd (East Kent, The Buffs) Regiment and the distribution of the British battalion was two companies

Site of New Battery, Mount Siloso, Singapore showing knoll after explosion. Courtesy of the Royal Engineers Museum.

Fort Siloso, Blakang Mati, showing underground stores and ammunition chambers.
Photographs by Mrs J. Harfield, 1977.

Fort Siloso, Blakang Mati. One of the 13-inch mortars, now on display at the Singapore Gun Museum (at Fort Siloso).

at Penang, a detachment at Malacca, and another at Labuan with the Headquarters and the remainder of the battalion at Singapore. In 1883 the garrison troops again changed and the Royal Inniskilling Fusiliers relieved the 2nd Battalion of The Buffs. The duties consisted of providing guards on numerous public and military buildings and in carrying out normal routine drills, with plenty of social life for the officers and for the Senior NCOs. In January 1885 it was recommended that a '. . . Corps of Malay Submarine Miners . . .' be formed and the Government were urged to commence building a barracks '. . . at once on the island of Pulo Brani so as to have as much accommodation as possible ready on 1st April next . . .'[73]

In February it was ruled that the Colonial Engineer should superintend the defence works that were being planned and built on Singapore Island and in March 1885 designs and plans for the defence works were forwarded from the War Office, having been approved by that department. The new plans included batteries at the following locations, with the armaments as listed:

Mt Siloso	1	9.2-in B.L. guns	
	2	7-in 6½-ton M.L.	
	2	64-prs M.L.	
Pasir Panjang	2	9.2-in B.L.	
	2	64-prs M.L.	
Mt Palmer	2	10-in B.L.	
Blakan Mati East	2	9.2-in B.L.	
	3	7-in 6½-ton M.L.	
Mt Serapong	2	9.2-in B.L.	
Teregah Point	3	64-pr M.L.	
Tanjong Katong	3	7-in 6½-ton M.L.	
Range shown [for]		7-in guns	2,000 yards
		9.2-in guns	3,000 yards
		10-in guns	3,000 yards . . .'[74]

At this time there was also a discussion regarding the stationing of an 'Ironclad' Floating Battery at Singapore, but this suggestion appears to have never reached any satisfactory conclusion and the construction of the new batteries continued.[75] On 25 June the question of the Corps of Malay Submarine Miners was again raised and it was considered '. . . most desirable that the Submarine Mining det. should be practiced in laying mines in the position which they would (normally) occupy . . .', but two days later orders were given for the suspension of the building of the barracks for the detachment of Submarine Miners pending further instructions.[76] The Government was, once again, having second thoughts on the expense of this additional unit being added to the garrison establishment.

The garrison battalion again changed in 1886 when the Royal Inniskilling Fusiliers were relieved by the 1st Battalion, The Buffs. The battalion, about 700 rank and file, moved from Malta and were notified as being due to arrive in a note dated 17 September 1885.[77] A decision was taken during the year that the armament for Fort Tanjong Katong would be changed from those guns already planned being exchanged for breech-loading guns at an estimated cost of £83,700, and in the following year, 1886, it was decided to reclaim an area adjacent to Fort Palmer so as to extend the defence works at that site.

During 1888, the 1st Battalion, The Buffs were relieved by the Prince of Wales' Volunteers (South Lancashire Regiment) and also during that year it was agreed by the Government that the Singapore Volunteers should be drilled on the same guns as those used by the Royal Artillery stationed on the Island of Singapore, which in this instance

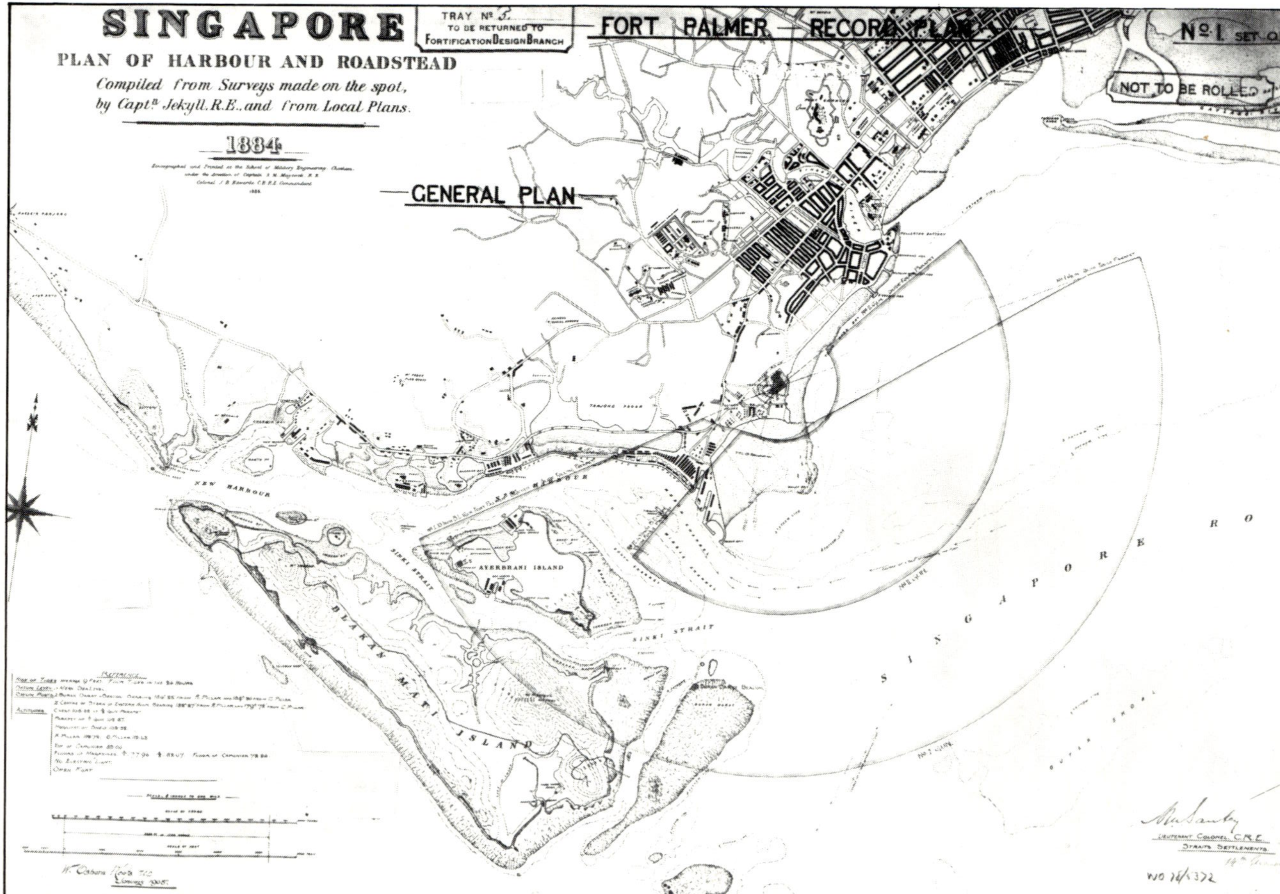

Plan of Singapore Harbour and defence positions, 1884. Courtesy of the Public Record Office, Kew.

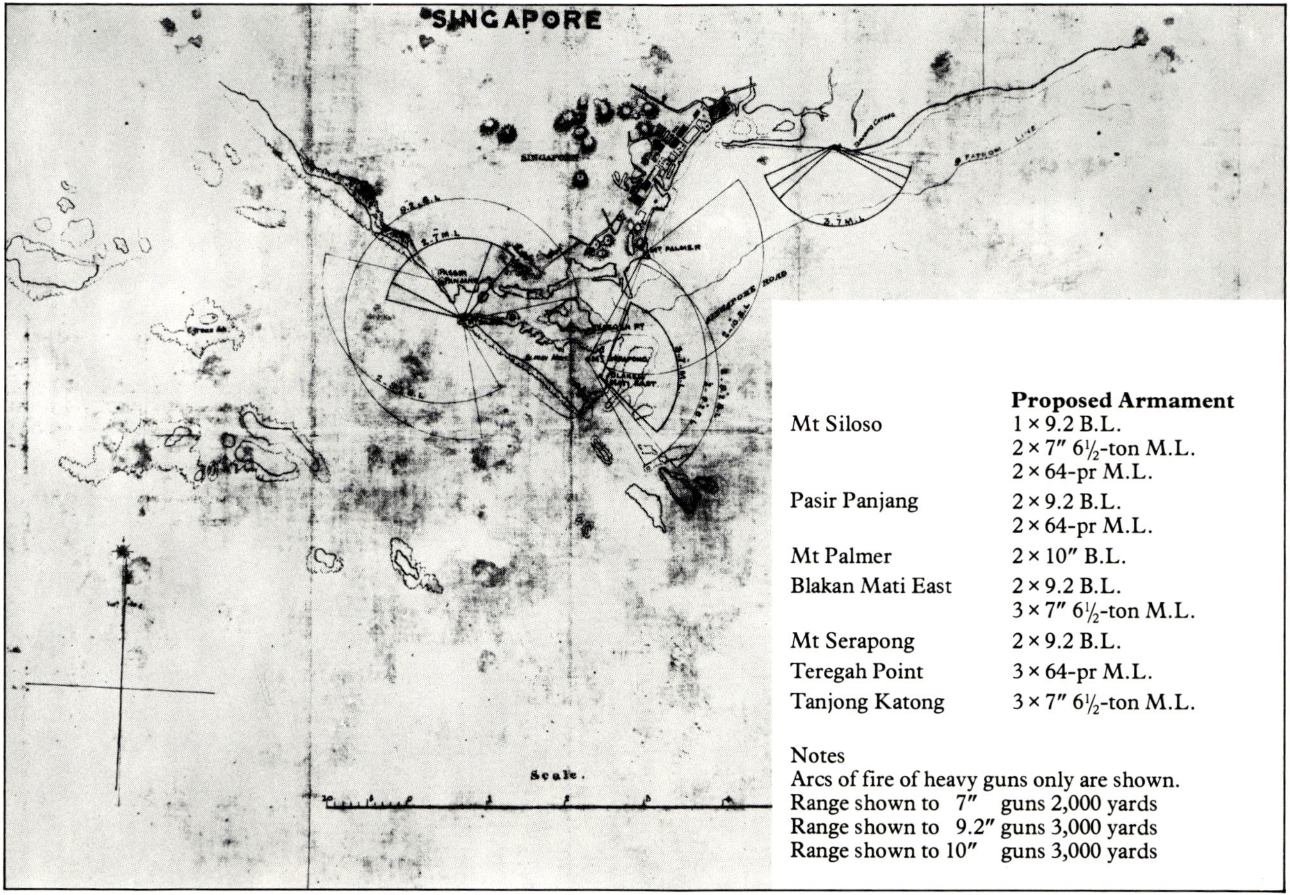

Plan of the proposed armament of Singapore harbour dated 1885. Courtesy of the National Archives, Singapore.

Fort Siloso, Blakang Mati. Armstrong guns now on view at the Singapore Gun Museum (located at Fort Siloso). Photograph by Mrs J. Harfield, 1977.

Fort Siloso, Blakang Mati. Close-up of Armstrong gun marking.

Photographs by Mrs J. Harfield.

Cannon on display outside the National Museum, Singapore, November 1971.

included the defence works on Blakan Mati.[78] As a result of an inspection of the battery that had been installed at Teregah Point on Pulo Brani it was found that the shell-store and the soldiers living accommodation were in the same building and it was ruled that this practice was to cease immediately and that a new shell-store was to be built. There was, of course, a distinct danger as not only were the soldiers living in the building but also the cookhouse was located in the same building and the resultant risk of fire was too great to ignore.[79]

The year 1890 saw another change in the garrison when the 2nd Battalion, The Northamptonshire Regiment relieved the Prince of Wales' Volunteers. Early in the year it was stated that although it was intended to move the troops stationed at Penang to the barracks being constructed on Blakan Mati the barracks were not ready for occupation. It was recommended that in order to speed up the removal of the troops from Penang instructions would be issued '... to erect attap barracks first of all as an experiment...'[80] With the withdrawal of the Royal Artillery from Penang authority was given for the dismounting of the guns at Fort Cornwallis and the Governor instructed the GOC to issue the necessary instructions.[81] At the end of the year 1890, the garrison received an increase to its establishment when half a Fortress Company, Royal Engineers was added. However, in 1891, the total expenditure for the previous year was said to have '... much exceeded £100,000 although the ½ Fortress Co RE did not reach Singapore until the end of the year the cost of raising and organising it was included during the past year...'[82] By the end of 1891 it was reported that although all the armament for the new fortifications had been received the actual number of troops required for the full establishment had yet to be met.

The question of the filling of the garrison establishment was discussed throughout 1892 at various times and in December of that year the following figures were given by the Governor, Sir Cecil C. Smith G.C.M.G.

'...

	Officers	*Warrant Officers, Non-Commissioned Officers and Men*
Royal Artillery	14	247
Native Artillery	2	118
Royal Engineers	6	89
Native Engineers	3	47
Infantry	28	985 (includes Penang)
Staff and Departments	12	29
Total	65	1515

From the last General Monthly Returns (Army Form A334) which I have received, the Garrison was on the 1st May last, 135 below its authorised strength, the Officers and 132 Men being deficient as follows:

	Deficient Officers	*Deficient Warrant Officers, Non-Commissioned Officers and Men*
Royal Artillery	1	35
Royal Engineers	3	9
Infantry		111
Total	4	155

But there were the following excesses:

	Officers	Warrant Officers, Non-Commissioned Officers and Men
Native Artillery		1
Native Engineers		13
Staff and Departments	1	9
Excesses Total	1	23

The nett number deficient was therefore:

Officers	Warrant Officers Non-Commissioned Officers and Men
3	132'[83]

The report included details of the various accommodation that was available for the troops and also details of the way in which the figures had been computed but in actual figures the amount of accommodation available for the garrison at Singapore was given as follows:

'...	Officers	All Other Ranks
Tanglin	26	661
Fort Canning	8	243
Pulau Brani		18
Ditto (for Natives)		90
Pearl's Hill		3
Forts		18
Penang	9	196
Total	43	1229

Thus there is a nominal deficiency in quarters for 22 Officers and 286 Non-Commissioned Officers and Men the authorised Garrison being: Officers 65, All other ranks 1,515.

The temporary Barracks, however, which have been built of unusual strength for Barracks of that class will accommodate 10 Officers, 1 Warrant Officer, and 269 Non-Commissioned Officers and Men. There is, therefore, an actual deficiency in accommodation in point of numbers for 7 Officers only, if the existing strengths only as distinct from the authorised strengths is considered ...'[84]

After further detailed recommendation for reorganisation of accommodation within the garrison the following request was submitted to the Secretary of State:

'... To sum up, I submit to your Lordship that the circumstances of the case, having regard to the basis on which your Lordship is now prepared to consider the matter, will be fully met by the following additional accommodation being provided:

	Officers	Non-Commissioned Officers and Men
Fort Canning	3	—
Pulau Brani	6	65
Blakang Mati		
(Native Artillery)	3	110
Ditto (Infantry)	2	100
Total	14	275

The full authorised strength of Royal Engineer Officers is never likely to be attained. To the expenditure for making this provision, will have to be added the re-appropriation services at Tanglin Barracks . . .'[85]

The question of the withdrawal of the two Companies of infantry from Penang Island was also discussed in the dispatch and it was pointed out that as the barracks at Penang had been adequately maintained it would be needless expense to withdraw the two companies as the figures given in respect of accommodation available, and required, at Singapore had assumed that the two infantry companies would remain at Penang.

A further development was proposed in May 1892 for the installation of a 'Defence Position Finding Station' to be constructed on Mount Wallich,[86] and in the following year the proposal was costed and the figure submitted was £24,102.[87] In the same year it was agreed that a new rifle range was needed at Tanglin Barracks and in November 1893 it was agreed that the erection of the Barracks, as proposed in December 1892, should be proceeded with and this included the work required on Pulau Brani, Blakang Mati as well as the barracks at Tanglin.

The whole question of defence and the manning of the defence position had been given a new emphasis in the Straits Settlements as in 1889 the forces within the settlements became a separate command with Command structure. The command included: '. . . Prince of Wales' Island, Malacca, Singapore and Christmas Island . . .' and the officers serving on the staff were given as:

'Governor and Commander-in-Chief	– Sir Cecil Clementi Smith KCMG
Aide de Camp	– Captain W. G. Massey RA
Colonel on the Staff (with rank of Major General)	– Colonel Sir Charles Warren, GCMG, KCB
Deputy Assistant Adjutant General	– Major P. H. Hammond RA
Garrison Adjutant	– Lieut G. A. Bramwell, Northampton Regt
Commanding Royal Artillery	– Lieut Colonel A. Burton-Brown
Commanding Royal Engineer	– Major A. C. Alexander
Senior Ordnance Store Officer	– Dep Assistant Commissary General of Ord. T. Heron
District Paymaster	– Staff Paymaster J. J. Tuck
Senior Medical Officer	– Surgeon Major J. G. Williamson'[88]

The year 1892 also brought about a general up-grading of the landforts that were to give protection to the harbour installations of the Colony of Singapore. The landforts on the main island of Singapore were, in 1892, (taking the forts in order from west to east) as follows:

Fort Pasir Panjang

Fort Pasir Panjang was first constructed about 1878 but the actual site had been selected as a part of the Colony's defence work some thirty-five years earlier. This is shown on the manuscript map drawn by Captain Best of the Madras Engineers[89] and dated 1843 that is now deposited in the National Archives of Singapore. The map shows his proposed new defence positions for '... the Town and Harbours of Singapore ...' The map gives details of the new harbour and lists thirteen forts or gun emplacements covering the whole water frontage of the town and harbour, including 'New Harbour' (this was later to be re-named Keppel Harbour after Admiral Sir Henry Keppel). On an un-named point, opposite the island of St George, as it was sometimes known, or Blakan Mati (also spelt Blakang Mati), there is shown on the map a proposed site for a battery which would give covering fire across the western entrance of the new harbour. In 1878 Captain Henry E. McCallum was given the task of designing and constructing the new defences for the Colony. The construction of the fort at Pasir Panjang commenced at that time and the western entrance of the harbour was secured by the crossfire of the guns of Fort Siloso, on Blakan Mati, and the two gun emplacements which then became known as Fort Pasir Panjang. The fort at this time consisted of a landing stage, a gateway with a portcullis, and a road leading up to the battery which consisted of two 7-inch muzzle-loading guns. Because of the hilly nature of the ground and the growth of secondary jungle, which at the time was uncleared, there was no overland route to the fort. Anyone visiting the fort during the present time will find it difficult to imagine the country as it was at that time. All supplies and ammunition to the fort had to be transported by boat to the landing stage and then carried up the hill to the gun platform. The defence system of Singapore was again reviewed in 1885 and the proposed plan showed that Fort Pasir Panjang was to be increased in size and upgraded from a two-gun fort to a six-gun defence work. The range of fire from the fort would be increased and the new proposal included two 9.2-inch rifled breech-loading guns and two 64-pounder muzzle-loading guns. The work was finally completed at Fort Pasir Panjang in 1892 and at that time the actual guns that were installed were listed as being: Two 9.2-inch RBL guns; two 7-inch ML guns and two 6-inch RBL guns. Fort Pasir Panjang was used extensively for training by the Singapore Volunteer Artillery as the accommodation had been extended and now included four new gun emplacements, accommodation for the artillerymen, underground storage and ammunition bunkers. A track was made to the fort in order that it could be reached by land and this permitted guns and limbers to be brought to the fort. Winsley describes the fort as follows '... it is charmingly situated. [There is] a portcullis at the Berlayer Point entrance, and an ammunition passage runs from the sea level, where the old Harbour Obelisk used to be, up to where the Signal Station now stands ...' The fort had, in 1892 an establishment of thirty-four men of the Royal Artillery who were accommodated 'on site'. The fort received further modification in 1901 but these did not alter the basic plan of the fort.

Fort Palmer

Fort Palmer was constructed during the period 1859–62 and was situated on the hill known as Mount Palmer, near Telok Ayer, and overlooking the eastern entrance to Keppel Harbour. The mount, and consequently the fort was named after Mr John Palmer who was a greatly respected member of the Indian merchant community of Singapore. He was associated with trading in Singapore from the early 1820s and died in Calcutta in 1836.

The armament at Fort Palmer was, in 1864, given as five 56-pounder guns and this armament was confirmed as being the same in 1867 by Winsley in his work *A History of the*

Fort Pasir Panjang: the Stable Block. Photograph by Mrs J. Harfield, 1971.

Fort Pasir Panjang, the Stable Block. Photograph by A.G. Harfield, 1973.

Fort Pasir Panjang, 1971. The Sea Entrance. Photograph by Mrs J. Harfield.

Fort Pasir Panjang, the central gun-platform, photographed in 1971 showing the dense undergrowth facing the sea (the direction of fire). Photograph by Mrs J. Harfield.

**Fort Pasir Panjang: entrance of the ammunition stores.
This section of the fortification was part of the 1892 rebuild.
Photograph by Mrs J. Harfield, 1971.**

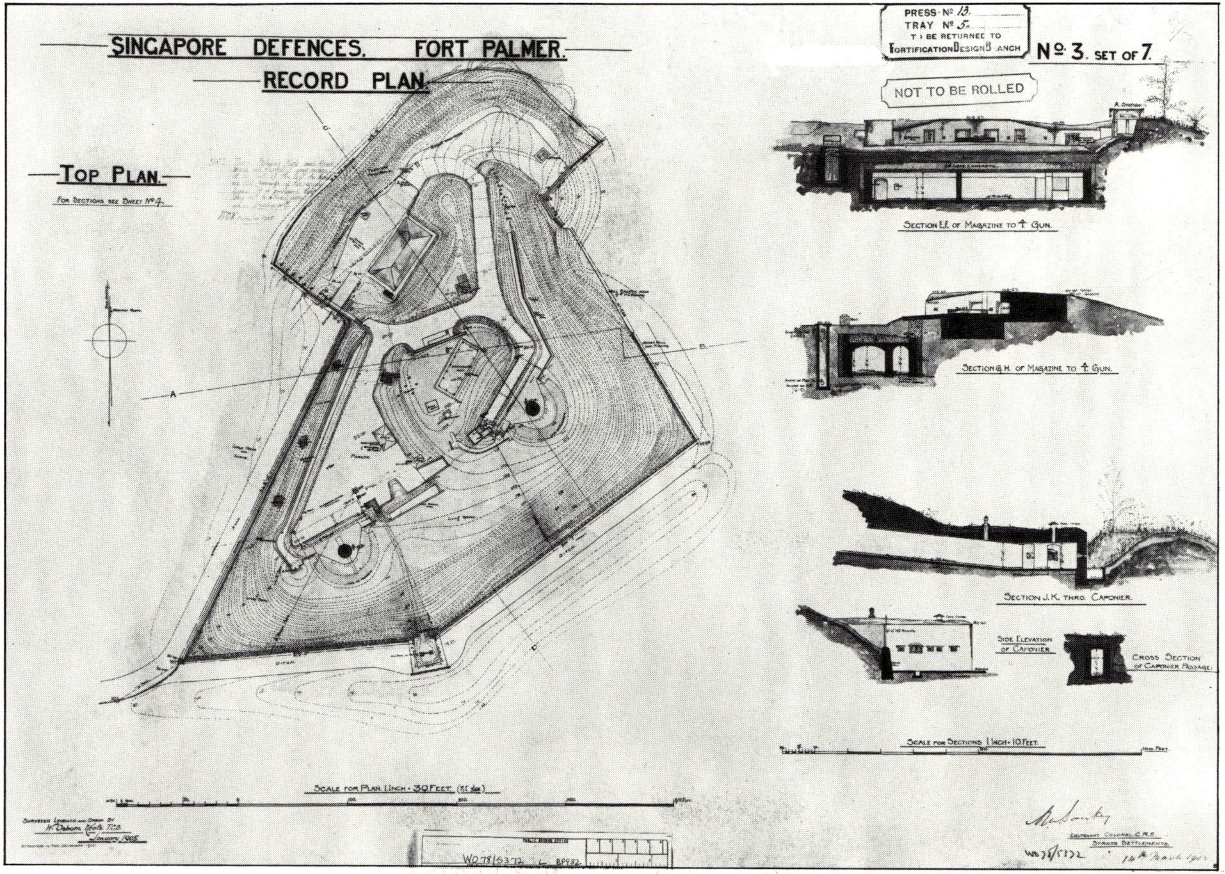

Plan of Fort Palmer, Singapore, dated 1907. Courtesy of the Public Record Office, Kew.

Singapore Volunteer Corps. The fort at that time consisted of earthworks, a magazine and a guardroom, etc. During the improvements to the defences of Singapore in 1878 the armament of Fort Palmer was changed to two 7-inch RML guns but these were later changed by the plan submitted and agreed in 1885 to two 10-inch BL guns. One of the 7-inch RML guns was removed from Fort Palmer and remounted at Fort Fullerton for use by the Singapore Volunteer Artillery Corps for gun drill practice.

In 1892 the fort was shown as having the two 10-inch BL guns and as having twenty-six men of the Royal Artillery stationed at the fort. The fort was again modernised in 1905 and at that time the defence plan shows Fort Palmer as having two 10-inch BL guns, two cartridge stores, two shell stores, accommodation for one non-commissioned officer and five men. In addition there was a general store, paint store and a small arms ammunition store. The living accommodation was quite basic as the plan allowed for water to be stored in a 700-gallon tank which was to be filled by Chinese coolies from the town supply. Lighting was by oil lamp and candles and the fort was bounded on the south and west sides by a concrete wall. The two guns were mounted in separate emplacements.[90]

The fort was later dismantled and part of the hill was used for the second Telok Ayer Reclamation project. At that time an Indian contractor purchased one of the 10-inch BL guns for $40 on the condition that he removed the gun or pay a penalty of $200.[91]

Fort Faber

Inland from Fort Pasir Panjang, and west of Fort Palmer, is Mount Faber which was named after Captain Charles Edward Faber, Madras Engineers. Captain Faber arrived in Singapore in September 1844 and became Superintending Engineer of the Settlement. The fort was built during the same period as Fort Palmer and was included in the overall defence plan of 1859–62. The original access to the fort was by a steep and narrow road leading off the Kampong Bahru Road and the site is described as being opposite the corner of Ayer Brani Island, where New Harbour and Sinki Strait meet. It was designed to cover both the New Harbour (Keppel Harbour) and the entrance between Ayer Brani Island (this island was later known as Pulau Brani) and Blakan Mati Island. The fort consisted of two gun emplacements situated on the top of Mount Faber which housed two 13-inch mortars. Fort Faber was used for practice firing during the 1860s and on 13 January 1865 a notice giving details of one of the practices was published in the *Government Gazette,* the details published were:

> '. . . There will be Shot and Shell firing from the Batteries in [*sic*] Mount Faber on the mornings of the 24th, 25th, 26th and 27th instant. The direction of firing – WSW by the entrance to New Harbour . . .'[92]

A report dated 1867 showed that the guns at Fort Faber had been changed and were at that time two 56-pounder guns. With the development of the harbour facilities adjoining New Harbour Fort Faber was dropped from the plan put forward in 1885. Winsley commented on Fort Faber unfavourably saying that it was '. . . never a really good fort . . .'

By 1892, Fort Fullerton had been converted and was no longer part of the defence plan. A Post Office had been built on part of the site and in 1890 a Drill Hall for the Singapore Volunteer Artillery Corps had been built on the eastern end of the old fort, and adjacent to the drill hall a 7-inch RML gun was mounted in order that gun drills could be exercised on training evenings. This was later removed during the 1907 reclamation programme and the final traces of Fort Fullerton were removed when the construction of Anderson Bridge was carried out in 1910.

Similarly Fort Canning had become, by the 1892 rebuild, more of a headquarters and

Fort Canning Gate, *c*1925. Courtesy of the Royal Signals Museum.

Fort Canning Gate, 1975. Photograph by Mrs J. Harfield.

Fort Canning Gate, viewed from the inside, 1977. Photograph by A. G. Harfield.

was the main station for the Royal Artillery based on Singapore Island. The artillery moved from Fort Canning in 1907 and from that date it became the Headquarters of the Army on Singapore Island and remained as such until the Second World War. Little of the old fort remains as during the years 1923 and 1927 the Singapore Municipality converted the top of Fort Canning Hill into a large underground municipal reservoir and in so doing demolished the old fort with the exception of the old gate and a small section of the original wall. The gate is an excellent example of the mid-nineteenth century fortress gateway and is still structurally sound with old double iron studded wooden doors, two sets, one at each end of the gate building, and outside the gateway two old cannon are embedded, muzzle foremost, into concrete and are now used as gateposts. The cannon are 9-pounders and date from the days of the early settlement at Singapore. One has a 'WD' stamp on it with 'P' and a figure '9', whilst below this is stamped a further letter 'P', '1800' and 'WG'. The second gun, on the left of the gateway, has been badly weathered and only an 'arrow' and a figure '44' are decipherable. No other traces of the fort's original armaments remain at the site. The existing buildings were used as the Headquarters of the British army between the two wars, and from the end of the war until the withdrawal of the British forces from Singapore in 1971. The office accommodation, accommodation block and certain married quarters were built during the period of the construction of the reservoir, and were occupied by the military in 1927. The native quarters were erected in 1929 and were then used continuously until the close of the 1960s.

Fort Tanjong Katong

At the eastern end of the defence line was Fort Tanjong Katong which was the only defence position built east of the Kallang River. It was built on the insistence of the merchants of Singapore who feared that the Colony might be subjected to an attack from the east. The fort was built during the 1878–9 build-up of the defences and in 1885 it was armed with two 7-inch, 6½-ton ML guns. During the following year it was planned that these two guns would be replaced by two 8-inch Armstrong guns, which were breech-loading. During 1888 the newly-formed Singapore Volunteer Artillery Corps carried out training at the fort and volunteer gunners who had classified on the 7-inch RML guns at Fort Siloso were then permitted to carry out voluntary training on the 8-inch BL guns at Fort Tanjong Katong. The course consisted of eight drills after which the volunteer was graded as having classified. The Singapore Volunteer Artillery Corps used the fort and the immediate area round the fort for their annual camp during the years 1890 and 1891. It was recorded that in 1892 thirty men of the Royal Artillery were stationed at Fort Tanjong Katong. However the fort was never popular and was regarded as a white elephant by the very same people, the civilian merchants, who had insisted on the fort being built at that site. In was known locally as 'The Wash-out Fort' as, due to the low level of the land at Katong the guns were unavoidably situated at sea level. A high tower was erected for the range-finding instruments but as this had been built on sand the firing of the guns during practice shook the tower to such an extent that the range-finding equipment had to be re-set after each salvo. Further problems arose when it was found that the 8-inch guns had insufficient range for the job that they were required to do, namely to defend the eastern sea approach to the city. In addition, as the guns had been obtained 'unofficially' (they had been diverted to Singapore during the time of the Franco-Chinese war), the supply of ammunition was always a great problem. It was later considered an impossible position to defend from a beach landing and in consequence the fort was abandoned and in later years the site became a 'People's Park'.

Also during the 1892 re-build, the three forts on Blakan Mati and Pulau Brani were brought up to recommended standard. The forts on Blakan Mati[93] were:

Fort Siloso

The fort was planned and constructed during the 1878–9 defence building and in 1885 had five guns mounted. These were one 9.2-inch BL gun, two 7-in, 6½-ton ML guns and two 64-pounder ML guns. In 1892 it was recorded that the guns had been mounted and that the fort was manned by eighteen men of the Royal Artillery. After 1907 the island became the home of the Royal Artillery based in the Colony and was used as an army training island.[94]

Fort Serapong

At the eastern end of the island, and on Mount Serapong, a small fort was constructed and is first mentioned in the plan of defence dated 1885. The fort was designed to hold two 9.2-inch BL guns and by 1892 the fort had been built and the guns mounted. The posted strength of the Royal Artillery personnel stationed at the fort during that year was thirty-six. Fort Serapong was, like the other landforts, completely self-contained and had facilities similar to those listed for Fort Palmer.

Fort Blakan Mati East

Fort Blakan Mati East is the fort at the eastern point of the island. Although this fort was not built until the 1878–9 period, Captain Best, on his map of 1843, marked in a site for a defence battery on Blakan Mati Island in the vicinity of the site of Blakan Mati East fort. The fort was armed with five guns and these are shown in the 1885 plan as being two 9.2-inch BL guns and three 7-in, 6½-ton ML guns. The fort was fully equipped in 1892 and then had a permanent Royal Artillery gunner strength of forty-four men. As a result of a visit of the Duke and Duchess of Connaught in 1890 the Fort was then renamed Fort Connaught.

The three forts continued to form a major part of the colony's defences and were armed and manned until the fall of Singapore during the Second World War.

The adjacent island of Pulau Brani had, during the nineteenth century, only one defence position and that was:

Fort Teregah

Fort Teregah was little more than a gun-platform. It was shown as a possible site for a battery in the Best plan of 1843 but was not built until the 1861 rebuild of the Singapore defences. At that time a gun-platform was built on the south-east corner of the island with the gun emplacements overlooking the entrance to the Singapore harbour area. In 1875 the living accommodation was built approximately a quarter-mile inland from the gun platform with a connecting path between the two. There was no attempt made to build a defence perimeter around these two sites, thus the term 'Fort' is, in this instance, rather misleading. The accommodation, mess-room, stores, etc had been constructed so that it was below ground level with a dry moat between the outer wall of the buildings and the adjacent ground. In 1888 the fort became the subject of a report, as the soldiers manning the site and the ammunition were housed in one building. As a result of this report a separate ammunition bunker was constructed. The self-contained living accommodation at this site, which was below ground level was extremely hot and uncomfortable but it was reported in 1892 that eight men were stationed at this site.

After the forts on Blakan Mati had been converted into training establishments the fort on Pulau Brani was abandoned. Unfortunately during the 1973 reclamation programme on Pulau Brani all traces of Fort Teregah were obliterated.

With the improvements to the land forts completed in 1892 and the proposals for bring-

Fort Teregah, Pulau Brani, 1971. Courtesy of J. W. Piper.

ing the strength of the Singapore, or Straits Settlements Command up to strength, the following year showed little change. The 2nd Battalion, The Lincolnshire Regiment moved to the Straits Settlements and took over the garrison duties.

In 1894 the question of the reduction of the European force was discussed with the emphasis being placed on the replacement of certain elements with Native troops. This proposal was rejected by the Colonial Defence Committee in February of that year.[95] At the end of the year there were further discussions with regard to the progress being made on the provisions of the new barracks on Pulau Brani and Blakan Mati. In the following year further improvements were required at Tanglin Barracks which included a proper water supply, and the improvement to the safety of the Tanglin Rifle range.[96] A proposal was also submitted for the construction of a hospital for the troops stationed on Blakan Mati island and later in the year, in December, the original proposal was '... agreed, with slight modifications ...'[97] and as a result the hospital was built during the following year.

There was a recommendation for the construction of a barrack block for the 'Asiatic Artillery' who were to be employed in the garrison and although the approval was requested in June it was not actually given until 29 November 1896.[98]

The garrison troops were again exchanged during 1896 and the 2nd Battalion, The Northumberland Fusiliers moved into the Colony and remained there until 1898 when they were, in turn, relieved by the 1st Battalion, The Prince of Wales' Own West Yorkshire Regiment.

An entry in the Governor's Diary for the year 1897 reveals that during that year the Volunteer Long Service Medal was approved for issue to those members of the Singapore Volunteer forces who had qualified. The medal had been instituted in 1894 for non-commissioned officers and men of the Volunteer forces of Great Britain who had completed twenty years. In 1896 it was extended to the Indian and Colonial Volunteer Forces and it was agreed that for Volunteer service in India the qualification period would be eighteen years.[99] The medal continued to be awarded for volunteer service until 1930.

With the outbreak of war in South Africa the British battalion in the Straits Settlements changed over quickly, the 1st Battalion, Prince of Wales' Own Regiment departed after having served only a little over a year in the Colony and was replaced by the 1st Battalion, King's Own Lancaster Regiment in February 1899 but they too departed in February 1900. It was agreed that due to the need to have British Army units available for service in the war in South Africa the garrison troops for the Straits Settlements would, once again, be provided from the Indian Army. Consequently, in March 1900, the 16th Madras Infantry arrived at Singapore and took over the garrison duties. This unit remained in the Colony until November 1901 when it was relieved by two regiments from the Indian Army, these being the 3rd Madras Infantry and the 13th Madras Infantry. During 1901 the state of the landforts was again reviewed and a number of modifications recommended and these were carried out over the period 1901–6. It was during this time that an additional gun emplacement was planned for Pulau Brani and was named Fort Silinsing, which was located on the high ground above Silinsing Point. The gun position was manned during the First World War and was modernised during the vast rebuilding programme of 1936.[100]

After the close of the war in South Africa the decision was made to increase the strength of the Straits Settlements Command from one battalion, as it had been prior to the outbreak of war, to two battalions, one from the British Army and the second battalion to be supplied by the Indian Army. As a result of this decision the 1st Battalion, The Manchester Regiment arrived in Singapore during April 1903 and relieved the 3rd Madras Infantry

Fort Silinsing, Pulau Brani. The gun emplacement (and right, one of the ammunition hoists). This includes the modifications made during the 1936 rebuilding programme. Photograph taken 1971. Courtesy of J. W. Piper.

who returned to India. The 13th Madras Infantry remained in the Colony until it was relieved by the 73rd Carnatic Infantry in January 1904. These two battalions served in the Colony until January 1905 when the 1st Battalion The Sherwood Foresters took over the duties of the British battalion.

During the period 1905 to 1914 there was a lull in the building activities in respect of the barracks and landforts; the only major change being the move of the Royal Artillery from Fort Canning to the barracks on Blakan Mati Island, and Fort Canning becoming the Army headquarters for the Straits Settlements. Although the education of European army children had been dealt with within each European infantry unit as it carried out its tour of duty in the Straits Settlements it was not until 1905 that the first written record of Army schools being in existence in Singapore appears and the *Register of Infants and Sewing Schools* for 1905 shows that such a school existed at Tanglin with seven pupils under Miss A. M. Wilkinson, the Schoolmistress. The record also shows that a school was in existence on Palau Brani in 1907 with '. . . 7 elder chidren and 9 infants . . . in the care of Mrs E. Ingram . . .'[101] By 1907 the army had four centres within the Colony and these were as follows:

Fort Canning – Headquarters of the army
Blakan Mati – Royal Artillery
Tanglin – Infantry garrison regiment
Pulau Brani – Royal Engineers and Ordnance Depot

It would appear that the school on Pulau Brani probably closed at the end of 1907 when the schoolmistress had departed to Hounslow.[102] It is probable that the pupils from Pulau Brani were then transferred to the Blakan Mati school where the number of pupils, in December 1908, was shown as being 6 elder children and 12 infants. An acting schoolmistress was appointed for the Blakan Mati school with a salary of £20 per annum and this appointment was filled by Mrs A. Bristow.[103] By the year 1910 the school at Tanglin was shown as having 2 elder girl pupils and 18 infants whilst the school on Blakan Mati had 5 elder girl pupils and 10 infants.

During the period leading up to the outbreak of the First World War the British battalion changed every two years and the Indian Army battalion at slightly less frequent intervals. The list of units that served on garrison duties in the Straits Settlements during the period 1805 to 1914 is given in Annexure A.

With the outbreak of the First World War in 1914, the British infantry battalion was withdrawn during the latter part of that year and the 3rd Brahmans, who were the Indian Army battalion doing duty in the Straits Settlements was relieved by the 5th Light Infantry. In February 1915, the 5th Light Infantry mutinied and the events leading to the mutiny and the story of the mutiny are dealt with in Chapter Fourteen.

With the mutiny duly quelled, the Straits Settlements settled back into normal routine by the beginning of April, but on 29 April a small disturbance broke out at Pasir Puteh, in the State of Kelantan, and resulted in one of the Police sergeants being killed. A small force was despatched from Singapore under the command of Lieutenant Colonel C. W. Brownlow consisting of 163 men of the 4th K.S.L.I (TF), fifty men of the Royal Garrison Artillery and twenty-one men from the Singapore Volunteer Infantry. The small Field Force arrived at Pasir Puteh on 5 May 1915 and were reinforced on 13 May by 140 men of the Malay States Guides, the latter being equipped with two machine guns. The force met with little opposition and were able to return to Singapore after only ten days leaving the

Cannon found at Changi and placed on display outside the RCT Squadron Office. Photograph by Mrs J. Harfield, 1971.

detachment of the Malay States Guides to 'police' the area. The army expedition was supplemented by a landing party from HMS *Cadmus* whose shore party had helped quell the mutiny in Singapore earlier in the year. The Malay States Guides remained in Kelantan until September when they departed for 'overseas' duties in Aden. They were relieved by a small force of one officer and thirty-eight men of the Malay Company, Penang Volunteers and two officers and twenty-four men of the Singapore Volunteer Infantry. This small force eventually departed from Kelantan in November 1915 and the maintenance of law and order was left in the hands of the small force of Special Constables who were later to be named the Kelantan Volunteer Rifles.[104]

The year 1915 saw the formation of two new volunteer units within the area, the first was formed in Malacca on 12 August 1915 and was named the Malacca Volunteer Rifles and was under the command of Major W. M. Sime. The unit had by 1918 increased in size and been renamed the Malacca Volunteer Corps with a strength of 390 all ranks.[105] The second unit to be formed was the Province Wellesley Volunteer Rifles which was formally constituted, as a purely volunteer force, on 3 November 1915. The unit only accepted Europeans in its establishment and never reached its full entitlement of volunteers which was six officers and 159 other ranks. It reached a total of 139 all ranks in 1917 and was ably commanded by Captain W. Duncan who was, in civilian life, the general manager of the Caledonia Group of Rubber Estates.[106]

NOTES TO CHAPTER ELEVEN

1. Walter Makepeace (ed), *One Hundred Years of Singapore,* (J. Murray, London, 1921), pp24 and 25.
2. Straits Settlements Government Gazette, No 3, Friday 21 January 1859, (National Library, Singapore: Film 994).
3. Ibid, No 20, Friday 20 May 1859.
4. Ibid, No 29, Friday 22 July 1859, and *Singapore Free Press* vol 26, No 30 of 28 July 1859.
5. Straits Settlements Government Gazette, Extraordinary Issue, Wednesday 10 August 1859.
6. Ibid, No 40, 7 October 1859.
7. Ibid, No 44, 4 November 1859.
8. Charles Burton Buckley, *An Anecdotal History of Old Times in Singapore 1819–1867,* (University of Malaya Press, 1965), p 675.
9. Ibid.
10. Ibid.
11. Governors Diary, entry No 2 of 12 May 1859, (National Library, Singapore).
12. Ibid, entry No 3 of 12 May 1859.
13. *Singapore Free Press,* vol 26, No 19, Thursday morning 12 May 1859.
14. Ibid.
15. Governors Diary, entry No 10 of 9 April 1860.
16. Ibid, entry No 10 of 19 September 1860.
17. Straits Settlements Government Gazette, vol 3, No 9, 2 March 1860.
18. Ibid, vol 3, No 16, 20 April 1860.
19. Ibid, vol 3, No 19, 4 May 1860. Lieutenant W. H. G. Palmer was appointed 4th Class, Executive Engineer from 4 May 1860.
20. Ibid, vol 3, No 6, Friday 10 February 1860. In fact Lieutenant Playfair resumed his appointment on 18 May 1860 under authority of SSGG vol 3, No 20 of 18 May 1860.
21. Ibid, vol 3, No 23, 8 June 1860.
22. Ibid, vol 3, No 31, 3 August 1860.
23. Governors Diary, entry No 21 of 23 April 1861.
24. Ibid, entry No 24 of 30 April 1861.
25. Ibid, entry No 25 of 15 June 1861.
26. Ibid, entry No 2 of 5 October 1861.

27. Ibid, entry No 3 of 5 October 1861.
28. Buckley, op cit, p686.
29. Governors Diary, entry No 13 of 23 January 1862.
30. Governors Military Diary, entry No 13 of 24 February 1862.
31. Ibid, entry No 20 of 14 May 1862.
32. Ibid, entry No 1 of 1 April 1862.
33. Ibid, entry No 2 of 1 April 1862.
34. Ibid, entry No 18 of 26 June 1862.
35. Ibid, entry No 19 of 26 June 1862.
36. Ibid, entry No 23 of 15 July 1862.
37. Ibid, entry No 24 of 15 July 1862.
38. Ibid, entry No 40 of 30 July 1862.
39. Ibid, entry No 7 of 16 April 1863.
40. Ibid, entry No 8 of 16 April 1863.
41. Ibid, entry No 16 of 18 July 1863.
42. Ibid, entry No 20 of 28 September 1863.
43. Buckley, op cit, p769.
44. Ibid, pp 769 and 770.
45. Madras General Order dated 8 February 1864, (National Library, Singapore. Film No 3308).
46. Ibid, dated 15 August 1865.
47. Ibid, dated 15 February 1864.
48. Ibid, dated 5 October 1864, (Film 615).
49. Ibid, dated 7 October 1864.
50. Ibid, dated 4 November 1864.
51. Ibid, No 116 dated 14 December 1864.
52. Governors Diary, entry No 1 of 3 March 1865. The entry in author's notebook of inscriptions of the military memorials in Northam Road Cemetery, Penang Island shows the memorial inscription on the grave of Major W. S. Mitchell as being:

'Sacred
to the memory of
Major William Somerville Mitchell
22nd Regiment
Aged 38'

53. Ibid, entry No 1 of 1 June 1866.
54. Ibid, entry No 2 of 1 June 1866.
55. Buckley, op cit, p769 and Governors Diary, entry No 6 of 10 July 1866.
56. Straits Settlements Government Gazette, No 35 of 1 September 1865.
57. Ibid, No 51 of 22 December 1865.
58. D. and J. Moore, *The First 150 years of Singapore,* (Donald Moore Press Ltd, Singapore, 1969), p327.
59. Governors Diary, CO 426/1. 5096 of 27 May 1867, (Public Record Office, Kew).
60. The New Cemetery, located adjacent to the Bukit Timah Road and was opened in April 1865 and was '. . . when first used, . . . a very dismal place, no sufficient drainage and water-logged within a few inches of the surface . . .' The cemetery was closed for burials when the Bidadari Cemetery was opened in 1903. The author had occasion to visit the cemetery in 1959 when commissioned to prepare a list of the military graves and memorials in the cemetery. At that time the eastern end of the cemetery was generally water-logged. The list produced at that time formed the basis of the entries in the 'Roll of Honour' which was prepared by the military authorities in the 1960s. The cemetery was converted into a garden in 1970 and with the exception of a few civilian memorial stones all the gravestones and memorials were destroyed at that time.
61. Alan Harfield, *Christian Cemeteries and Memorials in Malacca,* (British Association for Cemeteries in South Asia, 1984), p68.
62. Governors Diary, CO 426/1, 1970 of 27 July 1868.
63. Ibid, 7132 of 26 June 1869.
64. Ibid, 9812 of 7 September 1868.
65. Ibid, CO 426/2; 3874 of 20 April 1871; 10846 of 6 November 1871; 11538 of 27 November 1871; 11954 of 6 December 1871.

66. Digest of Service of 2nd Bn South Staffordshire Regiment, by courtesy of Regimental Headquarters, The Staffordshire Regiment (The Prince of Wales's).
67. Ibid.
68. Ibid.
69. Governors Diary, CO 426/3, 5718 of 6 June 1873.
70. Ibid, CO 426/4, 2354 of 29 February 1876.
71. Ibid, CO 426/6, entry dated 8 September 1879.
72. Ibid, CO 426/7, 18520 of 30 October 1882.
73. Ibid, CO 426/8, 842 of 15 January 1885.
74. Plan of Proposed Armament dated 31 March 1885, (National Archives, Singapore).
75. Governors Diary, CO 426/8, 8160 of 7 May 1885.
76. Ibid, 11118 of 25 June 1885; 11207 of 27 June 1885.
77. Ibid, 16487 of 17 September 1885.
78. Ibid, 14968 of 26 July 1886.
79. Ibid, 23790 of 3 December 1886.
80. Ibid, CO 426/10, 2760 of 12 February 1890.
81. Ibid, 2768 of 9 February 1891.
82. Ibid, 3883 of 24 February 1891.
83. *The Straits Times,* Friday 2 December 1892.
84. Ibid.
85. Ibid.
86. Governors Diary CO 426/11, 11392 of 6 May 1892.
87. Ibid, 5317 of 1 April 1893.
88. Harts Army List, 1891, p104.
89. C. A. Gibson-Hill states, in his book *Singapore Old Strait and New Harbour*, that the 1843 plan of the Singapore Defences was the work of 'Captain James Best'. My own notes also recorded the name as 'James Best'. A check of the East India Company Register for 1843, the Army List and Harts Army List for that year failed to identify a 'James Best' in either the British Army or in the Company service. Further investigation revealed that the officer was, in fact, Captain Samuel Best of the Madras Engineers. On re-examination of the map the confusion appears to have been caused by the writing of the forename which is written as 'Sam'el' which was obviously misread as 'James'. Samuel Best was the fourth and youngest son of George Best, a wine merchant of 22 Spring Gardens, Greenwich, and his wife Martha and was born on 31 October 1808. He was at Addiscombe during 1824–1825 and became an ensign on 16 October 1825, was promoted to the rank of Lieutenant on 17 December 1825 and Captain on 9 May 1842. He is shown as serving with the 'Board of Revenue in the PWD', as Secretary, in 1842. In the following year, 1843, he was responsible for the planning of the fortifications of Singapore. In 1845 he was made Superintendant of Roads in Madras and was responsible for the construction of the Southern Trunk Road and the Goolcheroo Pass. He died of jungle fever at Chitore (Chittoor) on 5 October 1852 on his return from the hills. The Hodson Index, (National Army Museum); C. C. Buckland, CIE, *Dictionary of Indian Biography,* (Swan, Sonnenschien & Co, 1906); and Frank Boase, *Modern English Biography,* (Frank Cass & Co, 1965).
90. Public Record Office, Kew. WO 78/5392.
91. Makepeace, op cit.
92. Straits Settlements Government Gazette dated 13 January 1865.
93. Blakan Mati was known by a variety of names these, and their sources, being:
Pulo Panjang – map drawn by J. N. Belin, 1755.
Po Panjang (Pulau – Island) – map attributed to Captain James Franklin, BNI, dated 1822.
Po. Panjan – Lithographed map by J. B. Tassin, published in Calcutta, 1836.
Isle of St George – map by Captain Samuel Best, Madras Engineers, dated 1843.
Blakan Mati Island – map by John Turnbull Thomson dated 1 January 1851, published by HM Admiralty.
Pulau Blakang Mati – map of Singapore by HM War Office dated 1911.
Sentosa – Singapore Guide and Street Directory for 1975. The island was renamed and converted into a 'holiday island'.
94. A 'Gun Museum' has now been established on Blakan Mati (Sentosa) at Fort Siloso.
95. Governors Diary, CO 426/11, 2888 of 17 February 1894.
96. Ibid, 1255 of 23 January 1895.

97. Ibid; 11754 of 6 July 1895; 23143 of 28 December 1895.
98. Ibid, CO 426/12; 12065 of 6 June 1896; 2161 of 29 November 1896.
99. Ibid, 3287 of 15 February 1897 and Winsley, op cit, p 171.
100. The author visited Fort Silinsing in 1971 and was able to photograph the main gun emplacement there. It would appear to have been a battery installed to cover the entrance to Singapore Harbour and to assist the guns installed at Fort Teregah. The whole fortification disappeared during the 1973–1974 reclamation programme on Pulau Brani.
101. R. P. Gaskell, 'British Army Children's Schools in Singapore', (Thesis presented to the University of Singapore, March 1971), pp 34 and 35.
102. Gaskell, op cit, p 35.
103. Register of Infant & Sewing Schools 1905–1910. Royal Army Educational Corps Museum, Beaconsfield.
104. Winsley, op cit, p 72, PRO Kew, CO 273/435.
105. Ibid, p 73.
106. Ibid, pp 73 and 74.

CHAPTER TWELVE

The Singapore Volunteer Units, 1854–1914

The Volunteer movement in the United Kingdom has been linked with times of stress and threat and in the mid-nineteenth century there was a great Volunteer revival which commenced in 1859; likewise in the Colony of Singapore local events also brought about an enthusiasm for Volunteer soldiering amongst the European population. The formation of a volunteer unit was first mooted in 1846 after the settlement had been disturbed by Chinese riots; however the Volunteer Corps was not actually formed in Singapore until 1854 when there had been a further outbreak of Chinese riots, lasting ten or twelve days.[1] The outbreak arose between the Hokkiens, from the province of Hokkien in China, and the Teo Chews from the province of Quantung. It occurred because the former refused to join in a subscription to assist rebel Teo Chews who had been driven from Amoy by the Imperial Chinese Troops.

The Volunteer Corps in Singapore was formed for the purpose of providing an internal security force in the settlement which was constantly being disturbed by riots. These riots, and the news of the outbreak of the Crimean War, led to a public meeting, which was held on Saturday, 8 July 1854 with the concurrence and support of the Governor, Colonel W. J. Butterworth, CB.[2] Mr John Purvis was in the chair and the meeting decided to form a Volunteer Corps, which was to be named the 'Singapore Rifle Corps'. The Governor accepted the offer of service and forwarded the list of Volunteers to the Government at Fort St George, Madras. The entries in the Governor's Military Diary record the events as follows:

> 'A deputation waited on the Govr to offer the services of a local Volunteer Rifle Corps to the Govt, and to request His Honre to propose a set of Rules for the Guidance of the Corps . . .'
>
> The Governor '. . . accepts the services of the Inhabitants and furnishes them with a general outline of the Rules for the formation of the Rifle Corps . . .
>
> Informs them that a list of the Volunteers will be forwarded to the Govt of India, which with the Govt of Hong Kong will be moved to supply any Regulations having reference to Volunteer Corps that may be in Force in India or China.'[3]

Initially thirty-two volunteers placed their names on the roll but by the time the Corps was actually raised the total had reached sixty-one. The governor became its first Colonel and Captain Ronald Macpherson, Madras Artillery, its first Commandant. On Thursday, 8 March 1855 the newly-formed Corps paraded on Government Hill (Fort Canning) this event was reported in the *Free Press* as follows:

> 'The Singapore Rifle Corps paraded on Government Hill (Fort Canning) on the afternoon of Thursday last, and went through a pretty stiff drill, after which they were reviewed by the Hon'ble the Governor (Colonel Butterworth) who then addressed them

in complimentary terms on the efficiency they had attained, and assuring them how proud he should have been to have headed them in actual service. His Honour requested that his name might remain upon their roll, [this parade was just prior to his retirement] and concluded by reading some dispatches from the Court of Directors and the Supreme Government of India, noticing in terms of approbation the promptitude with which the Singapore Volunteers had come forward with the offer of their services and expressing the hope that their example might be followed in other parts of India. The Corps, although they turned out on this occasion in somewhat diminished numbers, appeared, as far as the unprofessional eye could judge, to go through their manoeuvres with steadiness and precision, and we have no doubt that they will highly distinguish themselves whenever they may be called upon to take the field.'[4]

Governor Edmund Augustus Blundall presented, on 14 February 1857, a set of colours to the Singapore Volunteer Rifle Corps. Brigadier McLeod authorised the parade to be held on the Esplanade and as a sign of mourning for the late Colonel Butterworth, whose widow had prepared the Colours, the Corps wore a band of crêpe on their arms. The Colours were presented to Lieutenant W. H. Read, the Senior Subaltern, who is reported to have made a speech which concluded with the following words:

'We seek not glory of the battlefield, nor to embroider the names of victories on these colours. Ours are less martial, more peaceful aims. Our object is to assist in protecting the lives and property of the public, and to show the evil disposed how readily Europeans will come forward in the maintenance of order and tranquility. Should we ever be called upon to act, we shall be found prepared to do our duty, contented with the approbation of the Government and the applause of our fellow citizens.'[5]

On 18 July 1857, the Singapore Volunteer Rifle Corps ceased to be a 'private Corps' when the Legislative Council of India passed Act No. XXIII. The Act provided for '. . . good order and discipline of certain Volunteer Corps, and to invest them with certain power . . .'[6] The Corps was then classed as a Rifle unit and as such was clothed in dark green (Rifleman Green). This dress is documented in the SVRC Rules of 1857, where Rule IX reads as follows:

'At Full Dress Parades the members shall wear a Green Uniform of the material and pattern originally agreed upon; and at all other times they shall appear, when called on duty, in White Jackets, White Trousers, and Cap covers, Black Neck Ties and Black Shoes.'[7]

In the early days of the Singapore Volunteers the staff consisted of a Captain Commandant (Captain R. Macpherson), an Adjutant (Captain Robert Church of the 47th Regiment, Madras Native Infantry), two Lieutenants (W. H. Read and C. H. Harrison), three Sergeants and two Corporals. There were changes in 1857 when the Volunteers were officially recognised by the Indian Government and over the next two decades there was, in addition to the Captain Commandant (who from December 1857 was always a volunteer officer), a Colonel Commandant. This appointment was normally filled by the senior regular Military Officer stationed in Singapore.[8]

Captain W. H. Read was promoted on 1 January 1858 to become the first Volunteer Captain Commandant and Brigadier McLeod of the Madras Native Infantry was the first Colonel Commandant. A 'second' Captain was added to the establishment and Mr J. Purvis was given the appointment. The work of Adjutant and Quartermaster was delegated to a Volunteer Officer and the officer strength increased to three Lieutenants. Dr R. Little, MD, was appointed Hon Surgeon to the Corps. The title of Sergeant, accorded to the

Senior NCOs, was abolished in favour of the title 'Sub Officer', of which there were now four. The establishment of Volunteer Corporals at this time was also increased to four.[9]

In January 1858, due to unrest in China, Singapore suffered another outbreak of rioting and Buckley records that '. . . the Military and Volunteer Rifles were in readiness, and large guns were mounted on Government Hill (Fort Canning) and Pearls Hill. An additional Regiment was soon afterwards sent from Madras.'[10]

Captain Winsley, in his history of the Corps, commented on the fact that the regimental colours were 'of the old Ensign design [ie of the type before 1868] which differed from present-day Colours in that it was of larger dimensions and had a Union Jack in the top corner near the pike, the main field being the same colour as the regimental facings, while the pike had a spear head.' The regimental facings were, at that time, green. The original colours lasted about twenty-five years and on becoming badly worn were replaced. The original Pikes were deposited at the Central Police Station and, at a later date, were unfortunately disposed of as 'rubbish'. No clear details were known of the original Colours until 1937 when, whilst Captain Winsley was hunting in Raffles Library for illustrations for his book, a very faded photograph, dating from 1870, came to light showing the design of the Colours.

The unit's terms of service allowed for it to be called out if the Town was attacked by any foreign enemies, and also in the event of any disturbances within the Colony which the Police Force could not handle. The Corps paraded once a month, or more frequently if it was considered necessary, and in addition, in the months of February, May, August and November, a regular Drill practice on three or more alternate days (Monday, Wednesday and Friday) was carried out by Recruits until they were, in the opinion of the Commanding Officer, fit to join the ranks. As was the case with Yeomanry units in the United Kingdom, the Volunteers were subject to a fine if absent from a Drill without a suitable excuse, acceptable to the Commanding Officer. The fine for a member failing to attend was five dollars.

On 1 September 1858, Her Majesty Queen Victoria assumed the direct government of her Indian Empire and on the morning of 19 November the Proclamation was read in Singapore. For this occasion the Singapore Volunteer Rifle Corps paraded at 7.00am, together with the 43rd Madras Native Infantry, Madras Artillery and Marines and seamen from HMS *Amethyst*. The Governor read the Proclamation in English, it was then read in Malay, after which a Royal Salute was fired by the Artillery and a *feu de joie* by the troops.[11]

Two Volunteer appointments were published in the Straits Settlements Government Gazette on subsequent weeks, they were:

> 'Lieutenant M. F. Davidson has been appointed Adjutant of the Singapore Volunteer Rifle Corps from the 11th instant, vice Lt J. C. Smith who has resigned that appointment.'[12]
>
> 'Lieutenant J. C. Smith will continue to perform the duties of Quartermaster of the Singapore Volunteer Rifle Corps.'[13]

The next parade of note was on Monday, 26 November 1860 when the Volunteers paraded in the evening and were reviewed by the Governor, Colonel Orfeur Cavanagh. The parade was held on the Esplanade and the Volunteers turned out in full force under their Commandant, Captain Read. The Governor arrived, accompanied by Brigadier Burn and his Staff, and on his arrival the Volunteers presented arms. They then proceeded to demonstrate their drill ability and it is recorded that they '. . . marched past in slow and quick time and went through a number of Light Infantry manoeuvres, advancing, firing, halting, changing front in one direction and then the other, forming [a] square, retiring, and finally, having fired two volleys with remarkable precision, they formed up in their original positions and again presented arms.'[14] The Governor was extremely pleased with the performance and expressed his pleasure in an address to the Corps, during which he congratulated them on having the honour of being the first Corps of Volunteers to be enrolled under the Indian Army and commented on the fact that they were entitled, because of this, to bear upon their colours the inscription 'Primus in Indis'. This motto was in fact retained by the Corps until the formation of the Singapore Volunteer Artillery in 1888. Martial music for the parade was provided by '... the excellent Band of Her Majesty's 40th Regiment, Madras Native Infantry ... by the kind permission of the officers of the Regiment, and added much to the gaiety of the scene ...'[15]

In late 1861, the question of the effectiveness of the defences of Singapore was raised. The entry in the Governor's Foreign and Military diary records that the Brigadier commanding the garrison placed on record the effective operational strength of the garrison as being '... scarcely 200 effective Native Infantry with the Singapore Volunteers are available for Field Service, the strength of the Artillery being only 40 European and 70 Golundauz, these are not sufficient men to man the heavy guns at present in position ...' At this time however the Volunteer movement was still thriving in Singapore.[16]

The next recorded major event in which the Volunteers participated was in 1864. In the intervening years normal weekly training was carried out.

On 24 May 1864, eighty Volunteers attended the Queen's Birthday Parade which was held on the Esplanade at a little after 5.00am. The parade for this review numbered about 1,000 and would have been a colourful sight. The unit serving in Singapore at this time was the 34th Regiment, Madras Light Infantry, under the command of Lieutenant Colonel F. F. Warden. Eight hundred Sepoys of this unit were on parade dressed in red turbans, redcoats, black trousers and sandals; 120 British Artillery were dressed in white pith helmets with blue trousers and jackets, and the eighty Singapore Volunteers were recorded as being dressed in white trousers and light-coloured hollands. The guns at Fort Canning fired a 21-gun salute and the parade finished at about 8.00am. At this time (the 1860s) one of the guns at Fort Canning, a 68-pounder, fired a blank round at 5.00 am each morning as a dawn signal.[17]

On Monday 1 April 1867, the inauguration of the new Government took place at the Town Hall. The administration was then transferred from the East India Company to the Colonial Office. Among the military present were '... Major General Studholme Hodgson, Brigadier Ireland, Colonel Grant RA, Colonel Lovell CB, RE, and Colonel Cooke ...'[18] The latter was Lieutenant Colonel W. J. Cooke, the Commanding Officer of the 8th Madras Native Infantry who had provided the guard of honour outside the building, whilst the Volunteers provided the guard of honour on the front verandah. The new Governor, Colonel Harry St George Ord, late of the Royal Engineers, was received by a 17-gun salute from Fort Canning and the two guards of honour.

The Volunteer Corps was present at the May 1867 Birthday Parade but had, by that time, somewhat diminished in numbers. Those on parade were:

Captain Commandant C. H. H. Wilsone
Lieutenant von der Heyde
Ensign A. Duff
Hon Surgeon Dr Little,
47 NCOs and men.

The corps was complimented on its drill at this parade by the Inspecting Officer.[19]

During 1867 military expenditure was again the subject of discussion by the merchants of the Colony who, remembering that in 1863 nearly half of the Colony's revenue had been expended on defence, considered that too much was being spent on military defence. The main armaments, in 1867, consisted of some thirty-six guns which were located at Fort Canning and Fort Fullerton, and at two small earthworks at Mount Palmer and Mount Faber. The Headquarters of the Artillery was at Fort Canning which had been the Artillery barracks since its completion in 1861.

In the early part of 1868 the Corps establishment was increased to include half a Battery of Field Artillery and for this purpose it was loaned, from the Regular Artillery in Singapore at that time, two 12-pounder Howitzers.[20]

The Officers and NCOs of this Battery had to provide themselves with horses and the men were required to find a pair of ponies for each of the gun and ammunition wagons. The provision of these field guns encouraged the Artillery section to be active for a number of years but due, eventually, to the lack of firing practice, interest gradually declined and in October 1875 the guns and horse equipment were handed over to the British Resident in Perak who was, at that time, James Wheeler Woodford Birch the first British Resident to be stationed in that State.[21]

There had been a general decline in recruiting and attendance during the period leading up to the formation of the Artillery Battery and in a further attempt to encourage volunteering Captain W. H. Read, who, in 1864, had resigned because of his other Public commitments, was reappointed Captain Commandant on 4 August 1868.[22]

Captain Read, having reassumed command of the SVRC, found that his predecessor, Captain Wilsone, had organised a band consisting of drums, fifes and a bugler. By September 1868, the Drummers had reached the state of efficiency whereby Captain Read applied to the Government for Drums. In his application he did, however, comment that whilst he '. . . had never heard of a Rifle Corps with a Drum and Fife Band . . . he did not wish . . . to disturb the arrangements already in hand'.[23] It would appear that the request was refused as the Band continued to use borrowed instruments and practise under the instruction of the Drum Major of the Madras Native Infantry. This arrangement continued until that regiment left Singapore, when the SVRC Band was disbanded.

The Volunteers provided a mounted escort for the visit of HRH The Duke of Edinburgh to Singapore in 1869 consisting of the following personnel:

R. Dunman	McPherson
J. C. Ker	R. W. Maxwell
C. A. Maclaverty	Bligh
Rae	C. E. Velge
O'Laughlin	C. Phillips

A Guard of Honour of 62 men was also provided.[24]

The year 1869 brought further changes to the Corps. The first Volunteer Ordinance in the Straits Settlements was made law and became the 'Volunteer Ordinance 1869'. In the same year the Corps changed from the Enfield Muzzle Loading (Percussion) Rifle to the short Snider carbine which was chosen for the Corps by Colonel Macpherson.[25]

The provision of this rifle created a new interest and, with the accent on rifle shooting, the Corps increased its strength to over one hundred men. Competitions were arranged on the Old Race Course Range (a 400-yard range) and prizes for shooting were donated by Captain W. H. Read, Lieutenant R. Duff, and Sergeant C. B. Buckley (the author of the

ment. At that time patterns were submitted to the Governor of the Settlement for approval but unfortunately no record of these patterns can now be traced. Winsley recorded that the late Sergeant Major Charles Phillips, who was the drill instructor to the SVRC from 1866 to 1887, reported that the Minutes of the Volunteer Committee Meetings (which would, presumably, have given details of the dress) were deposited with the pikes and disposed of as 'lumber' at the time of the disposal of the pikes from the Police Station. The headdress for the Corps was, during the first twenty-five years of its service, a shako. The SVRC shako had, fixed to the front rim, a drooping horsehair plume, about five inches in height and dark green in colour, similar to the Light Infantry or Fusilier Regiments of the Regular Army. It would therefore appear that the Corps was somewhat freelance in its dress in that, although it was a Rifle Corps, it had probably, at one period, a red tunic as part of its Full Dress. The Full Dress Jacket was originally green (for the period 1857–67) and after 1886 it again became green and at that time the Corps adopted a Fusilier-type shako plume. It is possible that the original badge may have been a Fusilier Grenade, although Winsley was unable to find any accurate record or copy of the original SVRC shako badge.[31]

In 1887, the Jubilee Year of the reign of Queen Victoria, the SVRC had yet again dwindled to less than half a company and Major W. R. Grey,[32] who was then Captain Commandant, turned out a 'weak company' (in strength) which had been assembled for the purpose of the Jubilee Parade. Lieutenant C. B. Buckley reappeared for duty on this parade.

Mr W. G. St Clair, who was Editor of the *Singapore Free Press*, used his influence to urge the formation of a new Corps – a Volunteer Artillery Corps – to serve as the local reserve force to the Garrison. As a result of a meeting on 1 December 1887 between the Governor, Sir Cecil Clementi-Smith, GCMG, and a small committee headed by Mr St Clair, the Governor accepted the services of a roll of Volunteers for a Volunteer Artillery Corps. The Committee was requested to proceed with the formation. On 16 December 1887 a proclamation was made disbanding the Singapore Volunteer Rifle Corps.

The Singapore Volunteer Artillery Corps was formed by proclamation on 22 February 1881.[33] At the time of formation it had ninety-six volunteers on its strength and from 1 February, 1890, had, as Honorary Colonel Sir Cecil Clementi-Smith, GCMG. With the formation of this new Corps in the Colony the motto of the old Volunteer Rifle Corps was adapted to suit the new Volunteer Gunners who used as their motto 'In Oriente Primus'. The reason for the adoption of the motto was, according to the Singapore Volunteer historian, Captain Winsley, that the unit wished it to be known that it was the first British Volunteer unit to be formed in the Far East.[34]

Command of the unit was given to Major H. E. McCallum, CMG, late Royal Engineers, who was at that time the Colonial Engineer of the Straits Settlements. Mr W. G. St Clair joined as a Sergeant, having refused a commission, even though he was one of the main instigators in the formation of the Corps. It was he who led the Sub-Committee and persuaded the Governor to disband the old SVRC.

During the early days of the formation of the Artillery unit there were no Drill Hall facilities available and the Headquarters of the unit was located in the Town Hall which was where the Signalling, Squad and Carbine drills took place. The SVA General Committee held a monthly meeting in the Committee Room of the Singapore Cricket Pavilion.[35]

Gun drills were carried out on 7-inch rifled muzzle-loading guns which were located at Fort Siloso on Blakang Mati Island (Pulo Blakang Mati)[36] and later, in order to obtain extra efficiency, Volunteers were trained on the 8-inch breech-loading guns located at Fort

The officers of the Singapore Volunteer Artillery, 1893. Left to right: standing– Captain G. Bruce Webster; 2/Lieutenant E. M. Merewether; 2/Lieutenant J. Fabris; Lieutenant W. G. St Clair; Captain R. Dunman. Sitting– Captain and Paymaster F. W. Barker; Major H. E. McCallum; Lieutenant and Adjutant W. Jennings, RGA. Sitting front– Captain and Surgeon T. C. Mugliston; Lieutenant C. J. Davis. Courtesy of the Director, National Archives, Singapore.

The Singapore Volunteer Artillery with .405 Maxim gun, behind the Drill Hall, at the Fort Fullerton site during 1893. The 7-inch RML gun, drill mounted, can be seen in the background. Courtesy of the Director, National Archives, Singapore.

Tanjong Katong.[37] A training gun mounted at Fort Palmer was moved to a site behind the Master Attendant's office (the present General Post Office site), as it was found that Forts Siloso and Tanjong Katong were too difficult to reach for evening drill practices. The gun moved was a 7-inch RML type and it remained at that site for a number of years.[38]

An SVA Drill Manual was compiled by Major W. Brooke-Hoggan of the Royal Artillery, in 1888. Part I of the Manual dealt with the gun drill for the 7-inch RML gun and Part II with the 8-inch BL gun drill.[39] Twelve drills on the 7-inch RML gun qualified a member of the SVA to commence training on the 8-inch BL gun which was at Fort Tanjong Katong. Eight drills on this gun were considered to be a qualification. Members were required to carry out ten drills per annum under instruction of the Regular Army gunners stationed at Fort Siloso. The General Officer Commanding China and Straits Settlements, General W. G. Cameron, was particularly keen that the volunteers should train on as many types of weapons as was possible within the Colony.

Volunteers were enlisted by election. A prospective volunteer had to be proposed, seconded and balloted for by a general committee and this method of recruitment remained in force until the reorganisation of the Singapore Volunteer Corps in 1921 when enrolment was brought into line with the method used by the Territorial Forces in the United Kingdom. Promotion in the SVA, to all ranks, was by qualifying examination and this scheme also remained in force until 1921.

The authorised dress of the SVA in 1888, which was applicable to both officers and other ranks, was:

Headdress

Khaki helmet - surmounted by the Artillery metal ball, chin chain, blue puggaree with the SVA badge on the right-hand side.

Khaki Glengarry cap - with badge on the front left-hand side.

Uniform

Khaki tunic - of the patrol type with roll collar, later officers wore a white stiff collar on the inside.

Khaki trousers.

Gaiters - short blue gaiters (of the Marine Artillery type).

Boots - Black.

Officers wore Sam Browne belts with khaki uniform and all ranks carried swords.

Badges of rank were worn on the shoulder strap by officers and in addition, in the early years of the Artillery Corps, decorations of blue braid were worn on the sleeve, commencing with one or more rings about two inches above the cuff and continuing up almost to the elbow with looped and decorated braid. The amount of rings and braid increased with the rank of the wearer. The other rank uniform was also decorated on the arm with blue braid which, in their case, culminated with a 'Crows Foot' at the point. Captain Winsley believed these decorations to have been handed down from the dress regulations of the Singapore Volunteer Rifle Corps as the pattern used was identical to that used by the Rifle Volunteer units in the United Kingdom.

The date that this arm decoration was abolished is not accurately known but it would have been in the period 1893–95. This is confirmed by the existence of two pictures of officers of the SVAC, one dated 1893 and the other 1895. In the former the arm decoration is in evidence but, in the latter, officers are wearing tunics with two button cuffs and no braid.

Buttons were all of the regulation Artillery type and officers wore the bullet-shaped

The Singapore Volunteer Artillery. Major H. E. McCallum in the centre. Courtesy of the Director, National Archives, Singapore.

The drill hall was designed by the Commandant of the SVAC, Major H. E. McCallum, who was at that time Executive Engineer, PWD, Straits Settlements. The building was eventually moved in 1906 to the new Volunteer Headquarters at Beach Road and it remained in use at that site until 1933. At the time of the construction of the new drill hall at the Fort Fullerton site a gun emplacement was constructed for a 7-inch muzzle-loading gun in order that drills could be carried out without Volunteers having to travel to one of the Regular Army forts. The gun was eventually removed during the reclamation programme of 1907.

In March 1891, the Commandant, Major McCallum, returned to the United Kingdom for six months and during his absence command of the Corps was given to Captain R. Dunman. Winsley in his *History of the Singapore Volunteer Corps* gives the camp dress for 1891 as being 'white clothing; gaiters; sun hats (forage caps to be taken), regimental shirts, waist belts, swords and rolled coats . . .'[48] Most of the camps were under canvas and it is recorded that at the camp held at Fort Tanjong Katong between 5 and 11 November a messing allowance would be paid at the rate of 75 cents per diem for all personnel living in camp. Personnel not living in camp were required to pay their own messing charges.

On 4 April 1891, the four Maxim guns which were bought by public subscription finally arrived in the Colony on the SS *Glenshire* and Winsley comments that the acquisition made the Singapore Volunteer Artillery Corps the first unit in the British Army, both Regular and Volunteer, to form a Maxim Gun Company. The four guns were officially presented to the Corps on 29 May 1891 when the Corps paraded on the Esplanade to mark the Queen's Birthday. This was the first Birthday parade in which the regular units in Singapore had not been represented. The *Singapore Free Press* of 9 June commented unfavourably, saying '. . . Singapore celebrated the Queen's Birthday sporadically owing to the danger of letting British soldiers go out in the sun. It is fortunate we have so considerate a G.O.C. for, if the customary review had taken place, the whole battalion might have had to go to hospital, and left us defenceless . . .'[49]

The crews for the four Maxim guns were picked from members of the Corps who were already proficient in heavy gunnery and the drivers for the Company consisted, in this early stage, of personnel who possessed the proper-sized galloways or ponies in their private stables.

The Corps was inspected by Lieutenant Colonel Alexander Burton Brown, the CRA, in the absence of Major General Sir Charles Warren. The inspection was held at Fort Pasir Panjang and included gun drills. Little else is recorded for that year except that the Corps undertook several route marches during one of which they were accompanied by the Drum and Fife Band of the 41st Battery, Southern Division who were stationed at Fort Canning.

Camps for the year 1892 were again mainly under canvas, commencing with one on 19 January in the New Harbour area. Drills continued to be held on the 7-inch guns at Fort Siloso and when gun drills were held at Fort Pasir Panjang Winsley recorded that the volunteers '. . . previously made up the cartridges themselves at Fort Canning . . .'[50] The Lee-Metford rifle was brought into service by the SVAC during that year.

News of the Pahang rebellion first reached Singapore on 23 December, 1891 and Major H. E. McCallum departed for the 'front' in his Singapore Volunteer's uniform on 20 January 1892 in order to assist in the suppression of this small uprising. In April 1892, Lieutenant W. G. St Clair, who happened to be in Kuantan at the time, offered his services to John Pickersgill Rodger, the British Resident in Pahang, and was given command of '. . . a strong detachment of Perak Sikhs . . . with one 7-pr gun and its Pathan detachment . . .'

St Clair appears to have travelled to Kuantan with his SVAC uniform as in his memoirs he says '. . . Rodger asked me to go up [the Pahang River] in charge, which I did in SVA khaki, naturally . . .'[51] The cause of the Pahang uprising was the introduction of new taxes in the State to pay for the administration, and the law which banned the illegal tax collection by local chiefs. The first outbreak occurred when Bahman, a Chief in the District of Sungai Semantan, ambushed a Collector (Mr C. E. M. Desborough) and his party of fifteen Sikhs and six Malay policemen, on the Semantan River, at Lubok Trua on 15 December, 1891. Generally the uprising was over by the end of 1892, however a certain amount of unrest continued throughout 1893.[52] The war in Pahang was mainly confined to fighting small actions in the jungle and on the river with the Government forces pursuing the rebels. The rebels retreated into Trengganu and Kelantan States where they had sympathisers and fighting once again broke out in June 1894 when the rebels attacked and killed Mr E. A. Wise (Superintendent of the Ulu Pahang District) and wounded Captain H. Talbot of the Perak Sikhs. Four Sikhs were killed and four were wounded in this action. The situation remained serious until late 1895 when the rebel leader and some of his followers surrendered to the Siamese authorities and were taken to Siam. Two other rebels were reported as having been killed in Trengganu, these were To' Gajah and his son Mat Kilau.[53] Winsley recorded that only two Volunteers were involved in this campaign and commented that the Pahang campaign would give the Volunteers of 1938 a good grounding for 'defence operations on the East Coast of Malaya'. It is unfortunate that within a few years he was to be proved wrong as the punative Pahang campaign of 1892–95 was to be no blueprint for the operations against the Japanese army during its advance in 1941.

During 1893 the SVAC continued to carry out their normal volunteer training and in the following year, 1894, a new detachment was added to the Corps. This was the SVA Cyclist detachment whose dress was given as follows:

Khaki helmet
Khaki tunic
Khaki knickerbockers
Dark blue stockings
Brown canvas boots

The working dress of the unit, other than the Cyclists, was:

Khaki helmet, or field service cap
Khaki tunic, or regimental shirt (when shirt sleeve order was authorised)
Khaki trousers
Brown gaiters, Marine pattern
Brown boots.

Sun hats of the light pith beehive type were worn for drill and outside work when short sleeve order was authorised.[54]

Training routine continued unaltered during 1894 and 1895 but in 1896 various changes were made when Major H. E. McCallum departed for England and handed over command to Captain R. Dunman, who was promoted to Major; Lieutenant W. G. St Clair was promoted, in turn, to Captain. A farewell parade was held on 2 March for Major McCallum who in fact never returned to the Colony as whilst he was in the United Kingdom he was given another appointment.[55]

The unit was re-equipped during 1896 with six 2.5 RML 7-pounder mountain (screw) guns. Prior to his departure from the Colony Major McCallum had a land range constructed at Bukit Panjang on the Bukit Timah Road but this range was closed shortly after

The Old Gaol Bras-Basah Road, *c*1937, headquarters of 'F' Malay Company, Singapore Volunteer Corps. Courtesy of the Director, National Archives, Singapore.

The Old Gaol, Bras-Basah Road, 1977. Photograph by Mrs J. Harfield.

required to wait until 1911 before they could enlist. The establishment was increased to a Company of 120 men and it was then officially known as the Malay Company, Singapore Volunteer Infantry. The first Officer Commanding was Lieutenant G. S. Carver. The Governor allocated the old Gaol in Bras Basah Road, which had previously been allocated to the Chinese and Eurasian Companies, for use as the Company Headquarters.[64]

On 5 May 1911 it was proposed that the Maxim Gun Company SVA should be amalgamated with the SVA and for it to revert to its original role of a section of that unit. The move was defeated and it remained an independent company within the Corps. The following year the establishment of the SVA was reviewed and reduced from 107 to eighty-five all ranks. In fact at that time it was well under strength but as a result of a recruiting drive its strength increased from sixty-six to seventy-three all ranks.[65]

Another revised and up-dated edition of the SVA Handbook was published in 1912 and this covered not only training but administrative information as well. It contained detail of which items of clothing and equipment were provided free by the Government and also gave the various forms of dress, such as Drill Order, Review Order, Manoeuvre Order and Mess Dress. The training side of the book gave information on the drill for the 6-inch QF gun, how to use the Depression Range Finder, a section on sights and one on ammunition. It was, in the words of Winsley, '. . . Altogether a very handy little book of 39 pages . . .' In the same year training facilities were improved for the SVA in that a 6-inch QF gun was installed in a mounting within the compound of the Drill Hall at Beach Road.

New Regulations for the SVC were gazetted in December 1912 and in 1913 the Bearer Section SVA which had an establishment of one officer and ten other ranks was reorganised and became the Medical Company SVC.

During the period 1908 to 1912 there had been several changes in the dress of the various units of the Corps. An order dated 18 September 1908 authorised the issue of a new cap badge for the SVA. This badge was a gun with two scrolls, one above the gun bearing the motto 'In Oriente Primus' and the one below the gun 'Singapore Volunteer Artillery'. This badge was worn from 1908 until 1915 when a 'King's' crown was added above the top scroll. The new 'gun' badge of 1908 was worn only by the volunteers serving in the SVA. The old SVA Cypher badge continued to be worn by the Maxim Gun Company SVA until 1921. Members of the company also wore the 'SVA(M)' shoulder titles and SVI buttons.

With the introduction of the khaki field service cap the Corps introduced its own embellishments. The headdress worn was the khaki cap, with a black patent leather peak and chin strap with the following distinctions:

SVA and SVA(M) – Red band around cap with blue piping around the rim of the crown.

SRE(V) – Blue band around cap with red piping around rim of the crown.

Eurasian Company SVI – continued to wear the khaki forage cap.

Chinese Company SVI – Khaki field service cap, without peak with a segment of green cloth as a background for the badge.

Malay Company SVI – which was formed in 1910, wore the khaki field service cap, without peak, but with a band of red cloth.

With the outbreak of the First World War the Singapore Volunteer Corps was mobilised on 6 August 1914 and the strength at that time was given as being 450 all ranks. The units that were actually mobilised were:

Headquarters, Singapore Volunteer Corps

Singapore Volunteer Artillery

Maxim Gun Company SVA

HEADDRESS
SINGAPORE VOLUNTEER CORPS

The Slouch Hat
worn by the SVI

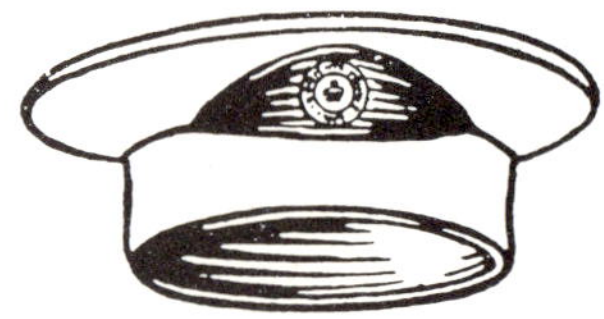

The Chinese Cap
Singapore Volunteer Infantry

The Malaya Company Cap
Singapore Volunteer Infantry

The Armoured Car Beret

Compiled from *A History of the Singapore Volunteer Corps* by Captain T. M. Winsley, SVC, published Singapore, 1938.

Singapore Royal Engineers (Volunteers)
Singapore Volunteer Infantry (Chinese Company)
Singapore Volunteer Infantry (Malay Company)
Medical Company, SVC
Singapore Volunteer Corps Band

The day after mobilisation a further company was formed which was the Cyclist Company, which was made up of two sections, a Motor-cyclist Section and a Cyclist Section.

One of the outposts from Singapore but under the administration of the Straits Settlements was the island of Labuan. The few Europeans residing on the island felt the need to have some official means of defending themselves and their property and it was agreed that a Detachment under the control of the Singapore Volunteer Headquarters could be formed. The new detachment, although they regarded themselves as a unit, was given the title of Labuan Defence Corps. It was probably the smallest volunteer unit to have been formed in the Far East. Winsley records that it was armed with Long Enfield rifles and the total number issued to the unit, on the basis of one per volunteer was twelve.

It became evident, during the latter part of 1914, that there were insufficient volunteers to carry out all the duties that were required of the Corps and on 23 October 1914 it was recommended that the old Singapore Volunteer Rifles, which had been disbanded in 1903, should be reformed. A European Company was formed under the command of Captain C. W. Darbyshire and later a Veterans Company was also formed. At the end of 1914 the total strength of the Singapore Volunteer Corps had risen to 687 all ranks.

NOTES TO CHAPTER TWELVE

1. Captain T. M. Winsley, SVC, *A History of the Singapore Volunteer Corps*, (Government Printing Office, Singapore, 1938).
2. Charles Burton Buckley, *An Anecdotal History of Old Times in Singapore 1819–1867*, (published 1902, reprinted by University of Malaya Press, 1963).
3. Governor's Diary Military, 1852–57, entry Nos 4 and 5 of 10 July 1854, (National Library, Singapore).
4. Buckley, op cit, p619.
5. Winsley, op cit, p5.
6. Ibid, p6.
7. Ibid, p7.
8. Ibid, p4.
9. Ibid, p5.
10. Buckley, op cit, p645.
11. Ibid, p665.
12. Straits Settlements Government Gazette, No47, Friday 19 November, 1858, (National Library, Singapore).
13. Ibid, No48, 26 November, 1858.
14. Winsley, op cit, p8.
15. Buckley, op cit, p682.
16. Governor's Diary, Military Foreign and Military, entry No2 of 5 October 1861, (National Library, Singapore).
17. Walter Makepeace, (ed), *One Hundred Years of Singapore*, vol II, (J. Murray, 1921).
18. Buckley, op cit, p787.
19. Walter Makepeace, (ed), *One Hundred Years of Singapore*, vol I, (J. Murray, 1921).
20. Makepeace, vol I, op cit, pp 385 and 403.
21. James Wheeler Woodford Birch, British Resident at Perak was later assassinated by Malays at Passir Salek on 2 November 1875.
22. Winsley, op cit, p11.
23. Ibid, p12.

bungalows damaged. In the meantime, Charles Adair Crymble, who was in command of the fort, defended the place against a large mob of Chinese. Crymble defended the fort with his few Fortmen and only abandoned the fort when he was the only survivor. An account of his actions records that '. . . although badly wounded, he then leaped over the parapet, miraculously avoided numerous sword and spear thrusts from the insurgents and escaped into the river . . .'[17]

The Chinese were eventually put to flight by the arrival of the Borneo Company's vessel *Sir James Brooke* which fired a few shots and the Chinese dispersed. Many of the Chinese were tracked down by Malays and Dayaks and were killed. Sir James Brooke was able to restore order in the town and rebuilt his residence and the fort.

In 1861 he eventually succeeded in defeating Rentap, a rebel Dayak chief, who had defeated various Government forces sent to subdue him and his followers. As a result of this fairly lengthy campaign, Rajah Brooke realising that he had a need for a stronger force than his Fortmen, raised in 1862 a new and larger military unit which he named 'Sarawak Rangers'. The new force was made up of 120 men, almost all of whom were Dayaks, split into two groups, one stationed at Kuching and the other recruited and used in the coastal area.[18]

Sir James Brooke, the first White Rajah, died at his home in Devon on 11 June 1868 and was succeeded by his nephew Charles Brooke who had been Rajah in everything but name for five years before the death of his uncle. Charles Brooke married in October 1869 and returned to Sarawak with Ranee Margaret Brooke early in 1870 and took up residence in the new palace which was described as being a '. . . handsome brick structure, more befitting the residence of a Rajah and worthy of the name Astana (Palace) . . .'[19]

During the reign of Rajah James Brooke a few small fortifications had been constructed away from Kuching. A fort was built at Skrang in 1849 and was known as Fort James, after the first Rajah, although no details of the type of construction have been found. In 1864 Fort James was considered to be redundant in that location and was dismantled and the majority of the material moved five miles down river, to Simanggang where it was re-erected. It remained unnamed for many years but after the arrival of Ranee Margaret in 1870 it was named after her, using her second Christian name Alice. Fort Alice was the only fort of its type in Sarawak, and in fact in Borneo. It was square in shape with small towers on the inland corners. The construction was entirely of 'belian' timber (ironwood) and had an open courtyard in the centre with steps in the form of a drawbridge which could be raised and which were drawn up at night or in the event of an emergency. It was garrisoned by a strong detachment of Sarawak Rangers, although the records do not show exactly the strength of the detachment.

The original Fort James at Skrang was moved as it was found that the site chosen was on low-lying ground and was liable to flooding. The site for the second fort, Fort Alice, was on a small hill rising from the river bank which afforded a good view of the country and also commanded a long stretch of the river which was the main highway of Sarawak.[20]

The reason for the location of forts in this area was the suppression of piracy and in an attempt by Rajah James Brooke to stop the Dayaks carrying out headhunting attacks on their weaker neighbours the Balau. In fact, Rajah Brooke attacked the rebels in that area on the night of 31 July 1849 and with the help of three British warships subdued the pirates, immediately after this a Mr Crookshank was sent to build the fort at Skrang which he did with the help of the friendly tribes. The fort was placed in care of Sherip Matusain, a Malay trusted by the Rajah, but he did not have the confidence of the local tribes and as a result a

Mr Brereton was sent to take charge. He was later assisted by a Mr Alan Lee. Brereton died in 1854 and for a while Charles Brooke, who was to become the second Rajah, replaced him and he in turn was relieved by Mr Watson who remained in charge of the fort until it was moved to Simanggang.[21]

There is an interesting legend relating to Skrang which pre-dates the fort. It is believed that the site of the fort was attacked by Captain the Hon Henry Keppel, and a landing party from HMS *Dido* which included some redcoated soldiers (Royal Marines). The attack was made in 1844 and during the fight with the Skrang pirates some of '. . . his Redcoat British soldiers were killed there. This later led to a belief among the Dayaks that the long-nosed monkeys are the spirits of these soldiers . . .'[22]

Fort Alice celebrated its hundredth year in 1964 and it was recorded, at that time, that the old tradition lived on in that '. . . a call [was] made by a policeman after he had struck the eight o'clock gong. This was the evening call when the Resident used to have his dinner and the drawbridge drawn up for the night . . .'[23] Originally the call would have been made by a Ranger and not a policeman. The translation of the call is as follows:

'Oh Ha! Oh Ha! Oh Ha!
The time is now eight o'clock,
The steps have been drawn up,
The door is closed
People from up-river,
People from down-river,
Are not allowed to come up to the fort any more . . .'

During 1871 Rajah Charles Brooke introduced a uniform for his unit the Sarawak Rangers which consisted of:

White jacket, single fronted, with stand collar, epaulettes, three rows of black braiding across the front of the jacket and with black braiding with a single loop above the cuff.
White drill trousers, plain.
Black cap, (worn by the Kuching company)
Black head handkerchief (worn by the coastal based company)

The walking-out dress at this time was national dress, which included the sarong and the parang. It is recorded that after a party of Rangers had raided an orchard in Kuching and the owner attempted to stop them the Rangers used their parangs and 'took his head'. Later the walking out dress precluded the wearing of the parang. The new approved dress was of the same design as the working dress except that it was of blue and red in place of black and white. No footwear was worn, but a plain belt was worn around the waist of the jacket.

In 1877 a Major Rodway was appointed as Commandant of the Sarawak Rangers and in the following year he personally supervised the planning and the construction of a new Fort at Kuching. The fort was completed in 1879 and was named Fort Margherita, after the Ranee. In the same year the Rajah issued an order amalgamating the Sarawak Rangers and the Fortmen into one force which was then given the title '*The* Sarawak Rangers'. They retained the dress that had been introduced in 1871 and this continued to be worn until the disbandment of the force in 1932 when the Rangers and the Police were disbanded and re-formed into a smaller, combined force which then became known as the Sarawak Constabulary.

At the time that Fort Margherita was built, a smaller fort was constructed across the Sarawak River, on the town side, at Pankalan Bato. This fort was a small box-shaped building without a courtyard or outer defences and was rather similar to a Norman keep.

Eastern rampart and tower, Fort Margherita.

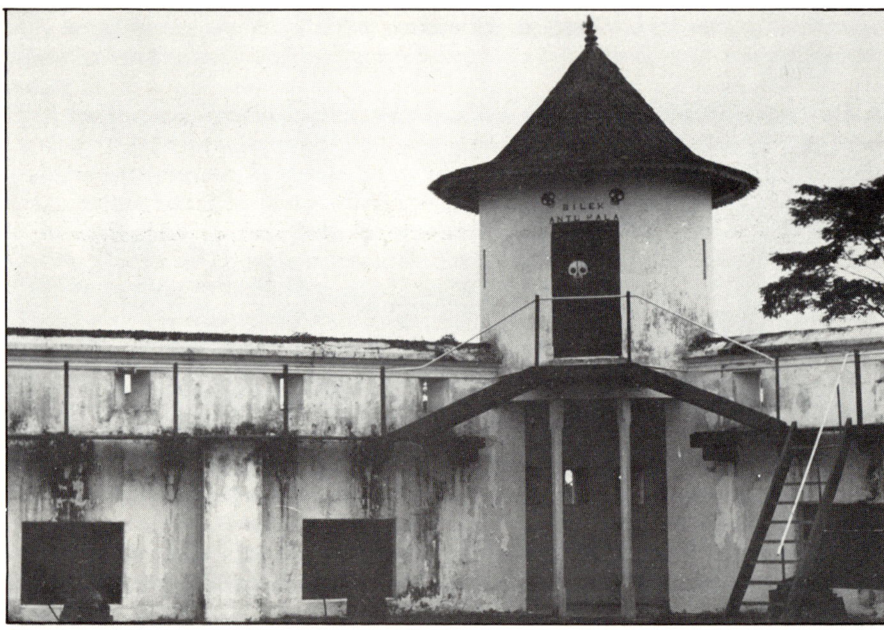

Eastern rampart and tower, showing the courtyard and gun ports, Fort Margherita. Photographs by Mrs J. Harfield.

Fort Margherita, south-western corner showing ramparts and overlooking the Sarawak River.

Fort Margherita, showing fire-step and gun ports.
Photograph by Mrs J. Harfield.

the forts were provided with armouries and quarters for the Fortmen on the ground floor and on the first floor the office and quarters for the District Officer. In the early days wives of officers were discouraged from living 'up-country' and remained in Kuching. The District forts are listed showing the location, name and the person after whom it had been named:

Location	*Fort*	*Namesake*
Bentong	Fort Lili	Ranee Margaret Alice Lili Brooke, wife of the second Rajah.
Kabong	Fort Charles	Rajah Charles Brooke. Built in 1878 it was destroyed in 1893 by sea encroachment and was then rebuilt further inland.
Belaga on the Rejang River	Fort Vyner	Son of the second Rajah, Charles Vyner Brooke was to become the third Rajah. The fort was started in November 1883 and was completed in six weeks and ready for occupation.
Kanowit	Fort Emma	Rajah Charles Brooke's mother. Other than a reference to the name of the fort and the location no other details have been found.
Muka	Fort Burdett	Named after Baroness Angela Burdett-Coutts, long-time friend and benefactress of the first Rajah.
Bintulu	Fort Keppel	Named after Captain, The Hon Henry Keppel, captain of HMS *Dido*, and long-time friend of Rajah James Brooke, later to become Admiral Keppel.
Baram River	Fort Rose	Named after Doctor Rose, the second Resident of the Baram River administration centre.
Trusan River	Fort Florence	Named after Mrs Florence Maxwell, wife of F. R. O. Maxwell who was the Rajah's deputy at the time that the Trusan district was transferred to Sarawak in 1885.

Rajah Charles Brooke left Sarawak for the last time on 5 December 1916 and died the following year on 17 May just a few weeks before his eighty-eighth birthday. News of his death reached Kuching by wire on Friday, 18 May and flags were immediately flown at half-mast.[28]

The Sarawak Ranger Band achieved its crowning glory in 1922 when it was present at the Borneo–Malaya Exhibition, which was attended by HRH The Prince of Wales. The exhibition was held at Singapore and the Band entered a band competition which had entries from British Army and Police bands which were serving in Singapore and Malaya. The first prize was won by a British Army band but the Sarawak Rangers gained second prize and were presented with a silver medal, three inches in diameter, bearing the head of the Prince of Wales on one side and an inscription on the other. The medal was proudly displayed in the office of the Commandant until the disbandment. Afterwards it was in the Office of the Commissioner of Constabulary but was lost during the Japanese occupation of Kuching.[29]

The Sarawak Rangers continued to be employed on guard and ceremonial duties within the state during the years following the First World War. The War Office publication *Notes on the Land & Air Forces of the British Overseas Dominions, Colonies, Protectorates & Mandated Territories* published in 1930 shows the military force of the State of Sarawak as follows:

'. . . His Highness the Rajah's Regiment – The Sarawak Rangers. This is a regular regiment of the State of Sarawak, in the service and under the entire control of H.H. the Rajah . . .'[30]

It also gives details of the constitution, etc:

'Constitution – The Regiment was constituted by order of His Highness the Rajah in 1862.

Enlistment – All enlistment is voluntary. A soldier enlists for a first period of five years, with re-engagements up to twenty-one years' service. The regiment is recruited entirely from the Dayak tribes of the State.

Terms of Service – The Regiment is liable for service within or beyond the limits of the State of Sarawak as His Highness the Rajah may be pleased to direct.

Discipline – All officers, N.C.Os and men are subject to military law and regulations as far as is compatible with local conditions . . .'[31]

Details of the establishment and armament were also given:

'. . . Establishment (War Strength)

Officers	4
Other ranks	608
	612

All officers are British Europeans, and are selected and commissioned by His Highness the Rajah.

Regimental headquarters and depot are at Fort Margherita, Kuching.

Armament – Fort Margherita consists of two batteries small M.L. guns, and one 3-pdr Hotchkiss for saluting purposes only.

The machine gun platoon is armed with Lewis and machine (German pattern) guns.

All ranks are armed with the S.M.L.E. rifle . . .'[32]

The training of the regiment was specified as being:

'All recruits are trained at the Depot, Fort Margherita, Kuching.

The annual training programme for the Regiment commences as from 1st January each year. All training is carried out on identically the same lines as in British units as far as local conditions permit. An annual range course is fired by all ranks, as laid down in Small Arms Training [pamphlet] . . .'[33]

The Commandant was always an ex-British Army officer and the instructor an ex-member of either the Army or Navy but was usually an ex-Sergeant of the Army.

It is recorded that the Rangers took part in over twenty expeditions, quelled a riot in Miri in 1923 and were actively engaged in an expedition against a rebel named Asun, in the Entabai district in 1932. However, before this expedition could be brought to a successful conclusion, Sarawak was hit by a slump and there was a need for major economies and as a result of this the Sarawak Rangers and Police were disbanded and a small force named the Sarawak Constabulary was formed. It had been hoped that many of the Rangers would volunteer to serve in this new combined military–police force but few would volunteer as they felt that to do so would be lowering their status and so the majority of the Rangers returned to their longhouses in the jungle. As a result of this a number of the ex-Rangers joined forces with Asun and then harassed the very Government that they had previously served.[34]

Fort Margherita was then taken over by the Sarawak Constabulary. After the Second World War the Fort was used as the Police Officers' Mess. By 1977 it had been converted

Sarawak Military Long Service Medal.
Photographs loaned by H. C. Pownall Esq, QC.

Cannon on display at Fort Margherita, note the gun is incorrectly mounted on the carriage. Photograph by Mrs J. Harfield.

into a Ranger and Police Museum and housed a good collection of weapons, swords, uniforms, and equipment of both forces. The prisoners' cells had been reconstructed as they had been when they were used as 'condemned prisoners' cells' and executions had been carried out at the fort.

A book entitled *Old Sarawak a Pictorial Study* by Craig A. Lockwood and Graham E. Saunders was published by the Borneo Literature Bureau in 1972. This book contained two illustrations which are of interest to the military historian but unfortunately they are undated. One is of the inside of Baram Fort at Marudi and the other shows the Sarawak Rangers on parade. Regrettably the latter portion of the book lacks pagination and although it would appear that both photographs are from the Sarawak Museum collection this is not made clear in the acknowledgement section of the book.

The Sarawak Rangers, unlike the Singapore Volunteer Corps, was a 'Regular' unit and as such the men serving in the regiment were, from 1908, able to qualify for the award of the Good Conduct Medal. The Sarawak Government Gazette dated 16 June 1908 published an order 'No VII, 1908' which authorised '. . . All men of 18 years service, provided that during that period they shall have had no serious crime entered against them, shall be entitled to wear a Good Conduct Medal . . .' They would also receive an increase of pay.[35] The Government Gazette dated 16 January 1919 stated that '. . . After 1st January 1919 all men irrespective of rank of the Sarawak Rangers and the Sarawak Police Force shall be eligible for a Good Conduct Medal provided they have served 18 years and that during that period they have had no serious crime entered against them . . .'[36]

From 1924 the system changed and the military, the Sarawak Rangers, were awarded a Military Long Service Medal. The award was authorised in Government Gazette dated 16 December 1924 and set out the regulations as follows:

> '. . . A Military Long Service Medal, the award of which shall be confined to members of the Sarawak Military Force.
>
> . . . Award of the Long Service Medals shall only be made after eighteen years . . . and shall be subject to production of a certificate of good conduct, clean record and satisfactory service.
>
> . . . A bar shall be granted for every further period of ten years of approved service.
>
> . . . [The Medals] shall be presented once a year only, on parade on His Highness the Rajah's Birthday . . .'

The award of the Military Long Service Medal existed for only a short period as the Sarawak Rangers were disbanded in 1932. The only record of the actual number of new medals awarded is contained in the Sarawak Government Gazette. The Gazette authorises the following awards on the dates shown: 24 September 1925 – 1; 1 October 1930 – 3; 1 October 1931 – 6.[37]

NOTES TO CHAPTER THIRTEEN

1. Aristide J. G. Papineau (ed), *Papineau's Guide to Malaysia*, (Andre Publications, Singapore, 1975).
2. Elizabeth Pollard, *Kuching Past and Present*, (Borneo Literature Bureau, 1972), p3.
3. Dr S. Santoso, *Standard Malay–English Dictionary of National Language*, (Phoenix Co, Singapore, 1972), 'Kuching – Cat'.
4. John Crawfurd, *A Descriptive Dictionary of the Indian Islands & Adjacent Countries*, (Oxford University Press, Kuala Lumpur, 1971), p377.
5. Steven Runciman, *The White Rajahs*, (Cambridge University Press, 1960), pp 46 and 47.
6. East-India Register and Directory for 1820, pp 74 and 75.

7. Ibid, for 1822, pp 50 and 51.
8. Runciman, op cit, p47.
9. The Hodson Index, (National Army Museum, London).
10. Runciman, op cit, pp 47 and 48, and
Robert Payne, *The White Rajahs of Sarawak,* (Robert Hale Ltd, London, 1960), p20.
11. East-India Register and Directory for 1833, (Bengal) p273.
12. Pollard, op cit, p4.
13. Captain, The Hon Henry Keppel was to become Admiral Sir Henry Keppel and his many connections with the East Indies resulted in the New Harbour at Singapore being renamed 'Keppel Harbour' in honour of this well-known naval officer.
14. Pollard, op cit, p8.
15. W. J. Chater, *Sarawak Long Ago,* (Borneo Literature Bureau, 1969), p8.
16. Pollard, op cit, p20. The letter was written by Harriette McDougall wife of the Reverend Francis McDougall, Bishop of Labuan and Sarawak, to her son Charley.
17. Chater, op cit, p94. Charles Adair Crymble assumed command of the fort in 1851 and died at Kuching in 1862. The stone on his grave gives the following dedication by the Rajah.

'THIS STONE WAS LAID BY
THE RAJAH AS A MEMORIAL
OF HIS PUBLIC SERVICE,
PRIVATE WORTH AND TRIED
FIDELITY'

18. Ibid, pp 11 and 12.
19. Ibid, p21.
20. Ibid, pp 28 and 29.
21. Ibid, p30.
22. Ibid, p29.
23. Ibid, p32.
24. Pollard, op cit, p37.
25. Ibid, p37. The map of Kuching dated 1870 illustrated on this page shows the Fort marked at the end of the 'Main Bazaar' at the junction of Rock Road. The Ranger Barracks are shown as being on the west side of Rock Road south of Carpenter Street.
26. Chater, op cit, p15.
27. Ibid, p17.
28. Pollard, op cit, pp 70 and 72.
29. Chater, op cit, p18.
30. *Notes on the Land & Air Forces of the British Overseas Dominions, Colonies, Protectorates & Mandated Territories (Exclusive of India and Iraq),* War Office Publication 26/Publications/1691, (HMSO, London, 1930), p92.
31. Ibid.
32. Ibid, p93.
33. Ibid.
34. Runciman, op cit, p240. By the end of the year 1932, Asun gave himself up to the authorities and was sent to a comfortable exile at Lundu. In 1933 he was allowed to return to Kuching but confined to the left-hand bank. As a result of the rebellion a group of young Dayaks (Dyaks) broke away and became head-hunters. They were outlawed but not entirely rounded up until 1941.
35. J. M. A. Tamplin, 'The Miscellany of Honours - Long Service Awards in Sarawak', *The Orders and Medals Research Society Journal,* No4 1982, p49.
36. Ibid, p50.
37. Ibid, p54.

CHAPTER FOURTEEN

The Mutiny in Singapore

The outbreak of the war in Europe in 1914 had, at first, little effect on the daily life of the Europeans within the Straits Settlements. Some of the younger men who were working for trading companies returned home to the United Kingdom so as to volunteer for Kitchener's army. The first indication of trouble in the East Indies came on 21 September 1914 when the German cruiser *Emden* first appeared in the Bay of Bengal and carried out a series of lightning raids on shipping and coastal installations in that area. Madras was shelled on 26 September and then the *Emden* moved further east and on 28 October 1914 entered Penang harbour and sank a Russian cruiser and a French destroyer.[1] A landing party was put on shore but this was captured and the German sailors became the first prisoners of war in the East Indies.

A few days before the attack on Penang Harbour, it had been decided that the Singapore Volunteer Rifles should be embodied and about 200 men of the Volunteers paraded and then marched to Fort Canning to take up duties within the garrison. Some of the Volunteers were required as guards for the camp which had been set up to hold those German nationals that had been placed under restriction shortly after the outbreak of war. All German nationals had been taken into custody, German ships in the harbour had been seized and the crews also taken into custody.

Prior to the outbreak of war the Singapore garrison had consisted of two regular regiments, one British and one Indian army regiment. There were, of course, Royal Artillery personnel manning the various forts. The British regiment, the 1st Battalion, King's Own Yorkshire Light Infantry, departed under orders for France towards the end of 1914 so that the garrison, in February 1915, consisted of the one Indian Regiment, the 5th Indian Light Infantry, a regiment of long-standing. It had been raised at Cawnpore, India in 1803 by Lieutenant F. M. Johnson, as the 2nd Battalion, 21st Regiment. It became the 42nd Regiment in 1824 and the 42nd Light Infantry in 1842. It was redesignated 5th Bengal Light Infantry in 1861.[2] In addition, troops of the Royal Garrison Artillery, Royal Engineers, together with small detachments of the 36th Sikh Regiment, Malay States Guides and Malay States Volunteer Rifles were assisting in carrying out the garrison duties. To help ease the position 200 men of the Johore State Forces were posted to Singapore on 14 February and were stationed at Tanglin Barracks. The detachment was under command of Captain H. Cullimore and Captain Abdul Jaffar. The 100 men of the Malay States Guides and Malay States Volunteer Rifles were under the command of Captain Sydney Smith and were billeted at a training camp at Normanton, about a quarter of a mile from Alexandra and adjacent to the Ayer Rajah Road. The Prisoner of War and Enemy Subjects Internment Camp was located at Tanglin Barracks, near Pierce Road. The guard for this camp was provided by the Singapore Volunteer Rifles under command of Second Lieutenant J. Love-Montgomerie.

Singapore was in a relaxed mood on Monday 15 February as it was New Year's Day in

Quarterguard of the 5th Light Infantry at Singapore, circa January 1915.
Courtesy of the *Changi Informer,* November / December 1971.

Singapore Gaol, *c*1915. Courtesy of the Director, National Archives, Singapore.

the Chinese lunar calendar and the whole weekend, including the Monday, was considered to be a holiday. The senior and elderly European population were either relaxing at home or at their club. The day that had started bright and sunny was by the afternoon dull with occasional light drizzle. It was hot and humid and the combination of the weather and holiday had lulled the guards into a false sense of security.

The 5th Light Infantry had been inspected during the morning by Brigadier-General D. H. Ridout as on the following day the regiment were to leave Alexandra Barracks and embark on the troopship *Nore* for Hong Kong. General Ridout had been pleased with the inspection and had complimented Lieutenant Colonel E. V. Martin on the turnout and bearing of his battalion. He addressed the sepoys and praised them for their past service and wished them good luck in their new posting, although he did not actually mention that the '5th' were destined for Hong Kong.[3] At this time Singapore was regarded as a military backwater and Regular troops were anxious to move westwards to Mesopotamia, or even France and whilst the British officers were keen to get into action the sepoys were not keen to go to Mesopotamia as they believed that the fight against the Turks was the start of a war against Islam. There were, within the regiment, a number of sepoys who stirred up this discontent. The fact that Brigadier General Ridout failed to mention that the regiment was bound for Hong Kong meant, as far as the rebellious sepoys were concerned, a clear indication that the regiment was bound for the Middle East.

In addition to this unrest, there was further discontent in the regiment as one havildar had expected to be granted a Viceroy Commission but had failed to be appointed. The men split into factions, those that supported the havildar in his claim and were bitter about his failure to become a VCO and another faction who were pleased that this ambitious soldier had been passed over.

The machine guns and ammunition had been sent to the *Nore* prior to the embarkation of the regiment and after the parade during the morning of 15 February the small arms ammunition had been collected at the Quarter Guard.[4]

At about 3.00 pm a shot was fired at the Quarter Guard of the 5th Light Infantry and the mutiny broke out. It was sudden and apparently completely unexpected. The mutineers took possession of all the ammunition which was, at that time, being packed into lorries under the supervision of Lieutenant H. S. Elliott. He and Captain P. Boyce attempted to quell the outbreak and were shot and the drivers of the lorries were chased away. The mutineers quickly gained possession of the whole camp as well as the ammunition. Captain M. F. A. MacLean, RGA, attached to the Mountain Battery of the Malay States Guides, was shot and killed by the mutineers whilst in his quarters. From this moment the mutineers left the immediate vicinity of the camp and commenced to terrorise the European residential area of Singapore.

Some of the British Officers of the 5th Light Infantry made their way from their quarters to the Indian Officers' Lines but were persuaded by the VCOs that it was impossible to stop the men. They then made their way to the camp of the Malay States Volunteer Rifles at Normanton. These three officers, Major W. L. Cotton, Captain L. P. Ball and Lieutenant W. D. Hall, together with the detachment of eighty-one men of the Malay State Volunteer Rifles under command of Captain Sydney Smith, withdrew from the advancing mutineers to the confines of the bungalow of the Commanding Officer of the 5th Light Infantry, Lieutenant Colonel E. V. Martin.[5]

The Civil Police were informed by Captain Ball and this information was quickly passed to the Colonial Secretary. Mrs Cotton, the wife of the second-in-command of the regiment had already taken refuge in the bungalow. The party proceeded to put the bungalow into a state of defence. The bungalow was quickly surrounded by the mutineers and telephone wires cut so that the besieged party were unable to keep in touch with the military Headquarters at Fort Canning. Colonel Martin had only been able to give a brief warning of the outbreak prior to the wires being cut. Eighty men of the 5th Light Infantry reported to the Commanding Officer and offered to join in the defence of the bungalow but they were directed to report to the nearest Police Station and declare themselves as being 'loyal', which they did.

The remainder of the mutineers divided themselves into two main bodies; one advanced towards the town area by the Pasir Panjang Road; the second party moved across country towards the barracks at Tanglin. The first party encountered four European civilians and murdered them; these were Mr and Mrs G. B. Woolacombe, Mr C. V. Dyson and Mr W. J. Marshall.[6] A section of this party turned away from the town and proceeded towards Pasir Panjang village. They then entered a bungalow and shot and killed Mr D. McGilvray, who was a corporal in the Singapore Volunteer Rifles but not on duty at the time, and in the same building found Mr E. O. Butterworth and Mr J. B. Dunn and killed them both. Meanwhile in Alexandra Road the mutineers met and killed three more Europeans, these were Mr N. F. Edwards, Mr H. B. Collins and Private W. H. Leigh, who was of the Singapore Volunteer Rifles but not on duty at the time of his death.

News of the mutiny had, by this time, alerted the Army Headquarters at Fort Canning and it was agreed with the naval authorities that a landing party of ninety officers and men was to be provided by HMS *Cadmus* which was in port at that time.[7] The landing party came ashore and intercepted the main body of mutineers in the city suburbs and prevented them from proceeding any further into the city.

The second party of about 100 mutineers reached Tanglin Barracks at about 4.00 pm and caught the garrison forces there completely unprepared. Although rifle shots had been fired, the guard had mistaken these for Chinese fire-crackers which were being let off frequently throughout the day as part of the Chinese New Year celebrations. They arrived at the camp just as the warning message was received by telephone. Second Lieutenant J. Love-Montgomerie, who was in charge of the section of the Singapore Volunteer Rifles guarding the POW and Internment Camp was killed as was Captain P. N. Gerrard of the Malay States Volunteer Rifles, who was serving as Commandant of the Camp.

The mutineers continued the attack on the camp and killed Captain H. Cullimore and Captain Abdul Jaffar of the Johore State Forces, and four soldiers of the Singapore Volunteer Rifles, namely Lance Corporal J. Harper, Privates F. S. Drysdale, A. J. G. Holt and B. C. Cameron. Private Jacob bin Salleh and Private Yakub bin Fali of the Johore State Forces were also killed in the defence of the camp. Two regular British servicemen were killed during this fierce but short action they were Sergeant E. H. Sexton, ASC, and Corporal J. R. V. Beagley, of the Royal Garrison Artillery.

The mutineers threw open the prisoner of war camp and then withdrew leaving Corporal G. O. Lawson, of the Singapore Volunteer Rifles, who was at the camp but not on duty, dead, three other soldiers of the Volunteers wounded and one German POW, E. F. Senftleben, dead. Seventeen prisoners-of-war escaped, although four were recaptured almost immediately. Four of the remaining eleven POWs who successfully escaped were from the German cruiser *Emden*. The landing party from the *Emden* that had been captured at Penang had, prior to February 1915, been transferred from Penang to the Internment Camp at Singapore. One of the prisoners who escaped was Lieutenant Jules Lauterbach, the former navigating officer of the *Emden*. He and some of his crew eluded capture, crossed the Straits into Johore, and eventually made their way to Shanghai.[8]

A small party of the mutineers penetrated the town near to the Sepoy Lines. In Outram Road they shot and killed Captain F. V. Izard and Major R. H. Galwey, both of the Royal Garrison Artillery. The party then saw and attacked Inspector Meredith of the Straits Settlements Police, however, although the mutineers killed his horse, he was able to make good his escape. At this time they shot and wounded two Chinese civilians who were unnamed. The mutineers then proceeded along New Bridge Road where they held up, shot at and killed three civilians and wounded a fourth. Dr E. D. Whittle, of the Government Medical Service, was killed and his wife wounded. Mr G. Wald, who was a Sergeant in the Singapore Royal Engineer Volunteers, was shot and killed as was Mr L. P. C. Smith and a car syce (driver) Hassan Kechil bin Hassan.[9] Near to the Singapore Jail, Warder J. Clarke was shot and killed and the party then attempted to obtain the keys of the Jail from the native warder but failed to do so. They then proceeded as far as the Central Police Station where they opened fire on the Sikh Guard. By this time it was almost dusk so the party then dispersed. The authorities were now quickly bringing order out of the initial chaos created by the mutiny. The Volunteer guard on King's Dock moved out and helped relieve the *Cadmus* Landing Party. The Sikh Police concentrated at the Central Police Station and the Singapore Volunteer Corps mobilized at their Drill Hall. Parties were organised to bring in women and children from the suburbs and accommodation was provided for them on the steamer *Ipoh*.[10]

Martial Law was proclaimed at 6.30 pm and pickets were set up at Tanglin Crossroads and at the end of Cluny Road. Troops of the Royal Garrison Artillery were brought in from the outlying forts and a headquarters was set up on the P&O Wharf to control the operation. The mail train from Penang arrived at Tank Road Station with 150 men from the Johore Military Forces under the personal command of His Highness, The Sultan of Johore.

Lieutenant Colonel C. W. Brownlow, of the Royal Garrison Artillery, with eighty men of the HMS *Cadmus* landing party, fifty men of the Singapore Volunteer Corps and twenty-one men of the RGA, proceeded to the junction of the Pasir Panjang and Alexandra Roads in an attempt to relieve Lieutenant Colonel Martin and his party who were still besieged in the bungalow. This relieving force was strengthened during the night of the 15th/16th by twenty-five armed civilians, which brought the total strength of the force to 176.

During the night two batches of men from the 5th Light Infantry surrendered at the Bukit Timah Road Police Station, the first batch was of sixty-four men and the second fifty-three men. The night was a restless one throughout Singapore with occasional sniping in the area of Orchard Road Police Station and Paterson Road. Two of the mutineers were killed during the night in exchanges of rifle fire between the police and mutineers.

Lieutenant Colonel Martin's bungalow remained in a state of siege, with Mrs Cotton providing tea and food for the defenders. The bungalow was illuminated by a searchlight from Blakang Mati which probably saved it from being rushed during the night. The mutineers attacked the bungalow at dawn, in the typical North West Frontier style, but Colonel Martin had expected this and had 'closed up' his defences and during the exchange of rifle fire about sixteen of the rebels were shot.[11]

Before dawn on 16 February, Colonel Brownlow's relief force advanced along Alexandra Road towards the barracks, the landing party from HMS *Cadmus* forming the firing line.

HMS *Cadmus* courtesy of the National Maritime Museum.

Contact was made with the mutineers at about 5.30 am and the Singapore Volunteer Corps advanced and occupied the barracks in support of the Naval contingent. Heavy firing on the left flank held up the attack for a short time but the armed civilians, under command of Captain E. A. Brown, SVC, occupying some of the barrack blocks drew the mutineers' fire whilst the HMS *Cadmus* party and the Singapore Volunteer Corps moved to the right and attacked from higher ground. The mutineers broke and the relieving force was again able to advance and reached the bungalow being held by Lieutenant Colonel Martin and his party. After the relief of the bungalow, Mrs Cotton was sent to Government House for safety. The combined forces were not considered strong enough to hold the position and the force then retired to Keppel Harbour and in doing so cleared the Keppel Golf Course of stragglers from the mutineer forces. The total loss during this relief operation was two killed and four wounded; Stoker C. F. Anscombe of HMS *Cadmus* and Special Constable A. Geddes being the two fatalities.[12] The mutineers' losses were never accurately known but it was estimated that about ninety prisoners were taken and eleven killed.

Overnight strong pickets were sent out to Tanglin and to the end of Cluny Road. During the morning of the 16th the Veteran Company SVC occupied Tanglin Barracks unopposed and took charge of the prisoners of war and the interned aliens, the former being mainly fifty prisoners from the *Emden*. The Volunteers were divided between the Drill Hall and Keppel Harbour, from which detachments were despatched and guards provided for Government House, the General Hospital and Fort Canning. A detachment of the Singapore Volunteer Artillery was based at Pasir Panjang Police Station under command of Lieutenant Colonel Scott RGA.

Whilst Lieutenant Colonel Martin's bungalow was being relieved, a party of the mutineers attacked the Orchard Road Police Station which, at that time, was being garrisoned by police and armed civilians. The attack was beaten off and two of the mutineers were killed. February 16 became a day of reorganisation and 200 European Special Constables were 'sworn in' and a force of 190 Japanese Special Constables raised by the Japanese Consul. All were supplied with arms by the military authorities.[13] All motorised vehicles were requisitioned and two lorries were converted into armoured cars. The defence forts on the islands were not neglected during this critical period and Winsley records that '. . . Fort Silinsing was manned throughout the mutiny by Lieut W. Murdoch, Corpl F. E. Dilley and 15 men, while Fort Siloso was manned by a mixed detachment of R.G.A. and S.V.A. . . .'[14] During the afternoon of the 16th ninety mutineers surrendered to the forces at Keppel Harbour.

The fighting diminished, although lively sniping continued for some days. The evacuation of women and children continued and they were accommodated on the Eastern Extension Telegraph Company's cable ship *Recorder*, the SS *Nile* and the SS *Penang*, all of which had been placed at the disposal of the military authorities. By the evening of the 16th, a complete line had been established from the P&O Wharf to Cluny Road which now successfully cut the mutineers off from the city. Motorised Forces were established ready to rush to any given trouble spot. The last European deaths were Lieutenant (Doctor) A. F. Legge, of the Singapore Volunteer Medical Company and Gunner J. Barry of the Royal Garrison Artillery, both of whom were killed near Alexandra Road Police Station during the sniping and skirmishes of the second day of the mutiny.[15]

A funeral service for the majority of the European Protestant victims of the mutiny was held at Bidadari Cemetery on the afternoon of 16 February, the service being conducted by the Bishop of Singapore. At a later date memorial headstones were erected over the individual graves.

The French cruiser *Montcalm* arrived on 17 February and landed two machine guns and 190 men; on the following day the Russian cruiser *Orel* landed a party of forty men.[16] On the same day the Japanese cruiser *Otawa* arrived and supplied a party of seventy-six men. The party of 190 men from the *Montcalm* were moved, by motor transport, to the Seletar district in the north of the island where some of the mutineers had been seen. Before the party arrived, the majority of the mutineers in that area crossed over into Johore territory and sixty-one of these surrendered to the Sultan.

The *Straits Times* published a statement to the effect that, of the 818 men of the 5th Light Infantry, 432 men had 'offered assistance', surrendered or had been captured. In the same edition Major W. L. Cotton, who had been appointed Commandant of the POW and Internment Camp, offered a reward for the capture of the missing POWs. A proclamation declaring Martial Law and officially putting the Volunteer Troops under the command of the 'Commander of Troops Straits Settlements', who was Brigadier General D. H. Ridout, was also published, together with a notice appointing Major A. M. Thomson as Provost Marshal with effect from 15 February 1915.

On 18 February, Lieutenant Colonel C. W. Brownlow RA led a force of seventy-six men out of Keppel Harbour with instructions to retake and occupy Alexandra Barracks. Six mutineers were captured at the barracks. The Japanese landing party came under fire as they approached Normanton Barracks. The sniper fire failed to halt the advance and twelve of the men of the 5th were taken but they proved to be hospital patients and followers belonging to the regiment..

On 19 February further reinforcements arrived when the Japanese cruiser *Tsushima* supplied a landing party of seventy-five men and on the following day six Companies of the 4th (Territorial) Battalion of the King's Own Shropshire Light Infantry arrived on the SS *Edavana* from Rangoon.[17]

On 20 February the Provost Marshal issued a public notice offering rewards as follows:

'$200 for the marking down of large bodies of mutineers consisting of ten and upwards.
$20 for the capture or surrender of mutineers up to ten in mumber.
$500 for the recovery of the body of Captain Boyce 5th L.I.
$5 for each rifle which can be proved to have been the property of native infantry or the Malay States Guides.
$1,000 for the capture of any German prisoner.
$50 for information leading to the capture of any German prisoner.

A. M. THOMPSON Major
Provost Marshal'[18]

Sixty of the mutineers were captured in and around Johore Bahru on 21 February and the *Straits Times* of 23 February reported that 614 mutineers were in custody or had surrendered and that fifty-two had been killed, wounded or drowned by the evening of 22 February making the total accounted for 666. A number of the men of the 5th Light Infantry who remained loyal and surrendered to the authorities were confined on board the ship SS *Cheang Hock Kian* under a guard of Chinese volunteers of the Singapore Volunteer Corps.[19] On 23 February two of the sepoys who had been captured were tried by a summary Court Martial which consisted of Lieutenant Colonel C. W. Brownlow, Major Edge, 4th KSLI (TF) and Captain L. P. Ball, 5th LI. The two sepoys were convicted and sentenced to be shot. The sentence was carried out at the Gaol at Outram Road at 5.00 pm on that day.

A further proclamation appeared in the *Straits Times* on Thursday 25 February to the effect that the Regiment had mutinied due to grievances over promotion to the rank of VCO. On the same day a party from the Russian cruiser *Orel* made contact with a body of mutineers. The mutineers opened fire on the Russian landing party, wounded two of the sailors and then withdrew. From that day onwards widespread searches were carried out in the more remote parts of the west side of the island, as it was certain that the eastern side of Singapore Island had been cleared. On 27 February nine more mutineers were captured, leaving a total of 125 unaccounted for, the nine mutineers were in a bad condition due to exposure and lack of food. The official bulletin on the following day reported that only 113 were then unaccounted for and it was also admitted that during the outbreak on 15 February three Chinese and two Malay civilians had been killed by the mutineers. One was Omar bin Amat Kaptin and the other Malay the car syce previously mentioned.

Tension in the Settlement gradually eased and on 3 March only sixty-two of the mutineers were unaccounted for, with fifty-six having been killed during the previous sixteen days. Although the search for the remaining mutineers continued, the incident gradually dropped out of the newspapers as the details of the Court Martial took the public eye and the threat from the few remaining mutineers at large diminished.

The Court Martial continued throughout March and on 25 March the Singapore Volunteer Artillery Maxim Company and a section of the Singapore Volunteer Rifles, under command of Captain H. Tongue, provided a firing party of 110 men for the execution of twenty-two of the convicted mutineers. The sentence was carried out at the Gaol at Outram Road. On the same day eight sepoys were sentenced to life imprisonment and transportation, seven sentenced to 20 years imprisonment and transportation, a further seven to 10 years with transportation and one sepoy to 7 years and transportation.[20] The convicted mutineers received little or no sympathy from the local population.

The loyal section of the 5th Light Infantry left Singapore some months later and served in the Cameroons and then in other parts of the Empire without any trouble. The regiment finally disbanded during the 1922 reorganisation of the Indian Army

The Court Martial proceedings continued through most of April and were reported in varying degrees in the *Straits Times*. A typical report given in the paper is that of Sepoy Taz Mohamed Khan which was reported as follows:

> 'The business of the Court Martial was resumed this morning, [Wednesday 14 April] when a sepoy named TAZ MOHAMED KHAN was called upon to answer charges of joining in the mutiny and desertion. The constitution of the Court remained the same, Lieut Col C. W. Brownlow, presiding, but in place of Mr Roland Braddell, Capt H. E. Wace, of the 4th Shropshire Light Infantry, prepared to prosecute. Mr de Silva was council for the defence.
>
> Mr Cullen, Assistant Superintendent of Police, stated that the accused was handed into his custody at Bukit Panjang. As a result of information received from the man, witness and Mr Campbell went in search of two other mutineers. Witness recognised the prisoner as the man handed over to him.
>
> Special Constable Campbell said it was on March 19 that the accused was arrested. He asked him if he knew where there were any other mutineers and he replied in the affirmative, but a search of the neighbourhood the man mentioned was not successful. Later, however, the two men he named were arrested.
>
> This was all the evidence put forward by the prosecution.
>
> The accused, making his statement, told the usual story. When the alarm sounded he

Execution of the first two convicted mutineers, Subedar Dunde Khan and Jemedar Chiste, outside the prison at Singapore. Courtesy of the Imperial War Musuem.

Execution of mutineers of the 5th Light Infantry.
Courtesy of the Imperial War Museum.

took his rifle and supplied himself with ammunition, but when he saw that mutiny was afoot he ran into the jungle where he joined up with two other men. He remained at large until March 19 when he was arrested by a Chinese. He stayed in the jungle so long because he was in fear of his life, but when arrested he was about to come into Singapore to give himself up. He rid himself of his rifle and ammunition at the very beginning and never once fired a shot.

This closed the case, the Court after retiring for a while to consider its finding, proceeded to deal with a batch of eleven men, also charged with joining in the mutiny and desertion . . .'[21]

The report carried in the paper on the following day read:

'The trial of the eleven men of the 5th Light Infantry, eight sepoys, a naik, lance naik and a bugler, occupied some time yesterday, as each accused went into the box to make his statement. The men were all arrested at Woodlands on February 18 by a party of the Johore Regular and Volunteer forces. Evidence was given as to the capture and subsequent examination of the rifles which were all black inside . . .'[22]

On Friday 16 April, the *Straits Times* published a notice to the effect that the promulgation of sentences on the mutineers who had recently been tried would be read out on Saturday 17 April at 5.00 pm '. . . outside the walls of the Criminal Jail . . .'[23]

The newspaper published on Monday 19 April 1915, carried the following account of the events of the preceding Saturday, under the heading of 'Large batch of Mutineers sentenced'.

'A large batch of men of the 5th Light Infantry, found guilty by the military authorities, of participating in the mutiny, were dealt with on Saturday afternoon, when the various sentences were promulgated outside the walls of the criminal prison. There was an attendance of some thousands of the general public, the ground being held by troops under command of Col GARRATT,[24] of the 4th Shropshire Light Infantry.

First of all sentence was passed on the men who have escaped the death penalty. They were led out and the terms of imprisonment which they have to undergo, ranging from life to one year, announced to the assembled spectators, after which they were taken back into the jail. Then two who were to be shot, Havildar SAMAND KHAN and Lance Naick [*sic*] FEROZ, were brought out under a guard of warders and executed with all despatch, the firing party consisting of the Royal Garrison Artillery. At the same time the other man, Lance Naik FAZAL ALI, was shot inside the prison walls, the order for the volley being a flag signal simultaneously with the firing outside. The timing was so accurate that it was practically impossible to tell that two volleys were fired. The man FAZAL ALI, who was executed in private, was unable to walk, he being the mutineer who was seriously wounded whilst firing at Mr L. A. Thomas, Assistant Superintendant of Police'[25]

In the same publication the list of other sentences was published and is summarised below:

Transportation for Life	1	Naik
	2	Lance Naik
	11	Sepoys
Transportation for 20 years	1	Sepoy
Transportation for 15 years	2	Lance Naik
	5	Sepoys
Transportation for 10 years	2	Lance Naik
	13	Sepoys

Governor addressing the crew of HMS *Cadmus* at Singapore, 1915.
Courtesy of the Imperial War Museum.

All the European casualties of the mutiny were buried in the Christian cemetery at Bidadari and later memorial plaques were placed in St Andrew's Cathedral. There were four plaques dedicated in the cathedral the details of which are as follows:

'In Memory of
Captain & 2nd in Command
HORACE CULLIMORE
Staff - Johore Military Forces
Late Royal Marine Artillery
Born 11 November 1869
Killed at Tanglin 15 Feby, 1915
Erected by Sergeants Mess R.M. Artillery'

'In Memory of
Stoker C. F. Anscombe
of HMS "Cadmus"
Who was killed in Action
during the mutiny of the 5th Light Infantry
16th February 1915
"Ready, aye Ready".
Erected by the Association of Engineers, Singapore'

'To the Glory of God
and
In Sacred Memory
of the undermentioned Officers,
Non-Commissioned Officers, and men,
of the Singapore Volunteer Corps
who lost their lives during the
Mutiny of the 5th Native Light
Infantry in February 1915.

	15th February	
2nd Lieutnt	J. Love Montgomerie	Rifles
Sergeant	G. Wald (Reserve)	Engineers
Corporal	D. McGilvray	Rifles
"	G. O. Lawson	Cyclist Scouts
Lce Corporal	J. G. E. Harper	Rifles
Private	B. C. Cameron	"
"	F. S. Drysdale	"
"	A. J. G. Holt	"
	16th February	
Lieutnt	A. F. Legge	Field Ambulance
	18th February	
Gunner	P. Walton	Artillery
	19th February	
Civilian	A. Geddes	Armed Civilians.

This Memorial is erected by their Comrades in the
Singapore Volunteer Corps'

'To the Glory of God
and
In Memory of
Major R. H. Galwey
Captain F. V. Izard
Captain M. F. A. MacLean
Corporal R. V. Beagley
Gunner J. Barry
All of The Royal Garrison Artillery
who were killed
in the Mutiny at Singapore
in February 1915.
This Tablet is Erected by their Comrades of the
Royal Garrison Artillery
in Singapore'

A summary of the casualties of the Singapore Mutiny is shown below:

	Killed		*Wounded*	
	Officers	*Soldiers and civilians*	*All*	*Total*
5th Light Infantry	2			2
Royal Garrison Artillery	3	2	3	8
ASC		1		1
Naval: HMS *Cadmus*		1		1
Volunteer Forces:				
SVA		1		1
SRE(V)		1	1	2
SVR	1	6	3	10
SVMC	1			1
MSVR	1	1	1	3
Johore Military Forces:				
European	1			1
Malay	1	2		3
European Civilians		13	1	14
Russian Sailors			2	2
Chinese Civilians		3		3
Malay Civilians		2		2
German Detainees		1	1	2
Totals	10	34	12	56

Of the mutineers the final total of casualties was given as 59 killed or drowned, 37 were executed for their part in the killings, 77 were transported for periods ranging from life to seven years, and 12 were imprisoned for periods ranging from one to five years.

All those Europeans buried in the Bidadari cemetery were given memorial headstones, provided either by their family or by various organisations or companies. It was unfortunate that, during 1959, a misguided decision was taken to remove the remains of the military casualties from Bidadari cemetery and re-inter the bodies at the Kranji War Cemetery. It is agreed that this cemetery is well maintained, but the decision called for the erection of 'standard' War Grave headstones, which were supplied. The original memorial headstones, which contained the details of the cause of death, etc, were left at Bidadari cemetery and were subsequently broken up and used as rubble.[33]

Much has been written on the cause of the mutiny but it is not within the scope of this work to record all the various theories, only to record the military events of those few violent days of 1915 when the war came to Singapore. Perhaps the most quoted of the theories will stand repetition to close this chapter and this is from the book *Turmoil and Tragedy in India* by General Sir George MacMunn, KCB, KCSI, DSO, which was published in 1935 and in this work he commented on the Singapore mutiny as follows:

> '. . . This strange, fierce outbreak, which might have been so disastrous, was believed to have been largely due to the influence of German agents but unconnected, to any great extent, with the wishes of the German residents. It found, however, a small cell of Moslim fanaticism to work with . . .'[34]

In addition to the men of the 5th Light Infantry there is conclusive evidence that a number of men of the Malay States Guides also joined the mutiny. The men of the Guides were generally recruited in India and were volunteers who agreed to serve in the Federated Malay States or in the Straits Settlements.[35] The Malay States Guides were officered by Indian Army Officers and were normally stationed at Taiping, Malaya.[36] With the outbreak of the First World War it became necessary for the unit to serve outside of its area and it was moved from Taiping to Singapore on standby ready to be moved to East Africa. There was apparently some confusion as to whether the sepoys had volunteered for overseas service outside Malaya and the Straits Settlements and it is generally agreed that the unit, as a whole, had not volunteered and did not wish to serve in East Africa.[37]

The unit was divided between Sikhs and Punjabi Muslim soldiers and it was the Punjabi soldiers that joined the mutiny although the exact details are not recorded in any great detail. The loyal portion of the unit in Singapore was moved to Johore and those soldiers at the Depot at Taiping were confined to barracks.[38]

Winsley records very little of the activities of the mutineers from the Malay States Guides but does show in his summary of the events of the mutiny that on 27 February there were 125 soldiers of the Malay States Guides '. . . still at large . . .'[39]

The soldiers of the Malay States Guides who were apprehended were dealt with separately from the men of the 5th Light Infantry and they did not receive the publicity that was given to the trial and sentences of the mutineers of the 5th Light Infantry.

Subsequently the Malay States Guides assisted with the garrisoning of Kelantan following the 1915 uprising and the unit then moved to Aden where it served until it was disbanded, at Aden, in September 1919.[40]

LIST OF PERSONS KILLED DURING THE MUTINY, FEBRUARY 1915

Name	*Unit/Employment*	*Where Buried*
Major R. H. Galwey	Royal Garrison Artillery	Bidadari*
Captain F. V. Izard	" " "	Bidadari*
Captain M. F. A. Maclean	" " "	Bidadari*
Corporal J. R. V. Beagley	" " "	Bidadari*
Gunner J. Barry	" " "	Bidadari*
Sergeant E. H. Sexton	Army Service Corps	Bidadari*
Captain P. Boyce	5th Light Infantry, Indian Army	Bidadari*
Lieutenant H. S. Elliott	5th Light Infantry, Indian Army	Bidadari*
Stoker C. F. Anscombe	HMS *Cadmus*	Bidadari*
Gunner P. Walton	Singapore Volunteer Artillery, SVC	Bidadari*
Sergeant G. Wald	Singapore Royal Engineers (V), SVC	Bidadari*
2nd Lieutenant J. Love-Montgomerie	Singapore Volunteer Rifles, SVC	Bidadari*
Corporal D. MacGilvray	" " "	Bidadari*
Private F. S. Drysdale	" " "	Bidadari*
Private J. Harper	" " "	Bidadari*
Private A. J. G. Holt	" " "	Bidadari*
Private G. O. Lawson	" " "	Bidadari*
Private B. C. Cameron	" " "	Bidadari*
Lieutenant A. F. Legge	Singapore Volunteer Medical Coy, SVC	Bidadari*
Captain P. M. Gerrard	Malay States Volunteer Rifles	Bidadari*
Private W. H. Leigh	" " " "	Bidadari*
Captain H. Cullimore	Johore Military Forces	Bidadari*
Captain Abdul Jaffar	Johore Military Forces	Not known
Corporal Jacob bin Salleh	Johore Military Forces	Not known
Private Yakub bin Fali	Johore Military Forces	Not known
Doctor E. D. Whittle	Straits Settlements Govt Service	Bidadari
Mr C. V. Dyson	" " " "	Bidadari
Mr J. Clarke	Warder, Singapore Prison	Bidadari
Mr G. B. Woolacombe	Eastern Extension Telegraph Company	Bidadari
Mrs G. B. Woolacombe	Wife of G. B. Woolacombe	Bidadari
Mr L. P. C. Smith	Eastern Extension Telegraph Company	Bidadari
Mr J. B. Dunn	Messrs Guthrie & Co	Bidadari
Mr E. O. Butterworth	Messrs Guthrie & Co	Bidadari
Mr W. J. Marshall	China Mutual Insurance Company	Bidadari
Mr A. R. Evans	The Borneo Company	Bidadari
Mr H. B Collins	*The Straits Bulletin*	Bidadari
Mr N. F. Edwards	Messrs Paterson, Simons & Co	Bidadari
Mr F. Geddes	Messrs Topham, Jones & Railton	Bidadari (Catholic Section)
Hassan Kechil bin Hassan	The Straits Motor Garage	Not known
Omar bin Ahmat Kaptin	of Pulo Brani	Not known
plus		
Three Chinese civilians	Names and occupation not known	Not known
Mr E. F. Senftleben	German detainee	Bidadari

*later transferred to Kranji

NOTES TO CHAPTER FOURTEEN

1. Donald Moore, *The Magic Dragon*, (Panther Books Ltd, 1975), p 188. The *Emden* was eventually destroyed by the Australian naval ship *Sydney* off the Cocos Islands.
2. Boris Mollo, *The Indian Army*, (Blandford Press, 1981), pp 65 and 124. Indian Army List, July 1914, p473.
3. Gerald H. Corr, *The War of the Springing Tiger*, (Osprey, 1975), p3.
4. Captain T. M. Winsley, SVC, *A History of the Singapore Volunteer Corps, 1854–1937*, (Government Press, Singapore, 1938), p61.
5. Ibid, p63.
6. Ibid, p64.
7. Ibid, p63.
8. Corr, op cit, p 13. Lieutenant Jules Lauterbach eventually returned to Germany and was given command of the sea raider *Mowe*.
9. Winsley, op cit, p64.
10. Ibid.
11. Ibid, p65.
12. Ibid, p66.
13. Ibid.
14. Ibid.
15. Ibid, p68.
16. Ibid, p66.
17. D. & J. Moore, *The First 150 years of Singapore*, (Singapore, 1969), p553.
18. Winsley, op cit, p68.
19. Song Ong Siang, *One Hundred Years History of the Chinese in Singapore*, (John Murray, London, 1923), p513.
20. Winsley, op cit, p 192.
21. *The Straits Times*, Wednesday 14 April 1915.
22. Ibid, Thursday 15 April 1915.
23. Ibid, Friday 16 April 1915.
24. Lieutenant Colonel A. N. R. Garrett, commanding 4th Battalion The King's Shropshire Light Infantry.
25. *The Straits Times*, Monday 19 April 1915.
26. Ibid.
27. Ibid, 19 and 20 April 1915.
28. Ibid, Tuesday 20 April 1915.
29. Ibid, Wednesday 21 April 1915.
30. Ibid, Thursday 22 April 1915.
31. Ibid, 17 May and 19 May 1915.
32. Winsley, op cit, p71.
33. *Author's note:* During a visit to the Bidadari Christian Cemetery in 1959 I found that the bodies of the military personnel killed during the mutiny had been exhumed and reburied at the Kranji War cemetery. A visit to the Kranji War Cemetery revealed that all had been buried and provided with the standard Commonwealth War Graves Commission headstone. Returning to Bidadari Cemetery I searched for the original headstones and found some of the broken stones in a heap near to the Protestant chapel. I photographed two, those of Captain H. Cullimore and Lieutenant F. V. Izard both of which were broken. When I returned later I was informed by the Chowkidar that the headstones were not required and had been broken up.

   A subsequent visit to the Cemetery by my wife and myself in 1971 revealed that all the civilian graves were still in good order with the majority of the stones unbroken. An examination of the burial register revealed that the pages covering the early months of 1915 had become badly damaged and were, in fact, almost destroyed by white ants.
34. Winsley, op cit, p71.
35. Public Record Office, Kew. File CO 273/43.
36. PRO, File CO 273/43, letter 298/2936.
37. PRO, File 273/435, letter 298/2936.

38. I am indebted to Lieutenant Colonel George H. Kelling, US Army, for allowing me to have access to his thesis entitled 'The British in Malaya' which gives details of movements of the Malay States Guides during February 1915.
39. Winsley, op cit, p 192.
40. PRO, File CO 273/477. Letter from C-in-C India to GOC Aden dated 26 August 1919 dealing with the 'disposition of the Malay States Guides'.

CHAPTER FIFTEEN

Between the Wars

By the close of the First World War, Singapore and Malaya had increased the size of the Volunteer forces from the three Corps that existed in 1914 to the following units that were in existence in 1918:

Singapore Volunteer Corps
Penang Volunteers
Malay States Volunteer Rifles
Malacca Volunteer Rifles
Province Wellesley Volunteer Rifles
Malayan Volunteer Infantry
Johore Volunteer Rifles
Labuan Volunteer Defence Detachment
Kelantan Volunteer Rifles

Although the war had ended and it was expected that the usual pattern for Volunteer soldiering would wane this did not immediately happen in Singapore and Malaya and in fact it was during 1919 that a new unit was formed and added to the Singapore Volunteer Corps, this was the Electic Light Section, SRE(V).[1] Singapore and Malaya settled down into a peaceful routine and it was at the end of 1921 that it was agreed that the Volunteer forces of the Straits Settlements should be reorganised and that they should be under one control. The existing Volunteer ordinances were cancelled and replaced by a new ordinance, The Volunteer Ordinance 1921.[2] The new ordinance came into force on 1 January 1922 and under this reorganisation the four existing Corps were disbanded and all commissions were '... called in ...'[3] The new combined force was to be known as the Straits Settlements Volunteer Force and Lieutenant Colonel F. E. Spencer, DSO, MC, RA was appointed as the first Commandant. The new Corps were each granted an adjutant from the regular forces and these appointments continued to be filled from the British Army until the Second World War. The establishments of the new Corps were as follows:

Singapore Volunteer Corps		*Officers*	*ORs*
Singapore Volunteer Artillery	(1 Company)	6	161
Singapore Royal Engineers (V)	(1 Company)*	5	87
Singapore Volunteer Rifles	(2 Companies)	10	316
Singapore Volunteer Infantry	(1 Eurasian Company)	5	158
Singapore Volunteer Infantry	(1 Malay Company)	5	158
Singapore Volunteer Infantry	(1 Chinese Company)	5	158
Singapore Field Ambulance	(1 Company)	2	124

*including a Signal platoon.

Attached to SVC

Johore Volunteer Rifles (European British Volunteers).
Labuan Volunteer Detachment.

Malacca Volunteer Rifles

Malacca Volunteer Rifles	1 Company European British
	1 Company Malay
	1 Company Chinese
Field Ambulance	1 Company

Penang and Province Wellesley Volunteer Corps

2 companies (European British) with additional machine-gun platoon
1 Chinese Company
1 Malay Company
Field Ambulance 1 Company[4]

The training was standard in all three Corps and was given as being

'Infantry and Field Ambulance	– 18 parades
RE & VA	– 24 parades
Reserve Service	– 6 parades'[5]

During the early years following the end of hostilities in 1918 the Regular forces returned to normal within the Straits Settlements, and in Singapore the British Infantry battalion was once again stationed at Tanglin Barracks; the Royal Artillery, of which there were two companies, were on Blakang Mati. The Royal Engineers, 41st Company (Fortress) RE, which had been increased from half a company to a full company in 1905, was stationed on Pulau Brani, together with the Royal Army Ordnance Corps. The headquarters was still stationed at Fort Canning. During the early 1920s it was recorded that twenty pupils in the school at Tanglin came from the families stationed at Fort Canning and Tanglin. On Blakang Mati the schoolmistress had twenty pupils also and these were from the families living in quarters on the island and also from Pulau Brani. The children attended school on Monday to Friday from 8.00 am until midday.[6] The school building at Tanglin was far from comfortable for the teacher and pupils as the building was an old church that had been adapted for use by the small school and was '. . . quite large with 2 small rooms at one end for children and a small space with an attap roof as a playground . . . Poor lighting was afforded by oil lamps and there were neither fans nor punkahs . . .'[7]

Generally the conditions in both the single soldiers quarters and the married quarters were poor at this time with mosquitoes, which were not brought under control until 1922/23, bugs, white ants, and ants infesting every barrack block and quarter. Entertainment was provided by barrack dances, whist drives, picnics on the beaches and various outdoor games and sports. There was a cinema but this only worked intermittently. By 1925 the regular troops in the Colony consisted of 2nd Battalion, the Royal Sussex Regiment, 32 Company RAMC stationed at Tanglin Barracks, Headquarters, 27th Heavy Battery RA and 3rd Battery Hong Kong and Singapore Artillery on Blakang Mati Island, and on Pulau Brani 41st Company (Fortress) RE, the Ordnance Stores Depot and a detachment of the Royal Corps of Signals. The Command headquarters remained at Fort Canning, although the Command Education Officer had established his office at Tanglin Barracks. In March 1926, the Royal Sussex Regiment were relieved by the Duke of Wellington's Regiment. It was during that year that a plan was evolved at Fort Canning for the provision of a base for Royal Artillery batteries to be sited at Changi so as to cover the eastern approaches to the Johore Strait. In early 1927 the Army Council sent a commission of three officers led by Major General Sir Webb Gillman, RA,[8] to prepare the detailed defence plan for the area. The Commission's terms of reference included the actual siting of guns, the distribution of

Fort Canning accommodation block as seen from the cemetery, *c*1971. Photograph by Mrs J. Harfield.

Fort Canning accommodation block. Photograph by A. G. Harfield, 1973.

troops and the agreement of the principles to be used when laying out the main Changi camp area.[9] By 1 July 1927 the commission had recommended a 15-inch battery to be sited at Bee Hoe; two 6-inch batteries on Battery Hill and at Telok Paku and two batteries, also of 6-inch guns, to be sited on Pulau Teking Besar, which was a small island opposite Changi. These sites all had adequate fields of fire and were, in respect of the mainland positions accessible for the heavy loads that would need to be moved into the area when constructing the gun emplacements. The commission also recommended and left detailed plans for two main barrack areas at Changi which when built became Kitchener and Roberts Barracks. The area allowed for ample scope for expansion and in addition to allowing for the artillery units there was also room for any additional infantry that might have been required. The Changi district chosen was an area comprising three main hills, one which became known as Battery Hill which was immediately to the west of the small kampong, Fairy Point Hill, and Temple, or Changi Hill. Between these hills were low-lying swamp areas which extended to the alluvial plain to the east of the Changi Road and as far as the sandstone hills of Selerang to the south. A description of the area at the time reads:

> 'The road from Singapore ended south of the mouth of the Changi River, near the Police Station, from where a narrow track entered a dark natural tunnel through dense jungle, eventually emerging into an area of young rubber trees and mangrove swamps around Fairy Point Hill ...'[10]

At this time there was in the area, apart from the small kampong, the Police Station, a private bungalow, a government bungalow owned by the Public Works Department and a Japanese hotel. This building was later, after considerable internal reconstruction, to become the first Officers' Mess at Changi.

Work was put in hand immediately and after considerable problems over land purchase the resident works staff moved on to the site in December 1927 and the jungle undergrowth at Fairy Point was cleared to make way for huts for the locally employed labour force. The first four officers' married quarters were built at Fairy Point and the Japanese hotel became the Officers' Mess for the Royal Engineers who were in charge of the construction of the new camps. By early 1928 the first four blocks of the married soldiers quarters had been built on Battery Hill. There is an interesting note to the effect that the roofing tiles for these quarters had, due to their non-availability in other areas, to be obtained from Marseilles. By this time the northern swamp had been reclaimed up to the high-water level and this part of the project was doubly necessary as not only did this clear the area of a breeding place for mosquitoes but it also provided the area for the main sports grounds.

Snakes were a constant source of danger during the construction in the Changi area and workers had to take special care when clearing the jungle. This work of clearing the swamp was followed by the construction of the first roads, the first road in fact being called 'New Road' but later when the Changi area had, after the Second World War, become RAF Changi, it was renamed Netheravon Road. Because of the weight of equipment and ammunition to be moved during the construction, and then later when the batteries were commissioned, it was decided that the ideal means of transport within the new War Department area would be a light railway. The plan was drawn up during the latter part of 1927 by the Royal Engineers and members of the Federated Malay States Railway. The railway line was 1½ miles long running from Changi pier to the 15-inch battery situated at Bee Hoe, with a loop-line to serve the ammunition depot which was located on the edge of the Selarang swamp. The line, with the exception of the loop, was built during 1928 and several hundred Chinese labourers were used during the construction of the railway. The railway was of standard gauge as it was considered that the narrow gauge used by the Federated Malay States Railway would not be adequate for the type of equipment that had to be moved. This therefore meant that the engine and the rolling stock had to be shipped out to Singapore from England. The locomotive was loaded from the freighter in Singapore Harbour onto a junk owned by a Chinese contractor. With the railway line in operation, work on the barracks progressed and it was necessary to provide another Officers' Mess as the original Mess in the converted Japanese hotel was considered to be in danger of collapse. Fairy Point House became the second Officers' Mess at Changi. Sites for five barrack blocks were prepared during 1928/29 on the hillside south-east of the reclaimed swamp and during 1930 the first two of these blocks were completed.[11] In 1934 the camp was connected to the Singapore water mains but reserve storage tanks were built on Changi Hill. After three years' continuous work the camp at Changi was beginning to take shape but then a halt was called on all building due to cuts in the defence spending by the British Government and one of the major casualties in the cut-backs was the defences of Singapore.

A certain amount of new planning was carried out in 1933 when the restrictions were lifted on the development of the Changi area.

Meanwhile the Volunteer forces had seen a number of changes and in 1922 it was recorded that the arms of the Straits Settlements Volunteer Force were:

All units - SMLE rifles, with the exception of P&PWVC who were still armed with older pattern rifles.

SVC - 6 Vickers guns and 20 Lewis guns.

P&PWVC - 4 Vickers guns and 12 Lewis guns.

MVR - 2 Vickers guns and 8 Lewis guns.[12]

On 9 February 1923 it was announced in the Straits Settlements Government Gazette[13] that His Majesty King George V had approved the grant to the reorganized Artillery and Engineer Volunteer units of the title 'Royal', and in the following month new establishment and equipment tables were issued for the Force.

During the year 1923, a new Volunteer Ordinance was promulgated.[14] The new establishment listed the following units as forming the Straits Settlements Volunteer Force:

Singapore Volunteer Corps	
Singapore Royal Artillery (V)	1 Battery
Singapore Royal Engineers (V)	1 Company, including a Signal Section
Singapore Volunteer Infantry	2 European Companies 1 Scottish Company 1 Eurasian Company 1 Chinese Company 1 Malay Company
Singapore Volunteer Field Ambulance	HQ and 2 sections
Machine Gun Platoon	
Labuan Volunteer Detachment	
Penang and Province Wellesley Volunteer Corps	
Penang Volunteer Infantry	1 European Company
Province Wellesley Volunteer Infantry	1 European Company

Penang Volunteer Infantry	1 Eurasian Company 1 Chinese Company 1 Malay Company
P&PW Volunteer Field Ambulance	HQ and 2 Sections
Independent Sections	1 Engineer, 1 Signal and 1 Machine Gun Section
Malacca Volunteer Corps	
Malacca Volunteer Infantry	1 European Company 1 Chinese Company 1 Malay Company 1 Eurasian Company
Malacca Volunteer Field Ambulance	HQ and 1 Section
Sections	1 Signal and 1 Machine Gun Section

The efficiency qualifications were set out in 1924 as being: Artillery, Engineers, Signallers and Machine Gunners – 24 parades; Infantry and Field Ambulance personnel – 18 parades per year.[15] It was during this year that a series of company camps were introduced and the SRE(V) held their camp on Pulau Brani with 41st Fortress Company RE and the SRA(V) camped on Blakang Mati. Two years later two new Asiatic units were formed in Singapore these were the Defence Electric Light Unit (Auxiliaries) and the second unit was the Singapore Royal Artillery (Volunteer) Asiatic Signal Unit, which was formed from twelve trained General Post Office operators.[16]

In 1927 there were minor changes in the dress of the Volunteer Force. Bandoliers were gradually replaced by webbing equipment and flannel shirts were also gradually replaced by cotton shirts and, for the Scottish Company, a superior and better fitting kilt was obtained and issued.

During the following year a new Corps Badge was introduced for the Singapore Volunteer Corps. The Royal Crown surmounted the whole, the circle had the words 'Singapore Volunteer Corps' with a double spray of oak leaves immediately above a scroll bearing the Corps' motto 'In Oriente Primus'. In the centre of the circle is the Singapore lion 'statant' with a palm tree.[17] Winsley records that:

> '. . . At the same time Collar Badges, hitherto only worn by Officers of the SRA(V) and SRE(V), were instituted for wear by all ranks throughout the Corps. A Corps collar badge was designed in the form of an oval garter surmounted by a Royal Crown with the SVC cypher in script, superimposed in white metal in the centre, the cut out design having been maintained . . .'[18]

Winsley also records that the SRA(V) and the SRE(V) continued to wear their own distinctive cap and collar badges, the other ranks being permitted to follow their Officers with regard to the collar badges. The Machine Gun Sections and Companies were also an exception in that they were permitted to wear a helmet 'flash' which consisted of '. . . crossed machine guns with the cypher S.V.C. below the centre . . .'[19]

It was during 1928 that the Straits Settlements Volunteer Force had yet another re-organisation and this saw the creation of the 'Battalion' organisation which was shown as being:

1st Battalion SSVF – Singapore Volunteer Corps

1st Battalion Signal Section.	Transferred from SRE(V) Signal Section.
A (MG) Company (European)	Amalgamating the European Company and the MG Platoon.
B Company (European)	Reformed.
C (Scottish) Company	originally formed 24 February 1922.
D (Eurasian) Company	originally formed 4th July 1918.

2nd Battalion SSVF – Singapore Volunteer Corps

E (Chinese) Company	originally formed 1902.
F (Malay) Company	originally formed 1910.
Machine Gun (Eurasian) Platoon	formed 1928.
2nd Battalion Signal Section	

3rd Battalion SSVF – Penang and Province Wellesley Volunteer Corps

4th Battalion SSVF – Malacca Volunteer Corps[20]

During this reorganisation the SRA(V), SRE (V) and the Field Ambulance units remained unchanged.

It was decided, during 1928, that the cotton shirts which had been introduced during the previous year were unsatisfactory and the order was issued to revert back to the flannel shirt.

The Johore Volunteer Rifles which had been reorganised during 1927 was, as a result of representations made to the Advisory Committee of the Straits Settlements Volunteer Force, renamed from 1 January 1928 the Johore Volunteer Engineers and this change was eventually ratified by the publication of a Johore Government Gazette in 1932.[21] The new unit retained the general format of its cap badge and merely substituted the word 'Engineer' for 'Rifles'. The collar badges and buttons were, however, changed.

It was during the years 1928 and 1929 that a limited number of officers and men of the Singapore Volunteer Corps joined with the Duke of Wellington's Regiment in the provision of special guards for seaborne duties. At this time there was an epidemic of piracy on the high seas between Singapore and Hong Kong, these attacks being mounted by pirates of the Bias Bay stronghold. The pirates would attempt to smuggle some of their own men on board ship as coolies and then attempt to take over the bridge when out of sight of land. The guards were provided to stop this act occurring. A guard of one officer and twelve other ranks was provided for the ships plying between Singapore and Hong Kong, normally by the regular garrison infantry unit at Singapore but on four occasions the Singapore Volunteer Corps provided the guards.[22] Only on one of the four occasions did the guard encounter any trouble and then it was only of a very minor nature.

During the years 1930, 1931 and 1932 both the Regular infantry battalion and the Volunteer forces were only occupied by training and the numerous social events that were by then popular in the Colony of Singapore. It was during September 1931 that the Armoured Car Section SVC was formed when a 4-ton six-wheel Albion lorry was delivered and the Singapore Harbour Board workshops built the body on this vehicle. Only one section was formed as there was, at that time, only one vehicle and the command of the section was given to 2nd Lieutenant C. E. Collinge, who in 1932 attended a suitable course whilst on leave in the United Kingdom.

The 1930 publication *Notes on Land and Air Forces of the British Oversea Dominions, Colonies, Protectorates & Mandated Territories* shows that the Singapore Royal Artillery

(Volunteer) battery was armed with four 3.7-inch howitzers and that the remainder of the arms for the Corps were SMLE rifles, Vickers and Lewis guns.[23] The colour service volunteers for the Straits Settlements Volunteer force were required to be over eighteen and under forty years of age and were to be enrolled for a first engagement of four years. After this initial period of enlistment the Commanding Officer was then empowered to extend the colour service for any period not exceeding four years, providing that the service shall '. . . come to an end when the Volunteer attains the age of 40 years . . .'[24]

In 1931 the Volunteer engineers were reorganised and became Singapore Royal Engineers (Volunteers) Fortress Company and as the designated Officer Commanding was absent from the Colony the unit was commanded during its first year in existence in its new form by Lieutenant (Acting Captain) F. Bedford. It was also during 1931 that the Corps had quartermasters written into the establishment and appointed and Winsley records that '. . . the appointment of quartermasters led to a marked improvement in organization and system (of stores accounting and procurement) . . .' In December 1931 it was recorded that in Singapore four Queen's Army Schoolmistresses were still being employed in the schools at Tanglin and Blakang Mati.[25] The garrison battalion stationed at Tanglin at the time was the 1st Battalion, The Gloucestershire Regiment. The Regular Army were then still located at Tanglin, Fort Canning on the main island and on the islands of Pulau Brani and Blakang Mati. The new military cantonment that was being developed at Changi already housed the artillerymen to man the completed gun emplacements and during the year 1932 elements of the Singapore Volunteer Corps.[26]

On Saturday 4 March 1933, His Excellency, The Governor, Sir Cecil Clementi, GCMG, opened the new Straits Settlements Volunteer Force Headquarters which had been built in Beach Road, Singapore. The opening ceremony was followed by a party and dance which was reputed to have had over 2,000 guests and hosts. The social life in the Straits Settlements in the early 1930s was important to both the Regular and the Volunteer units and dinners and dances, as well as sports days and picnics were frequently the order of the day.

The building of the new military camp continued at Changi and in 1935 it became necessary to close the granite quarry at Changi as it was getting close to some of the buildings and thereafter grantite had to be provided by contractors bringing it from the off-shore Pulau Ubin. By the close of 1935 the barrack blocks on Barrack Hill had been completed, as had the Royal Engineers Institute and the Sergeants' Mess, both of which had been built at the foot of Barrack Hill facing the Padang. The Royal Engineers Officers' Mess had also been completed at Fairy Point Hill.[27] The Royal Engineers established themselves at the northern end of Changi camp in the barracks which were then named Kitchener Barracks.

Roberts Barracks, which became the home of the Coast Artillery Regiment, was completed during the years 1934–36 and the Royal Artillery Officers' Mess was built on what had been known as Temple Hill. It was during the year 1934 that the first AA regiment arrived at Changi and two of the soldiers' barrack blocks within Roberts Barracks were used to accommodate the men, with the overflow being accommodated in a number of wooden huts which had been built on the eastern side of the Changi Road. This hutted camp, which was later used by Punjabi soldiers of the Hong Kong and Singapore Regiment, was to become known as India Barracks.[28]

The Regular battalions stationed in Singapore continued to train with the men of the Singapore Volunteer Corps and in 1933 it was reported that His Excellency, The Governor was a spectator at the GOC's Annual Inspection at Singapore which, for that year, took the

Blakang Mati accommodation block, now housing a museum display of 'The Surrender Chamber'. Photograph by Mrs J. Harfield, 1977.

Tanglin Barracks Officers' Mess, 1977.

form of a 'mock' attack in the Jurong area against the 1st Battalion, Straits Settlements Volunteer Force, with the Wiltshire Regiment and the Johore Military Force co-operating in the exercise.

The Volunteer units all participated in numerous rifle and shooting competitions during the years 1933 and 1934 and this, as always, formed a popular part of the annual training programme. The Wiltshire Regiment which had completed a two-year tour of duty in the Straits Settlements during the latter part of 1934 was given an extension of duty in the Colony for a further year. The regular forces were gradually building up in what was to become known during the pre-war period as 'Fortress Singapore'. The Straits Settlements Volunteer Year Book for 1934–1935 makes various comments on the build-up of the military force in the Colony with such comments as '. . . there have been many changes in the regular garrison, not so much, perhaps, in its composition as in its numbers . . .',[29] and '. . . possibly one of the biggest changes here [is] the way in which Blakang Mati has given place to Changi as the principal Gunner stronghold . . .'[30]

The presentation of Colours to the Singapore Volunteer Corps took place on the Padang on 26 May 1934. The Colours were presented to the Corps by His Excellency, The Governor, Andrew Caldecott Esq, CMG, CBE. The ceremony commenced at 5.30 pm with the arrival of the Governor who was accompanied by the General Officer Commanding, Major General E. O. Lewin, CB, CMG, DSO and after the ceremony a reception was held at the SVC Headquarters. Lieutenant Colonel J. H. M. Mee, and the officers of the 1st Battalion, The Wiltshire Regiment loaned the regimental band for the parade and the whole day was reported as having been a great success with the following comment being published:

> '. . . So ended a day of which the Singapore Volunteer Corps may well be proud. They possess now the great honour of carrying Colours and by their steadiness on parade, their soldierly bearing and their drill they showed they were worthy of the honour . . .'[31]

Five days prior to this ceremony the General Officer Commanding, Major General E. O. Lewin, had presented the King's and Regimental Colours to the Penang and Province Wellesley Volunteer Corps. This ceremony took place on the Polo Ground on Penang Island on 21 May 1934 at 9.00 am. The parade was complimented on similar lines to that given to the Singapore parade and it reflected a great deal of credit on the Commanding Officer of the P&PWVC, Lieutenant Colonel G. D. A. Fletcher, MC.[32] There was a third presentation within the Straits Settlements during 1934 and this was the presentation of the King's and Regimental Colours to the Malacca Volunteer Corps. This third ceremony took place at the High School Grounds at 5.00 pm on Saturday 7 July. The Parade was commanded by Lieutenant Colonel A. A. Lermit, OBE, VD, the Commander of the MVC and the presentation was made by His Excellency the Governor, Mr A. Caldecott.[33]

The year 1935 was the year in which the Jubilee Celebrations were held throughout the Empire to commemorate the Silver Jubilee of the reign of King George V. Singapore town (as it was then) was decorated and celebrated the event on 10 May. Fifty-two army personnel of the Straits Settlements Volunteer Force received the award of a 1935 Silver Jubilee Medal.

There was one major change during the year in the Volunteer forces in Malaya, which were not at this time part of the Straits Settlements force, when the Kelantan Volunteer Rifles were reconstituted as the Kelantan Volunteer Force.

Within the Straits Settlements defence was concentrated on the Colony of Singapore and with the building of new camps, the projected modernization of the land forts in the south of the island, and the build-up of the military forces there was a general mood of confidence and optimism. The life for the military was one of drills, duties and training, liberally punctuated with social events and sports. The sporting programme for the soldiers was varied and for the officers there were tennis parties, swimming parties, picnics across the Causeway in Southern Johore and no shortage of regimental dances and balls. For those who had free time there was always the afternoon tea-dance at the famous Raffles Hotel. There was, in fact, in 1935 little to show that in seven years time the whole of the Colony would be over-run and that the elaborately planned defences would prove to be completely inadequate.

NOTES TO CHAPTER FIFTEEN

1. Captain T. M. Winsley, SVC, *A History of the Singapore Volunteer Corps,* (Government Press, Singapore, 1938), p83.
2. Straits Settlements Government Gazette No1996 of 16 December 1921.
3. Winsley, op cit, pp 83 and 84.
4. Ibid, p86.
5. Ibid.
6. R. P. Gaskell, 'British Army Children's Schools in Singapore', (thesis presented to the University of Singapore, March 1971), p46.
7. Ibid.
8. During a later rebuilding programme a new barracks in the Alexandra area of Singapore was named 'Gillman Barracks', after Major General Sir Webb Gillman, RA.
9. Squadron Leader H. A. Probert, *History of Changi,* (Prison Industries, Changi Prison, 1970), p10.
10. Ibid, p8.
11. The five barracks blocks were used, during the 1960–1971 period by the Headquarters of the Far East Air Force as office accommodation and as part of the Royal Air Force Hospital, Changi.
12. Winsley, op cit, p89.
13. Straits Settlements Government Gazette No9 of 9 February 1923.
14. Ibid, No1997 of 21 December 1923.
15. Winsley, op cit, p95.
16. Ibid, p97.
17. The Singapore Lion and Palm Tree was adopted by Singapore District, and later by Singapore Base District, as the formation sign during the period 1946 to 1971. The lion and palm tree were in yellow on a green rectangle. Lt Col H. N. Cole, *Badges on Battledress,* Gale and Polden, 1953 and author's notes).
18. Winsley, op cit, p101.
19. Ibid.
20. Ibid, pp 102 and 103.
21. Johore Government Gazette No220 of 1932.
22. Winsley, op cit, p105.
23. *Notes on the Land and Air Forces of the British Oversea Dominions, Colonies, Protectorates & Mandated Territories,* (H.M.S.O., London, 1930), p74.
24. Ibid, p72.
25. The title Queen Alexandra Schoolmistresses was granted on 24 May 1928 with HM Queen Mary as Patron.
26. Winsley, op cit, p111.
27. Probert, op cit, p28.
28. Ibid, p30.
29. Straits Settlements Volunteer Force Year Book, 1934–1935, p163.
30. Ibid.
31. Ibid, p101.
32. Ibid, pp 122 to 125.
33. Ibid, pp 128 to 131.

APPENDIX A

List of Units who Served as Garrison Troops in the Settlements of Penang, Malacca, and Singapore during the period 1805 to 1915.

Records prior to 1805 do not show which of the Honorable East India Company units served in the two settlements of Prince of Wales Island (Penang) and Malacca.

Year	*Unit*	*Notes*
1805	20th Regiment Bengal Native Infantry	Prince of Wales Island and Malacca.
1806	”	”
1807	”	”
1808	20th (Marine) Regiment, Bengal Native Infantry	”
1809	”	”
1810	”	”
1811	”	”
1812	”	”
1813	”	”
1814	”	”
1815	”	”
1816	”	”
1817	”	”
1818	”	”
1819	”	Prince of Wales Island and Singapore
1820	”	”
1821	”	”
1822	”	”
1823	”	”
1824	”	Malacca, Penang and Singapore
1825	25th (Marine) Regiment, Bengal Native Infantry	”
1826	”	”
1827	”	Penang
	17th Regiment, Madras Native Infantry	Singapore and Malacca
1828	35th Regiment, Madras Native Infantry	Penang
	17th Regiment, Madras Native Infantry	Singapore and Malacca
1829	25th Regiment, Madras Native Infantry	Singapore and Malacca
	35th Regiment, Madras Native Infantry	Penang
1830	25th Regiment, Madras Native Infantry	Singapore and Malacca
	35th Regiment, Madras Native Infantry	Penang
1831	25th Regiment, Madras Native Infantry	Singapore and Malacca
	35th Regiment, Madras Native Infantry	Penang
1832	29th Regiment, Madras Native Infantry	Singapore and Malacca
	46th Regiment, Madras Native Infantry	Penang
1833	29th Regiment, Madras Native Infantry	Singapore
	46th Regiment, Madras Native Infantry	Penang
	5th Regiment, Madras Native Infantry	Malacca (for the Naning War)
1834	*East India Register – January*	
	29th Regiment, Madras Native Infantry	Singapore
	46th Regiment, Madras Native Infantry	Penang
	23rd Regiment, Madras Native Infantry	Malacca
	East India Register – May	
	15th Regiment, Madras Native Infantry	Penang (arrived February 1834)
	23rd Regiment, Madras Native Infantry	Malacca and Singapore
1835	15th Regiment, Madras Native Infantry	Penang and Malacca
	23rd Regiment, Madras Native Infantry	Malacca and Singapore
1836	15th Regiment, Madras Native Infantry	Penang and Malacca
	48th Regiment, Madras Native Infantry	Malacca and Singapore (arrived July 1835)
1837	15th Regiment, Madras Native Infantry	Penang
	48th Regiment, Madras Native Infantry	Malacca and Singapore

Year	*Unit*	*Notes*
1838	12th Regiment, Madras Native Infantry	Penang (arrived March 1837)
	48th Regiment, Madras Native Infantry	Malacca and Singapore
1839	8th Regiment, Madras Native Infantry	Singapore and Malacca
	12th Regiment, Madras Native Infantry	Penang
1840	8th Regiment, Madras Native Infantry	Singapore and Malacca
	24th Regiment, Madras Native Infantry	Penang
1841	39th Regiment, Madras Native Infantry	Singapore
	24th Regiment, Madras Native Infantry	Penang and Malacca
1842	39th Regiment, Madras Native Infantry	Singapore
	24th Regiment, Madras Native Infantry	Penang and Malacca
1843	4th Regiment, Madras Native Infantry	Singapore (arrived May 1943)
	24th Regiment, Madras Native Infantry	Penang and Malacca
1844	4th Regiment, Madras Native Infantry	Singapore
	27th Regiment, Madras Native Infantry	Penang and Malacca
1845	27th Regiment, Madras Native Infantry	Establishment reduced to 1 battalion to cover the three Settlements
1846	21st Regiment, Madras Native Infantry	Arrived May 1846 Headquarters based at Singapore
1847	21st Regiment, Madras Native Infantry	
1848	21st Regiment, Madras Native Infantry	
1849	21st Regiment, Madras Native Infantry	
	51st Regiment, Madras Native Infantry	Penang (arrived April 1849)
1850	51st Regiment, Madras Native Infantry	
1851	51st Regiment, Madras Native Infantry	
1852	43rd Regiment, Madras Native Infantry	Arrived April 1852
1853	43rd Regiment, Madras Native Infantry	
1854	43rd Regiment, Madras Native Infantry	
1855	38th Regiment, Madras Native Infantry	Arrived April 1855 (Singapore)
	29th Regiment, Madras Native Infantry	Arrived February 1855 (Penang)
1856	38th Regiment, Madras Native Infantry	Singapore and Malacca
	29th Regiment, Madras Native Infantry	Penang
1857	14th Regiment, Madras Native Infantry	Arrived Singapore February 1857
	29th Regiment, Madras Native Infantry	Penang
1858	14th Regiment, Madras Native Infantry	Singapore and Malacca
	29th Regiment, Madras Native Infantry	Penang
1859	14th Regiment, Madras Native Infantry	Singapore and Malacca
	22nd Regiment, Madras Native Infantry	Penang
1860	14th Regiment, Madras Native Infantry	Singapore and Malacca (see 1861)
	22nd Regiment, Madras Native Infantry	Penang
1861	40th Regiment, Madras Native Infantry	Singapore (arrived 22 January 1860)
	20th Regiment, Madras Native Infantry	Penang (arrived 18th April 1861)
1862	40th Regiment, Madras Native Infantry	Singapore and Malacca
	20th Regiment, Madras Native Infantry	Penang
1863	34th Regiment, Madras Native Infantry	Singapore from 29th April 1863
	20th Regiment, Madras Native Infantry	Penang
1864	34th Regiment, Madras Light Infantry	Singapore and Malacca
	35th Regiment, Madras Native Infantry	Penang (arrived 16th February 1864)
1865	34th Regiment, Madras Light Infantry	Singapore and Malacca
	35th Regiment, Madras Native Infantry	Penang
1866	8th Regiment, Madras Native Infantry	Singapore from 6th March 1866
	35th Regiment, Madras Native Infantry	Penang
1867	8th Regiment, Madras Native Infantry	Singapore
	35th Regiment, Madras Native Infantry	Penang
1868	8th Regiment, Madras Native Infantry	Singapore
	35th Regiment, Madras Native Infantry	Returned to Madras from Penang 1 April
	7th Regiment, Madras Native Infantry	Arrived Penang 17th March 1868

Year	*Unit*	*Notes*
1869	8th Regiment, Madras Native Infantry	Departed from Singapore May 1869
	7th Regiment, Madras Native Infantry	Moved from Penang to Singapore on 12th May 1869
1870	7th Regiment, Madras Native Infantry	Singapore
1871	7th Regiment, Madras Native Infantry	Departed from Singapore April 1871
	19th Regiment, Madras Native Infantry	Arrived at Singapore 8 March 1871
1872	19th Regiment, Madras Native Infantry	Departed from Singapore March 1872
	80th (Staffordshire Volunteers) Regt of Foot	Arrived at Singapore 17 March and departed 28 November 1872
	10th (North Lincolnshire) Regt of Foot (1st Bn)	Arrived at Singapore November 1872
1873	1st Bn, 10th (North Lincolnshire) Regt of Foot	Now referred to as Straits Settlements Garrison
1874	1st Bn, 10th (North Lincolnshire) Regt of Foot	
1875	1st Bn, 10th (North Lincolnshire) Regt of Foot	(does not include units drafted into the Straits Settlements for the Perak War)
1876	1st Bn, 10th (North Lincolnshire) Regt of Foot	
1877	1st Bn, 10th (North Lincolnshire) Regt of Foot	
1878	74th (Highlanders) Regt of Foot	
1879	28th (North Gloucestershire) Regt of Foot	
1880	2nd Bn, 3rd (East Kent The Buffs) Regt of Foot	
1881	2nd Bn, 3rd (East Kent The Buffs) Regt of Foot	
1882	2nd Bn, 3rd (East Kent The Buffs) Regt of Foot	
1883	Royal Inniskilling Fusiliers	
1884	Royal Inniskilling Fusiliers	
1885	Royal Inniskilling Fusiliers	
1886	1st Bn, The Buffs (East Kent)	
1887	1st Bn, The Buffs (East Kent)	
1888	Prince of Wales' Vols (South Lancashire Regiment)	
1889	Prince of Wales' Vols (South Lancashire Regiment)	
1890	2nd Bn, The Northamptonshire Regiment	
1891	2nd Bn, The Northamptonshire Regiment	
1892	2nd Bn, The Northamptonshire Regiment	
1893	2nd Bn, The Lincolnshire Regiment	
1894	2nd Bn, The Lincolnshire Regiment	
1895	2nd Bn, The Lincolnshire Regiment	
1896	2nd Bn, The Northumberland Fusiliers	
1897	2nd Bn, The Northumberland Fusiliers	
	1st Bn, The Rifle Brigade	From July 1897 to February 1898
1898	1st Bn, The Prince of Wales' Own (West Yorks Regt)	Arrived March 1898
1899	1st Bn, The Prince of Wales' Own (West Yorks Regt)	Departed to India January 1899
	1st Bn, The King's Own Royal Lancaster Regiment	Arrived February 1899
1900	1st Bn, The King's Own Royal Lancaster Regiment	Departed January 1900
	16th Madras Infantry	Arrived February 1900
1901	16th Madras Infantry	Departed October 1901
	3rd Madras Infantry	Arrived November 1901
	13th Madras Infantry	Arrived November 1901
1902	3rd Madras Infantry	
	13th Madras Infantry	
1903	3rd Madras Infantry	Departed November 1903
	13th Madras Infantry	Departed January 1904
	1st Bn, The Manchester Regiment	Arrived December 1903
1904	1st Bn, The Manchester Regiment	
	73rd Carnatic Infantry	Arrived February 1904
1905	1st Bn, The Sherwood Foresters	Arrived March 1905
	73rd Carnatic Infantry	Departed March 1905
	95th Russell's Infantry	Arrived April 1905

Year	Unit	Notes
1906	1st Bn, The Sherwood Foresters	
	95th Russell's Infantry	
	2nd Bn, The Queen's Own (Royal West Kent Regiment)	Arrived December 1906
1907	2nd Bn, The Queen's Own (Royal West Kent Regiment)	
	95th Russell's Infantry	
1908	2nd Bn, The Queen's Own (Royal West Kent Regiment)	Departed November 1908
	95th Russell's Infantry	Departed February 1908
	3rd Bn, The Middlesex Regiment	Arrived December 1908
	99th Deccan Infantry	Arrived February 1908
1909	3rd Bn, The Middlesex Regiment	
	99th Deccan Infantry	
1910	2nd Bn, The Buffs (East Kent Regiment)	
	99th Deccan Infantry	
1911	2nd Bn, The Buffs (East Kent Regiment)	
	99th Deccan Infantry	Departed January 1911
	3rd Brahmans	Arrived February 1911
1912	2nd Bn, The Buffs (East Kent Regiment)	
	3rd Brahmans	
1913	2nd Bn, The Buffs (East Kent Regiment)	Departed January/February 1913
	1st Bn, The King's Own (Yorkshire Light Infantry)	
	3rd Brahmans	
1914	1st Bn, The King's Own (Yorkshire Light Infantry)	
	3rd Brahmans	Departed August 1914
	5th Light Infantry	
1915	5th Light Infantry	(Regiment mutinied 15 February 1915)
	4th Bn (Territorial) The King's Shropshire Light Infantry	

The above list shows only those regiments that were sent as the Garrison infantry battalion to the Settlements. It does not show units that were sent to the Straits Settlements for specific campaigns, such as the Naning Campaign, Perak Campaign, etc.

There were, of course, other units stationed in the three Settlements in addition to the above, such as artillery and engineer units, or sections of such units, as well as ordnance, commissariat and medical personnel.

The list also precludes units 'in transit'.

Year and date	Casualty	Unit	Where buried	Notes
1809 21 February	Matross Mark Le Vache	Madras Artillery	Northam Road, Penang	No trace of grave
9 March	Pte John Panter	HM 30th Regt	,,	,,
14 March	Matross David McKinnon	Madras Artillery	,,	,,
13 April	Matross John Chadburn	Bengal Artillery	,,	,,
7 September	Pte James McGuiness	HM 47th Regt	,,	,,
19 September	Cpl John Watkyns	,,	,,	,,
26 September	Major John Jones (Aged 29 Years)	HM 25th Light Dragoons	,,	Grave exists
26 September	Pte William Bird	HM 47th Regt	,,	No trace of grave
10 October	Conductor John Grimwood	Ordnance Department	,,	,,
19 October	Cpl Thomas Shields	Bengal Artillery	,,	,,
12 November	Pte Thomas Cudworth	HM 47th Regt	,,	,,
27 November	Matross John Bennet	Bengal Artillery	,,	,,
1810 12 January	Matross Samuel Harris	,,	,,	,,
14 February	Surgeon John Sealey	20th Regt BNI	,,	,,
16 March	Matross Thomas Lune	Bengal Artillery	,,	,,
19 March	Matross James Matman	Madras Artillery	,,	,,
16 June	Sgt John Whitehouse	Bengal Artillery	,,	,,
19 June	Matross Joseph Kitzen	,,	,,	,,
12 November	Matross — Wooland	Madras Artillery	,,	,,
1811 21 May	Pte William O Neale	HM 14th Regt	,,	,,
21 May	Pte John Davis	HM 22 Light Dragoons	,,	,,
23 May	Pte Gabriel Napier	HM 59th Regt	,,	,,
30 May	Pte William Hill	,,	,,	,,
30 May	Pte William Brooks	,,	,,	,,
1 June	Pte Julius Smith	HM 14th Regt	,,	,,
12 June	Pte John Spencer	,,	,,	,,
19 June	Pte John Wood	HM 59th Regt	,,	,,
28 June	Matross William Dunn	Bombay Artillery	,,	,,
7 July	Pte Robert Hull	HM 59th Regt	,,	,,
5 August	Lieut John Shore	20th Regt BNI	,,	,,
14 August	Pte Hugh Carroll	HM 59th Regt	,,	,,
18 August	Pte Hugh McEvoy	,,	,,	,,
19 August	Lieutenant Edward Anderson	HM 33rd Regt	,,	,,
25 August	Gnr Samuel Gordon	Bengal Artillery	,,	,,
26 September	Pte John Lucas	HM 59th Regt	,,	,,
26 October	Name not known	Pte HM 14th Regt	,,	,,
3 November	Pte Johanies Hinday	Madras European Regt	,,	,,
6 November	Pte Peter Pursevant	,,	,,	,,
6 November	Pte John Pegneely	,,	,,	,,
8 November	Pte Henry Skipper	,,	,,	,,
9 November	Captain Henry Dunscombe	HM 59th Regt	,,	,,
10 November	Pte James Merrygold	HM 22nd Light Dragoons	,,	,,
14 November	Pte John Grovier	,,	,,	,,
14 November	Pte Frederick Baskie	Madras European Regt	,,	,,
14 November	Pte Anthony Abgarth	,,	,,	,,
30 November	Pte Andrew Peterson	HM 22nd Light Dragoons	,,	,,
30 November	Pte Mathew Isiak	Madras European Regt	,,	,,
15 December	Sgt William Bageley	HM 22nd Light Dragoons	,,	,,
15 December	Pte Joseph Brookie	Madras European Regt	,,	,,
21 December	Pte George Gaylor	HM 22nd Light Dragoons	,,	,,
1812 19 June	Pte James Giles	HM 59th Regt	,,	,,
5 November	Pte Andrew Dougherty	HM 59th Regt	,,	,,
1813 20 January	Sgt John King	Bengal Artillery	,,	,,
27 February	Matross Thomas Doogan	,,	,,	,,
30 April	Gnr William Davis	,,	,,	,,
23 May	Pte William Patten	HM 14th Regt	,,	,,
28 May	Matross William Burrows	Bengal Artillery	,,	,,
1 July	Col John Killick	,,	,,	,,
25 July	Lieut Henry Shadwell	7th Regt Bengal NI	,,	,,
14 August	Pte Thomas Jiffs	HM 14th Regt	,,	,,
14 August	Pte Thomas Kerr	,,	,,	,,
14 August	Pte Etheldred Gorse	,,	,,	,,
16 August	Pte Martin Spelling	,,	,,	,,
16 August	Pte William Judd	,,	,,	,,
23 August	Cpl Joseph Simpson	,,	,,	,,
23 August	Pte James Pryor	,,	,,	,,

Year and date	*Casualty*	*Unit*	*Where buried*	*Notes*
1813 28 August	Pte Simon Walker	HM 14th Regt	Northam Road, Penang	No trace of grave
3 September	Pte Richard Coliffe	"	"	"
6 October	Pte Patrick Walkens	"	"	"
12 October	Pte William Barrington	"	"	"
28 October	Pte Thomas Ball	"	"	"
18 November	Pte William Weldin	"	"	"
21 December	Pte Michael Ker	Bengal European Regt	"	"
24 December	Pte Alexander Smith	"	"	"
1814 21 June	Cpl John McCormick	Bengal Artillery	"	"
2 July	Matross John Hyland	"	"	"
7 August	Assistant Surgeon William Leslie	Bengal Medical Establishment	"	"
20 September	Deputy Commissary Thomas Robinson	Commissariat Dept	"	"
16 December	Matross William Bates	Bengal Artillery	"	"
1815 11 January	Matross John Wood	"	"	"
16 May	Lieut John Joliffe	21st Regt, BNI	Fort Road, Malacca	"
8 July	Matross Hugh Stewart	Bengal Artillery	Northam Road, Penang	"
17 August	Matross Charles Watt	"	"	"
6 September	Matross Thomas Kemp	"	"	"
16 September	Sgt John Dann	"	"	"
5 November	Matross John McLaughlan	"	"	"
12 November	Sgt John Brewer	"	"	"
1816 9 January	Matross John O'Niel	"	"	"
10 January	Matross Michael Horam	"	"	"
19 January	Matross Richard Mannington	"	"	"
29 July	Lieut Charles Leslie Walker	20th Regt, Bengal NI	Drowned at sea near Malacca	None
8 August	Matross John Brown	Bengal Artillery	Northam Road, Penang	No trace of grave
13 August	Lieut William Bruce	4th Regt, Madras NI	Fort Road, Malacca	"
8 September	Matross I. P. Bakewell	Bengal Artillery	Northam Road, Penang	"
5 November	Pte Richard Griegland	Bengal European Regt	"	"
1817 8 February	Mr Thomas Jeffs (Aged 44 years)	Commissary of Stores, Fort Cornwallis	Northam Road, Penang	Grave exists
26 March	Lieut Henry P. Short	20th Regt, Bengal NI	Fort Road, Malacca	No trace of grave
1818 18 February	Matross Thomas Tanner	Bengal Artillery	Northam Road, Penang	"
21 March	Matross William Lee	"	"	"
27 May	Captain George Holroyd Alley	20th Regt, Bengal NI.	"	"
22 December	Lieut Charles C. Nattes	Madras Engineers	"	"
22 December	Matross Alexander Low	Bengal Artillery	"	"
1820 11 August	Major James Scott	3rd Regt, Bengal NI	On board the *Castle Huntly*, off Penang	None
1823 13 May	Lieut M. Bell	not known	Fort Canning, Singapore	No trace of grave
23 May	Major Nicholas Manley	20th Regt, Bengal NI	Northam Road, Penang	"
1824 29 August	Conductor Richard B. Smith	Ordnance Dept	"	"
26 September	Lieut Samuel C. Cooke	40th Regt, Madras NI	"	"
1825 28 February	Lieut Bentley Buxton	Bengal Engineers	Fort Canning, Singapore	"
11 April	Brigadier General M. Shaw, CB	late HM 87th Regt	'. . . on board HM Ship Slaney . . .' Interred on 13 April	"
21 April	Lieut William Dalzell	35th Regt, Bengal NI	Fort Canning, Singapore	"
19 August	Ensign Robert Kings Mears	21st Regt, Bengal NI	Northam Road, Penang	"
30 August	Capt Hugh Davidson	30th Regt, Bengal NI	"	"
13 October	Lieut John Thomas Lowe (Aged 25 years)	65th Regt, Bengal NI	"	Grave exists
30 October	Conductor James Simons (Aged 33 years and 6 months)	Bengal Ordnance Dept	"	Grave exists
13 December	Assistant Surgeon William Henderson	Medical Estab	"	No trace of grave
1826 4 July	Lieut William Colquhoun	32nd Regt Bengal NI	Fort Road, Malacca	Grave exists
27 September	Captain John James Gamage	Madras Artillery	Northam Road, Penang	No trace of grave
20 November	Engign Henry Clarges Barrow	28th Regt Madras NI	"	Grave exists
1827 17 April	Lieut Christopher Keating	41st Regt Madras NI	"	Grave exists
5 August	Capt William Brookes	Madras Artillery	"	No trace of grave
1828 29 March	Sergeant Frapp	? Artillery	Fort Canning, Singapore	"

Year and date	Casualty	Unit	Where buried	Notes
1830 9 February	Lieut Francis Pelby	Bombay Engineers	Fort Road, Malacca	No trace of grave
1 March	Lieut William Edward Brooshooft	35th Regt Madras NI	Northam Road, Penang	,,
12 March	Ensign John Cheyne Wilkinson (Aged 23 years)	35th Regt Madras NI	,,	Grave exists
7 September	Assistant Surgeon John Black	Bengal Medical Estab	Fort Canning, Singapore	No trace of grave
10 September	Surgeon J. G. Sim	,,	,,	,,
30 September	Capt Thomas Culley	2 Regt Bengal NI	,,	,,
11 December	Ensign Alexander Dennistoun	11th Regt Madras NI (on furlough)	,,	,,
1831 20 August	Lieut James White	29th Regt Madras NI (Killed in action)	Fort Road, Malacca	Grave exists
30 September	Lieut James Iveson	7th Regt Bengal NI	Fort Canning, Singapore	No trace of grave
16 November	Captain Thomas Richard Coleman Mantell	48th Regt Madras NI	Northam Road, Penang	Grave exists
1832 7 February	Lieut John A. Shennan	46th Regt Madras NI	,,	No grave exists
29 March	Lieut Edmund Vincent Harding	29th Regt Madras NI (Killed in action)	Fort Road, Malacca	Grave exists
3 May	Ensign George Holford Walker	29th Regt Madras NI (Killed in action)	Alor Gajah	Grave exists — memorial plaque has been removed
31 May	Ensign William Pinsent Luscombe	36th Regt Madras NI	Northam Road, Penang	No trace of grave
1833 7 January	2nd Lieut Charles Anthony Sadler Bruere	Madras Artillery	,,	Grave exists
2 July	Lieut James Charles Collin Collins	25th Regt Bengal NI	Fort Road, Malacca	No trace of grave
1834 1 January	Conductor George Leith (Aged 40 years)	Madras Ordnance	Northam Road, Penang	Grave exists
1835 1 April	Lieut Colonel Gregory Jackson (buried with wife and son)	15th Regt Madras NI	,,	,,
1836 20 December	Lieut John Pinsent Germon	48th Regt Madras NI	Fort Canning, Singapore	No trace of grave
1840 26 December	Pte G. Axon	not known	,,	,,
1841 25 February	Assistant Surgeon William Shedden	Madras Medical Estab	Fort Road, Malacca	,,
28 October	Assistant Surgeon James Campbell Boswell	Bengal Medical Estab	Northam Road, Penang	Grave exists
1843 16 January	2 Lieut Thomas Bernard Cox	Madras Artillery	Fort Canning, Singapore	Memorial exists
26 April	Assistant Surgeon George Clendining Courtney	Madras Medical Estab	Malacca	Plaque in Church at Malacca
1844 3 November	Major General Gilbert Waugh (Aged 61 years 4 months)	Staff	Northam Road, Penang	Grave exists
13 November	Lieut Charles John Power	Madras Invalid Estab	Fort Canning, Singapore	No trace of grave
1845 9 February	Assistant Surgeon Charles Richardson	Madras Medical Estab	Fort Cemetery, Malacca	No trace of grave
1846 11 August	Overseer Robert James	Madras Commissariat	St Paul's, Malacca	Grave exists
1847 2 February	Lieut William Charles Law	21st Regt Madras NI	,,	,,
1 June	1st Lieut William Charles Wynne	Madras Artillery	Drowned in Straits of Malacca	—
1848 20 May	Lieut George John Henry Pauli (Aged 21 years)	Madras Artillery	Northam Road, Penang	Grave exists
1850 2 February	Lieut Colonel P. L. Pew	—	Fort Canning, Singapore	No trace of grave
22 February	Quartermaster Sergeant J. Harvey	51st Regt Madras NI	,,	,,
23 June	Assistant Surgeon John Wilson	Madras Medical Estab	Old Naval Cemetery, Labuan	Memorial now in Botanical Gardens
1854 24 July	Captain Ilay Ferrier	48th Regt Madras NI	Malacca	No trace of grave
20 December	Conductor James Craig	Madras Ordnance Estab	Fort Canning, Singapore	Memorial exists
1855 1 April	Sgt T. Lee	Ordnance Stores Dept	,,	No trace of grave
1856 15 September	Lieut Colonel James Richardson	2nd Madras European Regt	Northam Road, Penang	,,
1857 1 June	Lieut James Innes Gibson	14th Regt Madras NI	Fort Canning, Singapore	Memorial exists
1859 1 February	Conductor George Marchant	Madras Ordnance Estab	Northam Road, Penang	Grave exists
11 September	Deputy Assist Commissary Matthew B. Murray	Madras Commissariat Dept	,,	,,
8 December	Capt John Cross (Aged 83 years)	Bengal Artillery Veterans Estab	,,	,,

Year and date	*Casualty*	*Unit*	*Where buried*	*Notes*
1860 8 April	Pte J. Haynes	87th Regt (in transit)	Fort Canning, Singapore	No trace of grave
24 May	Major Robert J. Littlewood	Bombay Invalid Estab	”	
1862 14 August	Capt W. Lord	20th Regt, Madras NI	Old Naval Cemetery, Labuan	”
22 October	Quartermaster Sgt J. Coveney (Aged 43 years)	40th Regt Madras NI	Fort Canning, Singapore	Memorial exists
1863 21 February	Conductor Thomas Doyle (Aged 53 years 2 months)	Ordnance Dept	Northam Road, Penang	Grave exists
3 December	Assistant Surgeon James Thomas Fraser	Madras Medical Estab	Malacca	Plaque in church
1865 22 February	Major William Somerville Mitchell (Aged 38 years)	22nd Regt, Madras NI	Northam Road, Penang	Grave exists
20 June	Lieut J. G. Silva	not known	Bukit Timah Rd, Singapore	Cemetery reclaimed
1866 3 July	Gnr J. Harrington	Royal Artillery	”	”
1 August	Drummer W. Brown	35th Regt Madras NI	”	”
1867 24 March	Gnr F. Hulton	Royal Artillery	”	”
7 May	Gnr W. Barrett	”	”	”
1868 4 May	Gnr T. Smith	”	”	”
7 May	Gnr E. Collens	”	”	”
7 May	Gnr C. Cartwright	”	”	”
13 June	Gnr J. Bass	”	”	”
1869 4 January	Lieut H. S. Johnson	Madras Staff Corps	Northam Road, Penang	No trace of grave
16 February	Lieut Colonel B. Hodson	”	”	”
4 March	Lieut Colonel J. Curtis	”	”	”
28 March	Sgt James Dowse (Aged 40 years)	Madras Ordnance Dept	”	Grave exists
30 April	Gnr G. James	Royal Artillery	Bukit Timah Rd, Singapore	Cemetery reclaimed
29 June	Pte T. Cook	not known	”	”
29 October	Sgt — Pettegrew	Royal Artillery	”	”
6 December	Lieut Colonel Ronald Macpherson	Royal Artillery (Madras)	” (Memorial at Singapore and plaque in Christ Church, Malacca)	”
1870 31 May	Sgt R. Marsden	Army Hospital Corps	Bukit Timah Rd, Singapore	”
31 December	Sgt J. Eason	Royal Artillery	”	”
1871 13 February	Pte G. Thomas	75th Regt	”	”
13 March	Bdr J. Parker	Royal Artillery	”	”
12 April	Pte T. Wright	75th Regt	”	”
21 June	— A. Hodson	75th Regt	”	”
1872 3 January	Pte T. Bulter	10th Regt	”	”
3 January	Pte J. Powell	”	”	”
12 May	Pte J. Crosbie	80th Regt	”	”
18 May	Pte J. Fellows	”	”	”
4 June	Pte J. Donnelly	”	”	”
10 June	Pte J. Dempster	”	”	”
1875 —	Sgt F. Owen	10th Regt	) Combined grave at	Now abandoned –
	Pte J. Ball	”	) Sungei Ujong	to revert to
	Pte J. Newman	”	)	nature
	Pte H. Smith	”	)	
1876 4 January	Major Henry Lumsden Hawkins	Bengal Staff Corps (30th Regt MNI)	Bukit Chandran, Perak	Grave exists
4 January	Pte Jasper Ball	RMLI, HMS *Philomel*	”	”
4 January	Pte William C. Sone	”	”	”
17 January	Cpl John Clune	3rd Regt	”	”
29 January	Colour Sgt S. Julian	”	”	”
30 January	Pte William King	”	”	”
19 January	Pte Elijah Neal	”	”	”
7 February	Pte R. Killingbeck	”	”	”
8 February	Pte William Hobbs	”	”	”
29 February	Lieut & Adjt C. A. Colvill	”	”	”
17 March	Sgt J. H. Fogarthy	10th Regt	”	”
9 July	Pte Sam Watson	”	”	”
21 February	Pte W. Carter	74th Regt	”	”
22 March	Pte J. Tellford	”	”	”
28 March	Cpl G. Lenton	Royal Artillery	”	”
1877 19 July	Captain H. S. Phillips	not known	Northam Road, Penang	”
1882 4 June	Pte Robert Montgomery	R. Iniskilling Fus	”	No trace of grave
1883 13 April	Gnr Patrick Farrell	G. Bty 1st Bde RA	Bukit Timah Rd, Singapore	Cemetery reclaimed
18 November	Cpl S. Anderson	R. Inniskilling Fus	”	”

Year and date	Casualty	Unit	Where buried	Notes
1884 1 November	Bdr F. Fleming	Royal Artillery	Bukit Timah Rd, Singapore	Cemetery reclaimed
1885 20 February	Pte W. DeCourcey	R. Inniskilling Fus	”	”
27 July	Gnr W. Wiscombe	Royal Artillery	”	”
13 August	Capt J. A. Burnett	R. Inniskilling Fus	”	”
29 September	Pte P. McDowen	”	”	”
4 November	— W. Day	”	”	”
26 December	Pte W. Summerfield	1st Bn, The Buffs	”	”
1886 3 March	Pte A. Neat	”	”	”
4 June	Pte R. J. Obee	”	”	”
22 July	Gnr J. E. Barber	Royal Artillery	”	”
18 September	Gnr A. Dumbarton	”	”	”
22 October	Pte W. Burton	1st Bn, The Buffs	”	”
20 October	Drummer Guthrie	”	”	”
20 October	Lcpl E. Kelly	”	”	”
20 October	Pte C. Simpson	”	”	”
27 October	Gnr P. J. Madden	Royal Artillery	”	”
2 November	Pte F. E. Taylor	1st Bn, The Buffs	”	”
28 November	Pte H. Alborough	”	”	”
1887 14 October	Bdr Thomas Coon	Royal Artillery	”	”
not known	Lieut Alexander Spankie	S. Lancs Regt	Northam Road, Penang	Grave exists
1888 16 January	Pte James Finlay (Aged 37 years)	2 Northants Regt	”	”
8 March	Pte T. Banks	82nd Regt	Bukit Timah Rd, Singapore	Cemetery reclaimed
8 March	Pte W. Taylor	”	”	”
4 June	Pte W. Dennis	”	”	”
17 August	Gnr J. Randall	Royal Artillery	”	”
21 September	Pte J. Bimpson	S. Lancs Regt	”	”
17 November	Lieut T. T. Rowan	Royal Artillery	”	”
1889 26 January	— C. Watson	S. Lancs Regt	”	”
4 February	— J. Sharpe	”	”	”
8 October	Pte W. H. Aldryer	”	”	”
1890 12 February	Sgt — Senve	not known	”	”
17 February	Pte H. J. Luddington	not known	”	”
20 February	Pte J. Brook	not known	”	”
25 February	Pte G. Hoddle	not known	”	”
24 March	Pte E. Bucknell	2nd Northants Regt	”	”
1891 14 February	Pte J. Hodson	not known	”	”
5 March	Gnr G. E. Ferrall	Royal Artillery	”	”
8 May	Sgt R. Sewell	Royal Engineers	”	”
23 September	— D. Saunders	not known	”	”
1892 30 June	Gnr J. Moran	Royal Artillery	”	”
14 November	Gnr A. Morris	”	”	”
15 November	Pte W. Snell	not known	”	”
3 December	Pte F. Smith	not known	”	”
1893 17 February	Major General H. A. Gordon	Staff	”	”
27 June	— T. Martin	2nd Bn Lincolnshire Regt	”	”
1894 5 August	— H. Walker	not known	”	”
6 August	— A. Childs	Royal Artillery	”	”
29 September	Pte D. Starkings	2nd Bn Lincolnshire Regt	”	”
12 November	— W. Walker	not known	”	”
1895 18 March	Sgt D. Barnes	Royal Artillery	”	”
3 May	Sgt — Harvey	Royal Engineers	”	”
10 May	Pte T. Hughes	2nd Northumberland Fus	”	”
14 May	Cpl W. Donan	”	”	”
16 May	Pte P. Park	”	”	”
16 June	Pte H. Eastoe	”	”	”
20 June	Lieut C. V. M. Sarel	”	”	”
4 November	Pte G. Oliver	”	”	”
20 November	— S. Dawson	”	”	”
6 December	— L. J. D. Searle	”	”	”
1896 2 March	Gnr J. Mullen	Royal Artillery	”	”
29 April	Cpl R. Weallens	2nd Northumberland Fus	”	”
12 June	— J. Drew	”	”	”
1 August	— J. Newman	not known	”	”
21 September	Sgt Major J. Bannister	not known	”	”
22 September	Sgt W. O'Hanlon	2nd Northumberland Fus	”	”

Year and date	*Casualty*	*Unit*	*Where buried*	*Notes*
1896 1 October	Sgt J. Thompson	2nd Northumberland Fus	Bukit Timah Rd, Singapore	Cemetery reclaimed
21 November	Gnr J. Milner	"	"	"
1897 1 February	— — Seahill	not known	"	"
21 March	— C. Perry	not known	"	"
8 May	Gnr E. V. Wood (found drowned)	Royal Artillery	"	"
3 December	Pte — Hillman	not known	"	"
10 December	Pte — Sharratt	not known	"	"
1898 23 February	— F. G. Eels (4647)	1 West Yorks		"
15 March	— D. Murray	not known	"	"
26 April	— W. Yates	not known	"	"
30 April	— E. Canning	not known	"	"
13 May	Pte G. Gill	1 West Yorks	"	"
16 June	— C. Wells	not known	"	"
1899 10 April	Gnr J. Wilson	Royal Artillery	"	"
14 April	— W. Clark	The King's Own	"	"
15 August	Lieut S. R. Fletcher	East Surrey Regt	"	"
7 September	— A. Pierson	not known	"	"

Colonial Auxiliary Forces Long Service Medal. Medal loaned by J. M. A. Tamplin, Esq. Photograph by G. Parselle.

APPENDIX C

Medal Roll of the Straits Settlements Volunteer Force
for the Award of the Colonial Auxiliary Forces Long Service Medal
(The award became obsolete in 1930)

Ager, A. P., Private	Singapore Volunteer Corps
Abu Bakar bin Haji Arshad, Lieutenant	"
Abdul Aziz bin Masood, Lieutenant	"
Brown, E. A., 2nd Lieutenant	"
Baugh, C., Private	"
Biggs, L. A. C., Private	P&PWVC
Burt, C. G., Captain	Singapore Volunteer Corps
Buchannan, A., Private, MM	"
Coveney, A. E. G., Lieutenant (Quartermaster)	"
Craik, D. McLeod, Private	P&PWVC
Chan Tiang Seng, Private	Singapore Volunteer Corps
Chater, R. W., CQMS	"
Cheong Chin Heng, Private	"
Curran-Sharp, W. J., 2nd Lieutenant	"
Dawbarn, J. S., Captain	P&PWVC
Danson, J. R., CSM, DCM	"
Dandie, H., Private	Singapore Volunteer Corps
De Silva, R. E., Sergeant	"
De Buriatte, E. A., Captain	P&PWVC
Eusof bin Hussain Jewa, Lieutenant	Singapore Volunteer Corps
Fitt, J. F., RQMS	"
Fulcher, E. W. P., Lieutenant	"
Gilmour, H., Sergeant	"
Gallistan, E., Sergeant	"
Gan Hock Chuan, 2nd Lieutenant	unit not known
Grant, J. S., Corporal	P&PWVC
Hamilton, W., Private	"
Hay, T. C., Captain	Singapore Volunteer Corps
Holley, W. H., Private	"
Hodgins, J. F., Private	"
Haji Abas bin Omar, Lieutenant	"
Johnson, E. E., Private	"
Livingstone, A., Corporal	Malacca Volunteer Corps
Lesslar, T. S., RQMS	P&PWVC
Lobb, H. R. W., Private	Singapore Volunteer Corps
Lim Liang Quee, Private	"
Lewis, The Hon D. T., Gunner	"
Lee Kiah Wah, CQMS	"
Mayhew, T. O., 2nd Lieutenant	"
Makepeace, W., Captain & Hon Major	Singapore Volunteer Artillery
McLeod, K. M., Private	P&PWVC
Myles, G. S., Corporal	Singapore Volunteer Corps
Marsh, F. E., Sapper	"
Morrell, F. A., Captain	"
Mayger, F. J. L., 2nd Lieutenant	"
Mohamed bin Dol, Sergeant	"
Mukhtar bin Haji Hassan, L/Sergeant	"
Moncreieff, R. L., RQMS	"
Ogle, J. W. B., CSM	P&PWVC
Parsons, A. E., Private	Singapore Volunteer Corps
Perrott, R. T., Lance Corporal	Malacca Volunteer Corps

General Index

Note: The following abbreviations have been used in the General Index.
BNI - Bengal Native Infantry, Bombay NI - Bombay Native Infantry, MNI - Madras Native Infantry

Printed at Picton Print
keyboarding and
page make-up Veronica Pinniger
negatives Mike Simmons
platemaking Liz Wilkins
machining Neville Fox

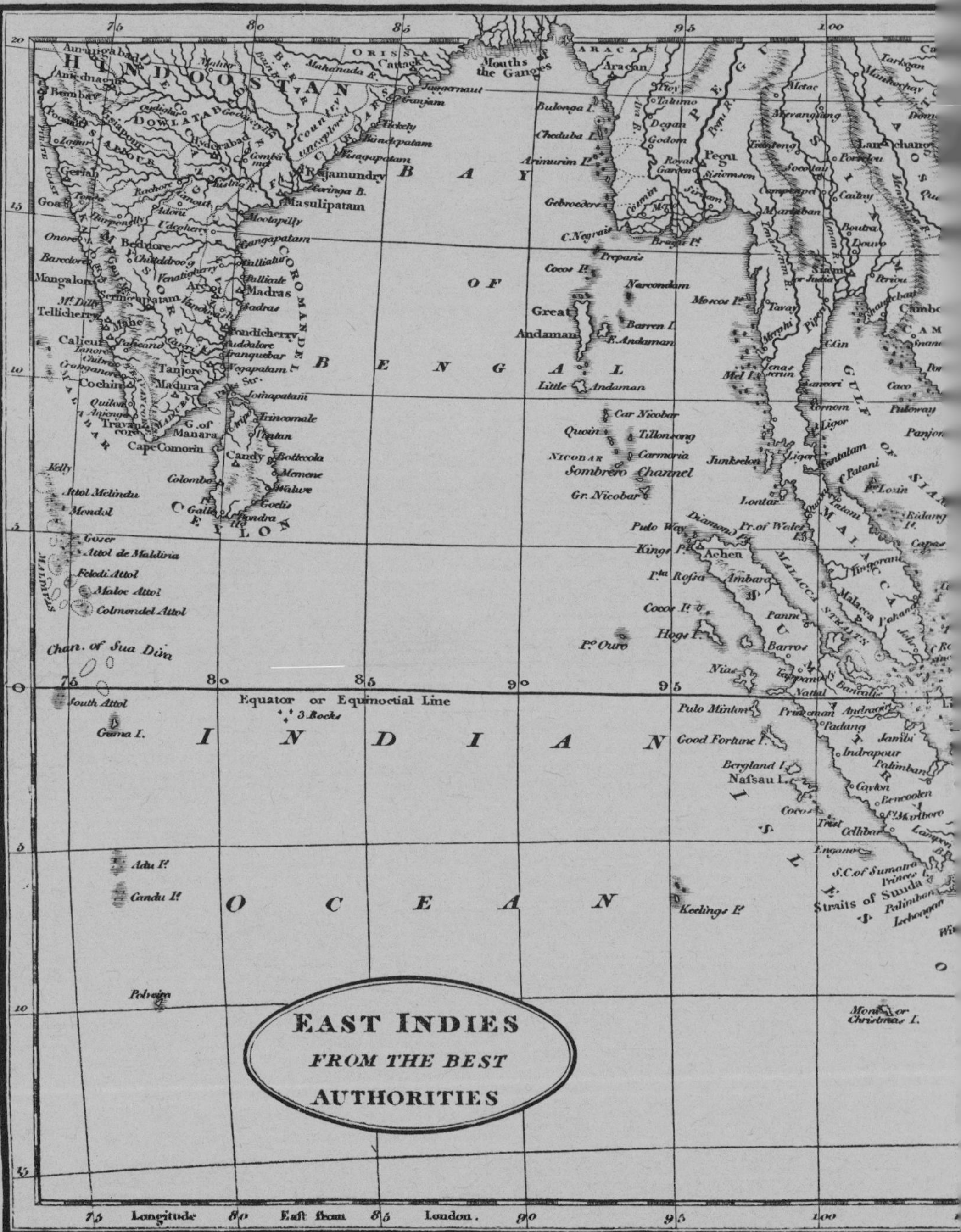
EAST INDIES
FROM THE BEST
AUTHORITIES
HINDOOSTAN
BAY OF BENGAL
INDIAN OCEAN
CEYLON
COROMANDEL
MALABAR
Equator or Equinoctial Line
Longitude East from London.
Bombay
Goa
Mangalore
Tellicherry
Calicut
Cochin
Quilon
Cape Comorin
G. of Manara
Madras
Pondicherry
Tanjore
Madura
Masulipatam
Rajamundry
Vizagapatam
Ganjam
Cattack
Mouths the Ganges
Aracan
Pegu
Martaban
Siam
Great Andaman
Little Andaman
Car Nicobar
Sombrero Channel
Gr. Nicobar
Candy
Colombo
Trincomale
Achen
Malacca
Malacca Straits
Bencoolen
Straits of Sunda
Keelings Is.
Christmas I.
Adu Is.
Candu Is.
South Attol
Chan. of Sua Dira
Colmondel Attol
Maloc Attol
Attol de Maldiria
Mondol
Attol Melindu
Kelly
Polveira
3 Rocks
Good Fortune I.
Nassau I.
Pulo Minton
Nias
Hogs I.
Pt. Ouro
Cocos I.
Pulo Way
Junkselon
Tavay
Mergui
Gulf of Siam
Patani
Pahang
Jambi
Indrapour
Padang
S.C. of Sumatra
Engano
75
80
85
90
95
100
20
15
10
5
0